NEUROMETHODS □ 23

Practical Cell Culture Techniques

NEUROMETHODS

Program Editors: Alan A. Boulton and Glen B. Baker

Practical
Cell Culture Techniques

NEUROMETHODS □ 23

Edited by

Alan A. Boulton

University of Saskatchewan, Saskatoon, Canada

Glen B. Baker

University of Alberta, Edmonton, Canada

and

Wolfgang Walz

University of Saskatchewan, Saskatoon, Canada

Humana Press • Totowa, New Jersey

© 1992 The Humana Press Inc.
999 Riverview Drive, Suite 208
Totowa, New Jersey 07512

This publication is printed on acid-free paper.∞
ANSI Z39.48-1984 (American National Standards Institute) Permanence of Paper for Printed Library Materials.

Photocopy Authorization Policy:
Authorization to photocopy items for internal or personal use, or the internal or personal use of specific clients, is granted by Humana Press Inc., provided that the base fee of US $4.00 per copy, plus US $00.20 per page is paid directly to the Copyright Clearance Center at 222 Rosewood Dr., Danvers, MA 01923. For those organizations that have been granted a photocopy license from the CCC, a separate system of payment has been arranged and is acceptable to Humana Press Inc. The fee code for users of the Transactional Reporting Service is: [0-89603-214-0/95 $4.00 + $00.20].

Printed in the United States of America. 9 8 7 6 5 4 3 2

Library of Congress Cataloging-in-Publication Data

Practical Cell Culture Techniques / edited by Alan A. Boulton, Glen B. Baker, and Wolfgang Walz.
    p. cm. — (Neuromethod ; 23)
    Includes bibliographical references and index.
    ISBN 0-89603-214-0 (cloth)
    ISBN 0-89603-348-1 (pbk.)
    1. Nervous system—Cultures and culture media. 2. Neurons.
I. Boulton, A. A. (Alan A.) II. Baker, Glen B., 1947– .
III. Walz, Wolfgang. IV. Series.
    [DNLM: 1. Neurons—cytology. 2. Tissue Culture—methods. W1
NE337G v. 23 / QS 525 T6155]
QP357.T57 1992
599'.0188—dc20
DNLM/DLC
for Library of Congress   92-1518
                                                        CIP

Preface to the Series

When the President of Humana Press first suggested that a series on methods in the neurosciences might be useful, one of us (AAB) was quite skeptical; only after discussions with GBB and some searching both of memory and library shelves did it seem that perhaps the publisher was right. Although some excellent methods books have recently appeared, notably in neuroanatomy, it is a fact that there is a dearth in this particular field, a fact attested to by the alacrity and enthusiasm with which most of the contributors to this series accepted our invitations and suggested additional topics and areas. After a somewhat hesitant start, essentially in the neurochemistry section, the series has grown and will encompass neurochemistry, neuropsychiatry, neurology, neuropathology, neurogenetics, neuroethology, molecular neurobiology, animal models of nervous disease, and no doubt many more "neuros." Although we have tried to include adequate methodological detail and in many cases detailed protocols, we have also tried to include wherever possible a short introductory review of the methods and/or related substances, comparisons with other methods, and the relationship of the substances being analyzed to neurological and psychiatric disorders. Recognizing our own limitations, we have invited a guest editor to join with us on most volumes in order to ensure complete coverage of the field. These editors will add their specialized knowledge and competencies. We anticipate that this series will fill a gap; we can only hope that it will be filled appropriately and with the right amount of expertise with respect to each method, substance or group of substances, and area treated.

Alan A. Boulton
Glen B. Baker

Preface

Most cells will survive removal from the natural microenvironment of their in vivo tissue and placement into a sterile culture dish under optimal conditions. Not only do they survive, but they also multiply and express differentiated properties in such a culture dish. A few cells do this in suspension, but most will need some kind of mechanical support substituting for their natural connections with other cells. The surface of a culture dish that might have to be coated is usually sufficient. The recent trend to standardization of conditions and the existence of commercial enterprises with adequate funds and specializing in the needs of scientists were responsible for the tremendous proliferation of cell culture techniques in all fields of research in the last 20 years. No longer does a scientist have to concentrate all his/her efforts on that technology; the new trends make it feasible to employ cell culture techniques as only one of the many methods available in a small corner of a larger research laboratory.

Some areas of research depend more heavily than others on cell culture techniques. Neuroscience is one of the areas that has developed hand in hand with the proliferation of cell culture methodology. Molecular biological aspects, cell differentiation and development, neurophysiological and neurochemical studies, as well as investigations into the nature of various diseases are now to a large extent dependent on the use of cell cultures. Some areas, such as glial cell biology and function, depend almost exclusively on cell culture-based research projects.

In general, most laboratories use cells that have been dissociated from the original tissue and that now form a monolayer or suspension in the culture medium; these are defined as primary cultures. Their properties are related closely to those of the cells from the tissue in vivo and can therefore be used to study cell function in isolation. It is also possible to select one cell type and obtain one homogeneous culture. If cells from these primary cultures are subcultured after repeated dispersion from the culture dish, and therefore undergo passages through different proliferation cycles, they are called cell lines, and may exhibit loss of some of their differentiated properties. It is then difficult to relate the cultured cells to functional cells in the tissue from which they were derived. Commercially available cell lines are, however, usually well defined, and often have advantages for molecular biological investigations and related problems. Thus, if one is interested in studying the interactions between two or more different cell types in culture, one can either coculture different primary cultures or can use an approach other than one involving dissociated cells. In organ cultures one obtains a three-dimensional culture of undissociated tissue. Here the characteristic architecture of the tissue is largely retained. In organotypic or primary explant cultures a fragment or slice of tissue is used, migration of the different cell types is promoted, and the cells recreate a three-dimensional relationship.

The advantages of the use of tissue culture techniques are obvious. They represent an economical means of obtaining large quantities of homogeneous cells. The isolation of a small population of homogeneous cells facilitates the observation and elaboration of interactions and mechanisms. The environment of the cells can be controlled readily, and all the more so now with the advent of chemically defined media. The cells can be easily defined and the interactions between different cell types can be studied by simply adding more than one cell type to a culture. Not only can specific chemical factors be studied in detail, but the in vitro nature of an experiment means that those can be varied to extreme values. The survival of the whole organism on which the vitality of the studied cell depends and which prevents the use of extreme single parameter settings is not a major problem in the artificial and standardized environment of a cell culture, and larger variations of such factors are possible. Because of increasing public sensitivity to animal rights issues, cell culture is destined to play an even greater future role in routine testing and toxicology tests. Physiological and biochemical mechanisms can be worked out in culture and subsequent in vivo work can concentrate on experiments of a confirmatory or modulatory nature. Cell culture methodology does not constitute a substitute for future animal research, but it should reduce the need for the use of animals. However, there are some disadvantages associated with the use of cell culture systems. The most important one is that the standardization of the techniques has not yet reached the point where results from different laboratories can be easily compared. We still lack an even rudimentary knowledge of how culture techniques and culture environment change the properties of cell cultures and create differences between in vivo and in vitro properties. In addition, some important parameters, such as cerebral blood flow, cannot be investigated very well in culture.

The present volume of Neuromethods concentrates on the preparation, maintenance, manipulation, and properties of tissue culture systems relevant for neuroscience research. References to the applications of such systems are minimal since many of the other volumes of this Neuromethods series focused on the use of cultured cells, but not on the creation of the cultured cells needed for the described applications. This book attempts to close that gap.

The book is organized into three parts. The first part consists of one chapter that addresses novices who wish to begin cell culturing. It provides valuable information about how to set up a cell culture laboratory, how to sterilize, how to check for contamination, and so on.

The second part of the book examines in detail the many environmental factors that play a role in tissue culturing. The reader is introduced to the use and necessity of cell markers to define the cell culture system used by individual laboratories. This question causes some major scientific controversies, and therefore a sound basis in defining and distinguishing cell types and subtypes is an absolute necessity. Another important issue is the optimal nutritional requirements of the different cell types. The development of chemically defined media for specific cells has opened up a complete new field and research possibilities. It is now possible to study development, expression, and interaction in culture by clearly defining all substances present, without any contributions from unknown factors. Related to this chapter is another one concentrating on growth factors. These are key factors in defining the final architecture of the CNS, synaptic as well as neuronal–glial interactions. Very important tools in cell culture are cell adhesion molecules substituting for the natural adhesion in vivo. They can also be used to select culture conditions for the attachment and therefore selection of specific cell types.

The third and last part of the book will concentrate on the establishment and the properties of some specific brain-derived cell types and systems in culture. This last portion has necessarily been selective. The reader is introduced to three-dimensional organotypic slice explant cultures, which are new and powerful tools with which to study interactions between different cells and cell types. The remainder of this third part of the volume concentrates on primary cultures from more or less homogeneous dissociated monolayer cultures. Neuronal cultures, with special emphasis on the hippocampus, are introduced; they are obviously the backbone of any interest in CNS function. But other cell types are becoming of increasing interest: For example, astrocytes and their subtypes are now of prime neurochemical interest. Oligodendrocytes and their role in myelin formation, as well as in proliferation and differentiation in vitro, are introduced to the reader. Finally, cultures of capillary endothelium derived from cerebral microvessels and their characterization are presented. We anticipate that this volume will suit the needs of the reader who is interested in this ever-expanding area of neuroscience research.

Wolfgang Walz

Contents

Environmental Influences on Cells in Culture
Jane E. Bottenstein

Detection and Analysis of Growth Factors
Affecting Neural Cells
Marston Manthorpe, David Muir, Brigitte Pettmann,
and Silvio Varon

The Role of Cell Adhesion Molecules
in Neurite Growth
Daren Ure and Ann Acheson

Three-Dimensional Organ Culture Systems
B. Rogister, J. M. Rigo, P. P. Lefebvre, P. Leprince,
P. Delree, D. Martin, J. Schoenen, and G. Moonen

Organotypic Slice Explant Roller-Tube Cultures
Susan Wray

Neurons
Jeffrey R. Buchhalter and Marc A. Dichter

Astrocytes
Bernhard H. J. Juurlink and Leif Hertz

Oligodendrocytes
Jean de Vellis and Araceli Espinosa de los Monteros

Brain Capillaries: *Cell Culture of Capillary Endothelium Derived from Cerebral Microvessels*
Nika V. Ketis

Contributors

ANN ACHESON • *Regeneron Pharmaceuticals Inc., Tarrytown, NY*

COLIN J. BARNSTABLE • *Department of Ophthalmology and Visual Science, Yale University, New Haven CT*

JANE E. BOTTENSTEIN • *Department of Human Biological Chemistry and Genetics, University of Texas Medical Branch, Galveston, TX*

JEFFREY R. BUCHHALTER • *Division of Neurology, The University of Pennsylvania, Philadelphia, PA*

P. DELREE • *Department of Human Physiology and Pathophysiology, University of Liege, Liege, Belgium*

JEAN DE VELLIS • *University of Southern California at Los Angeles, . Los Angeles, CA*

MARC A. DICHTER • *Department of Neurology, The University of Pennsylvania, Philadelphia, PA*

ARACELI ESPINOSA DE LOS MONTEROS •*University of Southern California at Los Angeles, Los Angeles, CA*

DIANE E. HAROLD • *Department of Physiology, University of Saskatchewan, Saskatoon, Saskatchewan, Canada*

LEIF HERTZ • *Departments of Pharmacology and Anaesthesia, University of Saskatchewan, Saskatoon, Saskatchewan, Canada*

BERNHARD H. J. JUURLINK • *Department of Anatomy, University of Saskatchewan, Saskatoon, Saskatchewan, Canada*

NIKA V. KETIS • *Department of Anatomy, Queen's University, Kingston, Ontario, Canada*

P. P. LEFEBVRE • *Department of Human Physiology and Pathophysiology, University of Liege, Liege, Belgium*

P. LEPRINCE • *Department of Human Physiology and Pathophysiology, University of Liege, Liege, Belgium*

MARSTON MANTHORPE • *Vical, Inc., San Diego, CA*

D. MARTIN • *Department of Human Physiology and Pathophysiology, University of Liege, Liege, Belgium*

G. MOONEN • *Department of Human Physiology and Pathophysiology, University of Liege, Liege, Belgium*

DAVID MUIR • *Department of Biology, University of California, San Diego, La Jolla, CA*

BRIGITTE PETTMANN • *Center for Neurochemistry, Strasbourg, France*

J. M. RIGO • *Department of Human Physiology and Pathophysiology, University of Liege, Liege, Belgium*

B. ROGISTER • *Department of Human Physiology and Pathophysiology, University of Liege, Liege, Belgium*

J. SCHOENEN • *Department of Human Physiology and Pathophysiology, University of Liege, Liege, Belgium*

DAREN URE • *Department of Anatomy and Cell Biology, University of Alberta, Edmonton, Alberta, Canada*

SILVIO VARON • *Department of Biology, University of California, San Diego, La Jolla, CA*

WOLFGANG WALZ • *Department of Physiology, University of Saskatchewan, Saskatoon, Saskatchewan, Canada*

SUSAN WRAY • *Laboratory of Neurochemistry, NINDS, NIH, Bethesda, MD*

Basic Techniques for Cell Culturing

Diane E. Harold and Wolfgang Walz

1. Introduction

Until recently, the cell culture technique was employed as the sole biological system within a laboratory, and major efforts were concentrated on the setting up and running of such a cell culture laboratory. However, recent developments in several areas, use on monoclonal antibodies and cell physiological techniques, for example, gave rise to a need for establishing a small cell culture laboratory. Being only one of several major methods or biological preparations used within the research group, there is a need for cost to be kept to a minimum. This chapter addresses all the major issues that should be considered by someone lacking previous cell culture experience, but wanting to use one of the many techniques introduced in the following ten chapters. All the basic problems, such as, design of the laboratory, sterilization of equipment, and basic techniques in handling cultured cells are introduced and referenced. It is our hope that the reader venturing into the area of cell culture for the first time will find this chapter a useful guide and a source of encouragement. At the end of this chapter is an appendix of equipment and supplies, along with the names and addresses of the companies from which they may be obtained.

From: *Neuromethods, Vol. 23: Practical Cell Culture Techniques*
Eds: A. Boulton, G. Baker, and W. Walz ©1992 The Humana Press Inc.

2. Design and Equiping
of a Cell Culture Laboratory

In setting up a cell culture laboratory, one must determine what facilities and equipment are already available, as well as the role tissue culture will play in the research program. It is easiest to make use of facilities that are already available, especially when space and funds are limited. The following is a list of basic needs:

1. Sterile work area;
2. Glassware, instruments, and the like;
3. Cleaning and sterilization facilities;
4. Storage for medium, serum, glassware, and so on;
5. Incubator(s);
6. Source of pure water;
7. Microscope(s) (Paul, 1975; Freshney, 1987).

As mentioned above, the role of tissue culture must also be considered as the needs for biochemical studies will differ from those for developmental biology, cell physiology, and the like. The number of personnel, culture techniques employed, and use of hazardous materials are also important factors to consider.

2.1. Sterile Work Area

If a separate room is not available, the work area may be placed in a low traffic area of the laboratory. Patterns of air flow and air quality are important considerations. A low traffic, low airflow area means less opportunity for dust particles and microorganisms to become airborn (Paul, 1975). A tissue culture room should be of higher pressure than adjoining hallways or rooms, i.e., air should flow out of the room. Direction of airflow is easily tested by holding a tissue suspended from one's fingertips while standing at the open doorway. The source of air entering the tissue culture area should be considered as contaminants may enter from adjacent animal facilities or from outside air intakes that are located too close to the ground. As well, animals and microbiological work should not be allowed in close proximity to tissue culture facilities (Paul, 1975).

If cultures are being maintained for only several days, a laminar flowhood in a separate room may be an unnecessary expense. A work surface with adequate lighting, shelter from dust from the sides and above, and easily cleaned surfaces located in a quiet area of the lab is sufficient (Paul, 1975). Longer term cultures present a greater opportunity for contamination, plus more time and effort has been invested in them. In this instance, having a tissue culture room equipped with a laminar flowhood is worth the expense. More details on flowhood will be presented in Section 2.5.

2.2. Glassware and Instruments

The type and amount of glassware required will, of course, depend on the size of the tissue culture operation. It should be stressed that tissue culture glassware be kept separate from general lab glassware. This ensures all glassware has been through the same cleaning process. Many items such as pipets, tissue culture dishes, and filters, are available in convenient sterile, disposable form. If glass pipets are being used, one must have canisters to store pipets in, both during and after sterilization. Much of the glassware may be similar to that in use in almost any lab. The following is a rough guide as to glassware needs (Paul, 1975):

> bottles for media and other solutions
> Erlenmeyer flasks
> graduated cylinders
> beakers of various sizes
> petri dishes (optional: tins for sterilizing and storage)
> centrifuge tubes with screw caps
> graduated pipets and canisters for sterilization and storage
> Pasteur pipets (straight and bent tip)
> microscope slides and coverslips
> funnels

A wide variety of tissue culture grade plastics are available, such as multiwell plates, tissue culture dishes, T-flasks, and disposable roller bottles. This is not to say that glass is no longer used. For example, techniques requiring large numbers of cells may make use of glass roller bottles and Erlenmeyer flasks may be utilized in cell aggregate culture.

Dissection instruments often required are (Paul, 1975):

knife handles and sterile scalpel blades
fine forceps: straight and curved tip
ultrafine forceps (e.g., Dumont brand)
surgical scissors:
  standard pattern, fine straight tip, and fine curved tip

Another essential piece of equipment is a phase contrast microscope for regular inspection of cultures. One may wish to purchase a microscope designed for use with photography equipment and fluorescent dyes. In addition, a dissection microscope may be necessary with certain types of tissue culture techniques.

2.3. Storage

The basic storage space requirements are for glassware, disposable plastics, media, serum, and other liquids (Freshney, 1987; Paul, 1975).

A commerical refrigerator with a large freezer compartment is ideal for storage of media and serum. Some may require an additional freezer. If so, it should be noted that an upright freezer permits easy access. A liquid N_2 freezer of appropriate size is necessary for those intending on preserving cells for later use. More information on low-temperature freezers may be found in Freshney (1987). Refrigerators with glass doors should not be used, as exposure to light may cause stored medium to deteriorate.

Laboratory space must be set aside for storage and preparation of equipment to be sterilized. This space must be clearly separated from the area reserved for storage of sterile equipment. Sterile glassware and plastics should be kept where they are readily accessible while the operator is performing tissue culture manipulations. Storage space may consist of shelving units, although cupboards do provide better protection from dust, and the like.

2.4. Incubator Facilities

Small scale tissue culture operations may have their incubator needs easily and economically fulfilled by the purchase of one or several waterjacketed incubators (Freshney, 1987; Paul, 1975). Having two incubators means one can be used while the other is being cleaned or repaired. Incubators are most conveniently located near the laminar flowhood or sterile work area. An inverted phase contrast microscope may be located nearby. Whether or not one requires a CO_2 incubator depends upon the buffering system being used in the medium. Zwitterionic buffers, HEPES, for example, and closed culture systems, Leighton tubes, for example, do not require 5% CO_2 in air. However, bicarbonate buffer systems require the presence of CO_2 (Douglas and Dell'Orco, 1979). Regardless of whether CO_2 is required, an incubator must be capable of maintaining temperature within the desired range. Some culture techniques require a certain type of incubator to meet their needs. Examples of this are cultures of cells in suspension and culture of cell aggregates.

2.5. Laminar Flow Hoods

Although a laminar flow hood is not a substitute for good aseptic technique, it does help ensure sterility when used properly. When choosing a flowhood, consideration must be given to the containment level required for product and personnel protection. This may be determined from MRC or NIH guidelines.

The flowhood should be large enough to accommodate necessary items for preparation or feeding of cultures and still have a comfortably sized work area. The flowhood should be easily cleaned, comfortable to work at, and have good lighting. Several types of lighting are available. Some workers use exposure to UV light to sterilize the air and work surfaces in the flowhood. However, UV light must make direct contact with microorganisms to kill them. As well, UV light does not reach into crevices, the tubes have a short lifespan, and they are expensive. Use of alcohol as a disinfectant is not only effective, but is far more economical than UV light. Beyond this, one has a choice of fluorescent light or yellow-filtered light. A 400–450 nm wavelength yellow filter has been suggested as a means of protecting media and other substances from the harmful, degenerating effect of fluorescent light.

Laminar flow hoods may be divided into two basic types, horizontal and vertical laminar flow. Both offer some degree of product protection via a constant flow of filtered air within the work area, but only the vertical laminar flowhood offers personnel protection. In both type of flowhoods, the room air is filtered through a high-efficiency particulate air (HEPA) filter, that removes particles of diameter greater than 0.3 μm (Douglas and Dell'Orco, 1979; Barkley, 1979). In a horizontal laminar flowhood, filtered air flows parallel to the work surface, toward the operator. Thus, this type of flowhood only provides product protection. Cell culture work, which falls into low risk categories (e.g., Medical Research Council level A) may by conducted using a horizontal laminar flowhood. Beyond the low risk level a vertical flow safety cabinet is a must. In vertical flow hoods, filtered air flows from the top of the cabinet toward the work surface. The drawing in of room air at the front of the work surface decreases the possibility of harmful agents leaving the work area. The different models of vertical laminar flowhoods vary in the level of protection they provide. In general, a class II biological (vertical flow) safety cabinet is suggested to be the most versatile (Barkley, 1979). When selecting a major piece of equipment, it can be useful to make use of literature supplied by various companies.

2.6. Source of Water

The location of water taps and sinks has a major influence on laboratory layout. Large sinks are required for the soaking, washing, and rinsing of glassware. These sinks require hot and cold water, as well as a pure source of water for rinsing. A deionizer, a single glass distillation apparatus, or reverse osmosis unit yields water of sufficient quality for rinsing (Freshney, 1987; Douglas and Dell'Orco, 1979). It cannot be forgotten that an autoclave's effectiveness is determined by the purity of its water source. Care should be taken to ensure that steam is not contaminated, for example, steam pipes are periodically cleaned with agents that are cytotoxic. Ideally, an autoclave capable of generating its own steam is used and the water for steam generation has been purified by distillation, for example (Perkins, 1969; Douglas and Dell'Orco, 1979).

Treated water is also necessary for preparation of media and other solutions. The water needs to be of higher purity than that for rinsing of glassware. There are a number of means by which this high quality water may be obtained. For those on a limited budget and/or setting up a small cell culture facility, it is best to make use of water purification systems that are already available. Triple-glass distilled or deionized-distilled water may prove to be the easiest to obtain. Another alternative is to subject water to reverse osmosis, then charcoal filtration and deionization followed by submicron filtration (Freshney, 1987; Douglas and Dell'Orco, 1979). When using charcoal filtration and/or deionization, one should inspect the filter cartridges regularly to make certain they are working properly.

The following is a summary of the properties of pure water:

1. Specific conductance, 0.1–0.06 $\mu\Omega$
2. Specific resistance, >16MΩ
3. Total matter, <0.1 mg/L
4. Silicates, <0.05 mg/L

Purified water should be stored for as short of a time as possible, since water quality does deteriorate over time. Although plastic carboys are cheaper and more durable, borosilicate glass carboys are the preferred storage vessel (Douglas and Dell'Orco, 1979).

2.7. Cleaning and Sterilization Facilities

Full advantage should be taken of cleaning and sterilization facilities that are already available. This includes the following items: pipet and bottle washers, autoclave, and a sterilizing oven. A pressure cooker capable of obtaining 15 lb pressure may be used to sterilize a small number of items (Freshney, 1987). There are also small sterilizers available that are designed for sterilization of dissection instruments (Inotech Steri).

The first requirement in cleaning are large sinks or tubs in which glassware may be soaked. Actual details as to steps taken in sterilization, decontamination, and product protection will be covered in a separate section.

With the respect to location, cleaning and sterilization facilities are best located in a room separate from the tissue culture facilities because of the heat and moisture they generate. At the very least, they should be far away from the actual tissue culture area in a spot with good ventilation.

3. Aseptic Techniques

Aseptic technique is a cornerstone of successful tissue culturing. This means the operator must work in such a manner to prevent contamination of material that is already sterile. The tissue culturist him-/or herself is the source of contamination most difficult to control. Aseptic technique is based on common sense more than anything else, but here are some basic rules to follow:

1. Limit access to tissue culturing area.
2. Decontaminate work surfaces with 70% ethanol before and after manipulations.
3. Permit no mouth pipeting, eating, drinking, chewing gum, or smoking.
4. Wear a lab coat that is used for nothing other than tissue culture.
5. Wash hands before and after.
6. Do not work closer than 15 cm from bench top and tie back long hair.
7. Keep arm motion away from work area, i.e., do not have equipment and materials arranged so you have to reach across work area. In a horizontal laminar air flowhood, do not work behind or over an open vessel. In a vertical laminar air flowhood, do not work over top of an open vessel (Freshney, 1987).
8. Control aerosis, e.g., wipe up spills with a cloth dampened with alcohol, do not talk, sneeze, or cough while facing work area.

Other points to consider are the layout of materials in the work area and proper technique in performing manipulations, such as, pipetting. These are best learned by observing an experienced tissue culturist and, of course, by practicing. The following is a brief look at several examples of proper technique.

Only after turning on the flowhood and decontaminating work surfaces with alcohol may required materials be put into the hood. Allow 10–15 min before doing any further work. Open vessels should be angled, as this allows the least possible opening for dust and other particles to fall in. For example, a sterile test tube is removed from its canister as follows:

1. Hold canister horizontally, remove lid, and place inside down.
2. Still holding canister horizontally, shake it gently until you can grasp onto one test tube.
3. Insert test tube at an angle into test tube rack.
4. Replace lid on tin.

Metal slanted racks are available for keeping bottles at an angle. Remember to wipe slant racks, bottles, and other similar items with alcohol as they are placed into the hood.

Pipeting devices, such as pipet-Aid (Bellco Glass, Inc., Vineland, NJ) allow for smooth drawing up and expelling of liquids from pipets. The pipet is held near the top while attaching the pipeting device to it. Grasping the pipet lower down introduces the possibility of contaminating solutions when the pipet is inserted into the solution. When feeding cultures the medium should be expelled without splattering. The pipet is held at an angle horizontal to the body and care is taken to not touch the tip of the pipet to the culture dish. When adding one liquid to another, serum to medium, for example, the liquid is expelled with the tip of the pipet below the surface of the liquid you are adding it to. Whenever practicable, using a pipet is preferable to pouring when transferring liquids.

4. Detection of Contamination

Cultures need to be inspected periodically for presence of contamination. A change in pH of medium, often an acidic change, is a good indicator of bacterial contamination. Microscopic investigation of suspect cultures will confirm whether or not bacterial contamination is present. A simple rule to follow is that precipitates of medium constituents and debris will only be subject to Brownian movement, whereas bacteria will exhibit directed movement. Yeasts will appear to be round or ovoid and buds may be visible under microscopic examination (Freshney, 1987). Under visual inspection, mold may appear as a whitish mass and is often found along the perimeter of a culture dish. Inspection under a microscope reveals the telltale filamentous mycelia. One should also watch out for deterioration or slowed growth of cultures and cloudiness of medium. Abnormal growth or deterioration of cultures may be indicative of mycoplasma contamination. This type of contamination can only be positively identified by specific techniques, such as fluorescent staining with Hoechst 33258. Using this technique, mycoplasma appear as fine fluorescent particles or filaments over the cytoplasm (Freshney, 1987).

As long as the contamination is an isolated recent occurrence, simply safely discard the culture and any materials that have come into contact with it, such as, medium and disposable pipets. Repeated, widespread contamination may have a variety of origins, depending on the type of contamination. Serum and the human mouth or throat are sources of mycoplasma (Adams, 1983). The possibility of mycoplasma infection may be avoided by purchasing serum certified to be mycoplasma-free (Douglas and Dell'Orco, 1979) and by avoiding mouth pipeting as well as sneezing, coughing, or talking over cultures. Yeasts and mites commonly are harbored in animal quarters. Maintenance of a clean work area and equipment, adherence to aseptic technique, and maintenance of laboratory equipment such as flowhoods should keep the possibility of contamination by mold and bacteria to a minimum. Widespread mold growth, for example, may indicate a mold-contaminated incubator. Regular cleaning of the incubator with detergent and 70% alcohol along with autoclaving of its shelving units helps prevent its contamination (Freshney, 1987).

Possible sources of contamination besides the operator to consider are:

1. Apparatus (dishes, flasks, and so on);
2. Biologicals (culture media, trypsin, bicarbonate);

3. Tissue (depending on source and dissection);
4. Work area (broken filter on flowhood, contaminated air, and so forth).

5. Sterilization and Decontamination

Getting rid of contamination may require sterilization. Sterilization may be achieved by:

1. Physical destruction of contaminants (dry heat, autoclaving, irradiation);
2. Physical removal (e.g., filtration of solutions such as medium);
3. Chemical destruction (e.g., antibiotics).

The availability of disposable tissue culture ware that has been sterilized by gamma-irradiation is a great convenience. As with many convenient items, however, disposables can be expensive. In addition, disposable items cannot completely replace glassware in everyday use in tissue culture laboratories.

During culture preparation or manipulation, a tub of water should be handy for immersion of used glassware. This prevents media and the like from drying onto the glassware. Prior to cleaning, cotton plugs are removed from pipets and all glassware is rinsed with water. After rinsing, remove labels and felt pen marks. This may require soaking in warm detergent for several hours then wiping off the marks with a cloth or abrasive cleaner.

Cleaning with detergents has the following advantages over cleaning with alkalines: Glassware does not have to be boiled; and glass surfaces upon which cells are to be grown do not have to be neutralized with HCl or H_2SO_4 (Paul, 1975). A detergent must be selected that is suitable for the water in your area, i.e., hard or soft water. The detergent must also be easy to rinse off and nontoxic (e.g., Microsolve, 7X, Haemosol, Stergene, RB S25, Decon 75) (Paul, J. 1975). Glassware is soaked overnight, rinsed well with tap water, then rinsed with distilled or deionized water and dried while inverted. Rinsing must be thorough to ensure complete removal of detergent.

Care must be taken with dissection instruments to prevent damage to their tips and cutting edges. Tissues should not be allowed to dry onto the instruments. When finished with, dissecting, instruments may be placed into a water-fillled beaker or pan that has its bottom lined with cotton gauze. Dropping instruments into an unlined container may cause damage to the tips. Instruments may then be cleaned by soaking in detergent, scrubbed with a toothbrush, then rinsed under tap water, followed by distilled or deionized water. Instruments have a final rinse in 95% alcohol prior to dry heat sterilization. Dry heat does not corrode cutting edges as does steam and metals are good heat conducters. Care must be taken to not expose instruments to excessive heat (≥160°C), which may damage cutting edges (Perkins, 1969).

5.1. Dry Heat Sterilization

The principle of dry heat sterilization is conduction of heat from the exterior to the interior of an object in the absence of moisture. This form of sterilization is destructive to and unsuitable for fabrics or rubber articles. However, it is ideal for oily substances, powders, metals, and most glassware (Perkins, 1969). The main disadvantage of dry heat is that it is not a rapid means of sterilization. All items to be sterilized in this manner should be well cleaned beforehand. The oven should not be overloaded and the load should be standardized to ensure equal heating of the load.

The advantages of dry heat sterilization are as follows:

1. Can be used for almost any type of glassware;
2. Eliminates the possibility of steam-related contamination;
3. Means of sterilizing items that cannot be sterilized any other way, e.g., powders and oils.

The disadvantages are:

1. Time consuming;
2. Unsuitable for certain items, such as fabrics and rubber;
3. Can be difficult to control because of the presence of cool spots in oven and some items are poor conducters of heat (Perkins, 1969).

A variety of ovens are available for use. Almost any oven will do, including a domestic model, however, an oven with forced air circulation is preferred. A temperature range of 160–180°C maintained for one to two hours is generally acceptable for sterilization of material in a forced air circulation type oven (Freshney, 1987). This time period does not take into account the time required for the materials to reach the sterilizing temperature—which may be quite considerable. Gravity convection type ovens are slower to heat and are subject to a less uniform temperature throughout the chamber. When using such an oven to sterilize glassware, care should be taken not to overload the oven, and a temperature of 200°C held for a longer period of time is suggested. One may find measuring the temperature and using sterilization indicators useful, especially if there is some doubt as to the efficiency of the oven. Regardless of the type of oven used, it must be remembered that the center of the load is least accessible and space must be left between items. Having two small ovens may prove to be advantageous, as they allow for more rapid, uniform heating, plus are economical when a small load is required.

5.1.1. Preparation of Materials

In preparing pipets for dry heat sterilization, the first step is to plug the back ends with cotton batting. Pasteur pipets are put into their glass containers and graduated pipets are sorted and put into their appropriately labeled metal canisters. Glass Petri dishes may be wrapped in foil or paper designed specifically for dry heat sterilization. Test tubes have their orifices covered in foil and are inserted into test tube racks. Alternatively, they may be sterilized, minus the foil, in metal canisters. Beakers and graduated cylinders likewise have their open ends covered with foil during sterilization and storage (Freshney, 1987). Commonly glass coverslips are sterilized in glass Petri dishes. The problem is many of the coverslips become fused together during heating. In our experience, we have found glass Columbia jars, often used in staining, are an ideal receptacle for sterilization and storage of glass coverslips.

5.2. Autoclaving

The effectiveness of autoclaving results from the combination of moisture with heat. Its advantages are:

1. Rapid sterilization;
2. May be used for sterilizing rubber and fabric;
3. Does not leave a toxic residue provided water used for production of steam is of good quality. Impurities or toxic residues such as copper may come from water pipes or may be a result of insufficient water purification.

The basis for sterilization in autoclaving is direct steam contact. Thus, the most isolated points require direct exposure at the correct temperature for the minimum amount of time for sterilization to occur. The standard is 115–120°C at 15 psi for 15–20 min with a pre- and postvacuum cycle (Freshney, 1987).

5.2.1. Preparation of Materials

Small items, such as rubber stoppers, may be placed into Petri dishes then wrapped in foil. Bottles with screw caps are autoclaved with the cap slack and covered with foil. The caps are tightened through the foil when the bottles have cooled. Solutions that are not heat sensitive, such as phenol red, may also be autoclaved. However, the evacuation step used to assist the drying process must be omitted in this instance. Beakers have their orifices covered in foil, as described previously (Freshney, 1987).

5.3. Sterilization Indicators

Sterilization indicators are a useful adjunct to autoclaving and dry heat sterilization. Autoclave tape is not a sterilization indicator. However, it does clearly indicate whether or not an item has been autoclaved. Some types of sterility indicators are glass tubes in which there is a liquid (Browne's sterilizer tubes, for example) (Adams, 1983) or a pellet of sulfur. These are not as reliable as paper sterilization indicators. Paper indicators are chemically treated such that they change color when exposed to the correct combination of temperature with moisture over time. Biological tests may be performed in which spores within a small vial are autoclaved then subsequently incubated. If there is no growth within a week, sterility is confirmed (Perkins, 1969). The major drawbacks of this indicator are the length of time required for confirmation of sterility and the possibility of breakage of the vial within the autoclave.

In general, autoclave tape obviates any doubts as to which articles have been autoclaved and which have not. Paper indicators are considered to be good indicators of sterility, plus they provide the results far more rapidly than biological tests (Perkins, 1969).

5.4. Physical Removal: Filtration

There is a variety of filters available, most designed for specific purposes. The question is then, how do you decide which is the appropriate type of filter? There are five basic points to consider:

1. Pore size of filter;
2. Chemical compatibility;
3. Hydrophobic vs hydrophilic filter;
4. Autoclave vs presterilized;
5. Toxicity of filter.

A pore size of 0.2 µm provides the sterilization necessary for tissue culture. Chemical compatibility refers to the compatibility of the filter with what is to be filtered. This becomes important in toxicological studies in which certain drugs may be dissolved in acids. If one is sterilizing media or other aqueous solutions, a hydrophilic filter is required. Hydrophobic filters are ideal for sterilizing gases, CO_2, for example. Although autoclave filters are less expensive, presterilized disposable filters are convenient and more reliable (Paul, 1975). If one is using filtration as a means of sterilization, it only makes sense to use the filter that provides the least possibility for failure, i.e., the presterilized disposable filter. Wetting agents and Triton detergent may be present on some filters. If in doubt as to their presence, water can be used to remove these toxic agents. The best way to avoid this possibility is to use a filter that is known not to contain these toxic agents.

5.5. Alcohol and Flaming

Swabbing with alcohol (ethanol) and flaming are two commonly used techniques for sterilization in performing tissue culture procedures. Alcohol is a surface disinfectant. Swabbing work surfaces with 70% alcohol is a means of removing dust and keeping the area clean. Flaming with a Bunsen burner fixes accumulated dust in place.

When wiping a work surface or various other items with alcohol, it may be useful to keep in mind that one is not scrubbing a floor. That is, simply wipe from one side to another working from the top to the bottom of the surface. Similarly, a bottle is wiped in a spiral fashion from the neck area toward the base. During tissue culture procedures, any spillage should be wiped up by a cloth dampened with alcohol. This keeps the area clean and prevents production of aerosols.

Ninety-five percent alcohol in conjunction with flaming is an effective means of sterilizing dissection instruments. First, clean instruments are soaked in a beaker of alcohol. Care must be taken to avoid getting alcohol onto the handles of the instruments. The alcohol is ignited by a brief passage of the instrument tip through the Bunsen burner flame. The now sterile instrument is kept in a sterile Petri dish such that the working area or the instrument is in the dish and the handle points outward. The lid of the Petri dish covers the sterile working area of the instruments.

The necks of vessels are flamed all the way around before and after removal of their caps. Culture vessels such as T-flasks are flamed before and after loosening the cap. Similarly, beakers, flasks, and cylinders are flamed around their orifices before use. The neck of a bottle or vessel is flamed once more before replacing the cap. One may want to give consideration as to the placement of the Bunsen burner. This is because heat from the burner may interfere with laminar air flow. It is for this reason that some workers prefer to have the burner just outside of the flowhood (Douglas and Dell'Orco, 1979).

5.6. Chemical Destruction: Use of Antibiotics

As with the use of a flowhood, use of antibiotics should be viewed as an adjunct and not a replacement for good technique. For cultures started from tissues obtained from slaughterhouses or human patients, a high initial concentration of antibiotics is necessary to eliminate bacteria that are already present. Consideration should be given to the following:

1. Whether antibiotic is against gram-positive and/or gram-negative, as well as whether it is cytolytic or cytostatic;
2. Possible interference of antibiotic with cell growth and metabolism;
3. Stability of antibiotic in medium.

Although there is a variety of antibiotics available, the combination of sodium penicillin G (100 µg/mL) with streptomycin (50–100 µg/mL) in media meets general antibiotic needs (Paul, 1975; Perlman, 1979). For tissues obtained from an abattoir, for example, an initial addition of Amphotericin B (Fungizone®) at a concentration of 2.5 µg/mL to serum-containing medium should be sufficient to prevent fungal and yeast contamination (Paul, 1975). If a serum-free medium or balanced salt solution is being used, the concentration of antimicrobial agents should be decreased by about one-third to avoid cytotoxic effects (Perlman, 1979). In general, antifungal, antiyeast, and antibacterial agents remain stable in media for 3–5 d at 37°C (Perlman, 1979).

6. Appendix

Antimicrobial agents
Gibco Laboratories, 3175 Staley Road, Grand Island, NY 14072
Sigma Chemical Company, PO Box 14508, St. Louis, MO 63178
Columbia staining jars
A.H. Thomas Company, PO Box 99, Swedesboro, NJ 08085-0099
Deionizers
Barnstead, 2555 Kerper Blvd., Dubuque, Iowa, 52001
Bellco Glass Inc., 340 Edrudo Road, Vineland, NJ 08360

Corning Glass Works, Corning, NY 14831
Elga Group, The, Lane End, High Wycombe, Bucks, UK
   Detergents
Decon Laboratories Ltd., Conway Street, Hove, E. Sussex, BN3 3LY UK
Diversey Ltd., Weston Favell Centre, Northhampton, NN3 4PD UK
Flow Laboratories Inc., 1710 Chapman Ave., Rockville, MD 20852
   Filters
Gelman Sciences, Inc., 600 S. Wagner Road, Ann Arbor, MI 48106
Millipore Corp., Ashby Road, Bedford, MA 01730
Nucleopore Corp., 7035 Commerce Circle, Pleasanton, CA 94566
Sartorius GmbH, PO Box 19, 3440 Gottingen, Germany
   Freezers
Fisher Scientific Co., 711 Forbes Ave., Pittsburgh, PA 15219
New Brunswick Scientific Co., Inc., 44 Talmadge Road, Edison, NJ 08817
   Glassware
Bellco Glass, Inc., 340 Edrudo Road, Vineland, NJ 08360
Corning Glass Works, Corning, NY 14831
   Incubators
Fisher Scientific Co., 711 Forbes Ave., Pittsburgh, PA 15219
Flow Laboratories, Inc., 1710 Chapman Ave., Rockville, MD 20852
Forma Scientific, PO Box 649, Marietta, OH 45750
Lab-Line Instruments Inc., 15 & Bloomingdale Ave., Melrose Park, IL 60160 (Burkhard Scientific Sales Ltd., UK)
Napco, 10855 S.W. Greenburg Rd., Portland, OR 97223
New Brunswick Scientific Co., Inc., 44 Talmadge Road, Edison, NJ 08817
Precision Scientific Co., 3737 West Cortland St., Chicago, IL 60647
Tekmar Co., P.O. Box 37202, Cincinnati, OH 45222 (W.C. Heraeus GmbH, Germany)
   Laminar Flow Hoods
Baker Co., Inc., Sanford Airport, Sanford, ME 04073
Flow Laboratories, Inc., 1710 Chapman Ave., Rockville, MD 20852
Gelman Sciences, Inc., 600 S. Wagner Road, Ann Arbor, MI 48106
   Media and Serum

Flow Laboratories, Inc. 1710 Chapman Ave., Rockville, MD 20852
Gibco Laboratories, 3175 Staley Road, Grand Island, NY 14072
Sigma Chemical Company, PO Box 14508, St. Louis, MO 63178
    Microscopes
E. Leitz, Inc., Link Drive, Rockleigh, NJ 07647
Nikon, Nippon Kogaku K.K., Fuji Bldg., 2-3 Marunouchi 3-chome. Chiyoda-Ku, Tokyo, Japan
Olympus Corporation of America, 4 Nevada Drive, New Hyde Park, NY 11042
Carl Zeiss, Inc. 444 Fifth Ave., New York, NY 10018
    Pipets
Bellco Glass Inc., 340 Edrudo Road, Vineland, NJ 08360
Corning Glassworks, Corning, NY 14831 USA
Fisher Scientific Co., 711 Forbes Ave., Pittsburgh, PA 15219
    Pipetting devices
Bellco Glass Inc., 340 Edrudo Road, Vineland, NJ 08360
    Reverse osmosis equipment
Barnstead, 2555 Kerper Blvd., Dubuque, Iowa, 52001
Elga Group,The, Lane End, High Wycombe, Bucks, UK
Fisons Scientific Apparatus, Bishop Meadow Road, Loughborough, LE11 ORG, UK
Millipore Corp, Ashby Road, Bedford, MA 01730
    Sterilization indicator tape
Fisher Scientific Co., 711 Forbes Ave., Pittsburgh, PA 15219
    Sterility indicators
Bennet & Co., Brempton Common, Nr. Reading, Berks, UK
    Stills
Corning Glassworks, Corning, NY 14831
Fisons Scientific Apparatus, Bishop Meadow Road, Loughborough, LE11 ORG, UK
    Tissue culture grade plastics
Bellco Glass Inc., 340 Edrudo Road, Vineland, NJ 08360
Corning Glassworks, Corning, NY 14831
Flacon (Becton Dickinson & Co.) Oxnard, CA 93030
Linbro Lux (Lux Scientific Corp.) 1157 Tourmaline Drive, Newbury Park, CA 91320
Nunc (Gibco Laboratories) 3175 Staley Road, Grand Island, NY 14072

Acknowledgment

D. E. H. is the recipient of a research traineeship from the Saskatchewan Heart and Stroke Foundation. W. W. is the recipient of a scientist award from the Medical Research Council of Canada.

References

Adams R. L. P. (1983) *Cell Culture for Biochemists,* Elsevier, New York.

Barkley W. E. (1979) *Safety Considerations in the Cell Culture Laboratory, in Methods Enzymology,* Vol. LVIII (Jakoby W. B. and Pastan I. H., eds.), Academic, New York, pp. 36–43.

Douglas W. H. J. and Dell'Orco R. T. (1979) *Physical Aspects of a Tissue Culture Laboratory,* in *Methods in Enzymology,* Vol. LVIII (Jakoby W. B. and Pastan I. H., eds.), Academic, New York, pp. 3–18.

Freshney R. I. (1987) *Culture of Animal Cells: A Manual of Basic Technique,* 2nd ed. Liss, New York.

Paul J. (1975) *Cell and Tissue Culture,* 5th ed., Churchill Livingstone, New York.

Perkins, J. J. (1969) *Principles and Methods of Sterilization in Health Sciences,* 2nd ed., Charles C. Thomas Publisher, Springfield, IL.

Perlman D. (1979) *Use of Antibiotics in Cell Culture Media,* in *Methods in Enzymology,* Vol. LVIII (Jakoby W. B. and Pastan I. H., eds), Academic, New York, pp. 110–116.

Identification
of Cell Types in Neural Cultures

Colin J. Barnstable

1. The Problem of Heterogeneity in Neural Cell Cultures

Tissue culture provides an opportunity to study the functions of the nervous system under strictly controlled conditions. As amply documented in other chapters in this volume, the culture medium, the surface upon which the cultures are growing, and various cellular and soluble factors can all modulate the growth and behavior of neural cells. A major variable in these cultures, however, remains the heterogeneity of the tissue used for the culture. Without knowing the relative proportions of neurons and glia, or the relative proportions of different subclasses of neurons, it can be difficult to interpret experimental results. If two types of electrophysiological responses are found under different growth conditions, is it because cell properties have become altered, or because different cell types are now present? If the amount of a particular neurotransmitter is altered by the addition of a certain growth factor, is it because the factor affected the synthesis, storage, or release of the transmitter, or because the proportion of cells utilizing this transmitter has altered?

Given such variables, it is not surprising that the most successful use of neural cell cultures has been from regions of the nervous system that do not show a lot of heterogeneity. The superior cervical ganglion of the sympathetic nervous system is

From: *Neuromethods, Vol. 23: Practical Cell Culture Techniques*
Eds: A. Boulton, G. Baker, and W. Walz ©1992 The Humana Press Inc.

almost entirely composed of a single type of neuron. Studies of the membrane and biochemical properties of these cells have shown differences under varied culture conditions (Furshpan et al., 1976; Patterson and Chun, 1977). Because only one cell type is present, these differences almost certainly reflect different behaviors of the same cell type. For most of the cultures discussed in this chapter, and elsewhere in this volume, such automatic identification of cells is not possible. The other ways by which cell types can be identified form the subject of this chapter.

Most successful cultures of neural cells are set up from embryonic or neonatal tissue. At these stages, many of the cell types have not fully matured. It is often not enough knowing what cell types are present, but it can also be important to know their state of differentiation.

2. Types of Cell Markers Available

2.1. Morphological Markers

Many cell types in the nervous system have a unique morphology, and there are clear examples where this is preserved in tissue culture. One of the most striking examples of this is in the culture of rod photoreceptors from mature salamander retina (Bader et al., 1978; MacLeish et al., 1983). These highly specialized cells can be isolated from enzymatically digested retinas and appear morphologically normal. They can maintain this normal morphology and remain functional, as judged by light sensitivity, for extended periods in culture. Another very characteristic type of neuron is the cerebellar Purkinje cell. In cultures of mouse cerebellum, these cells can be easily identified by their size and very dense network of dendritic processes (Messer et al., 1984).

Such unique morphological features are not common. Nevertheless there are several types of culture where morphology has proven to be an adequate way of identifying cell types. By careful dissection of the hippocampus, it is possible to obtain cell cultures that consist primarily of pyramidal cells. The size and shape of the cell bodies, the presence of distinct dendrites and axons all serve to identify these cells (Bartlett and Banker,

1984). Cultures of chick spinal cord are more typical of the mammalian CNS in that they contain a number of different cell types. Motor neurons can be distinguished from the other cell types by their size, but this is not an absolute criterion in culture. Some of the largest cells in these cultures were identified electrophysiologically as motor neurons, but the sizes of GABAergic interneurons showed considerable overlap (Farb et al., 1979).

Perhaps a more typical culture is shown in Fig. 1. Neonatal rat retina cells were plated onto a collagen substrate and grown for 7 d. As the figure shows, cell morphologies range from small cells without processes, through bi- and tripolar cells, to multipolar cells, all of which preferentially adhere to an underlying layer of flat cells. To use such cultures effectively, it is essential to have a way of unambiguously identifying each of the cell types present.

A common feature of the examples in which morphology is useful is that the cells have already differentiated before the tissue is dissociated. Cultures from early embryonic tissue that differentiate in vitro do not generally develop such characteristic features.

2.2. Enzyme Markers

There are numerous enzymes that can distinguish different types of neural cells. Examples of these are the enzymes involved in neurotransmitter metabolism to distinguish different subclasses of neurons, isozymes of common enzymes, such as enolase, that are found in neurons but not glia, and enzymes involved in myelin synthesis to distinguish oligodendrocytes and Schwann cells from other glia.

Although it is relatively easy to carry out biochemical assays for these enzymes, there are not many histochemical methods that can utilize enzymatic activity to localize a cell type in a mixed culture. One of the few successful examples of this is acetycholinesterase. By carrying out the reaction with an acetylthiocholine derivative, a dense insoluble product is formed in the cells expressing the enzyme. Because the enzyme is extremely active, the assay is very sensitive. Specific pharmacological methods are available to ensure that acetylcholinesterase

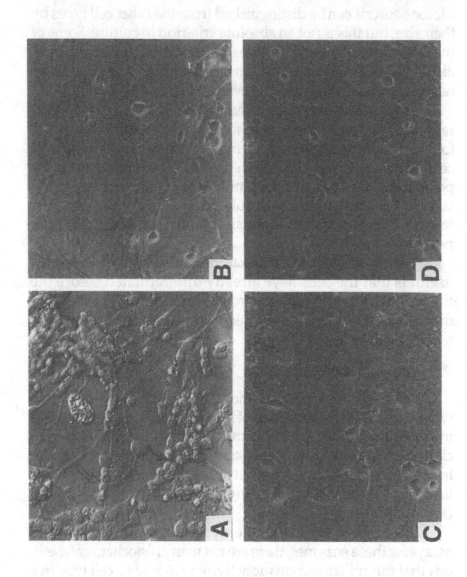

Fig. 1. Monolayer cultures of neonatal retinal cells. (A) View showing clusters of small cells (solid arrow) and such cells attached to processes (open arrow). (B–D) Typical bipolar and tripolar cells with phase-bright cell bodies and long processes. Occasionally, cells with more pyramidal morphology were observed (open arrows in B and D). (E) Multipolar cell with long processes. (F) Flat cells. Panels B–F were photographed from less dense areas of cultures to make it easier to see the processes from individual cells. Scale bars, 50 μm. Taken from Akagawa and Barnstable (1986).

is being visualized rather than a nonspecific esterase. In spite of this, the method does have two limitations. The first is that the enzyme is relatively ubiquitous. It is certainly not confined to cholinergic neurons, and in many culture systems several classes of neurons might express it. The second problem is that the enzyme can also be found in extracellular forms. This can give rise to labeling patterns in which cells might appear to express the enzyme, but are in fact, adjacent to enzyme secreted from other cells and adherent to extracellular matrix.

2.3. Pharmacological Markers

Three classes of pharmacological agents have had widespread use as cell markers. These are toxins, neurotransmitters, and transmitter receptor ligands.

Of the toxins, the most specific to be used widely has been α-bungarotoxin. This polypeptide binds with very high affinity and specificity to the nicotinic acetylcholine receptor, and it can be conjugated to fluorescent dyes or enzyme markers for microscopic identification of binding sites, or to radioactive tags for subsequent autoradiography (Burden et al., 1979; Jessell et al., 1979). Other toxins that have been used extensively include tetanus toxin and cholera toxin, both of which bind to gangliosides on the cell surface. In mature cell cultures, tetanus toxin labels all neurons, but not glia (Dimpfel et al., 1977). With immature cells the situation is less clear, and some glial cell precursors may also react. Cholera toxin, which reacts with GM1 ganglioside, has been used to monitor granule cell differentiation in the cerebellum, since mature granule cells express much higher levels of the ganglioside than the immature precursors (Willinger and Schachner, 1980).

This method works well in tissue sections where the two labeling levels can be compared in adjacent regions, but it is less easy to use in culture where it can be harder to measure absolute levels of binding to individual cells.

The most effective way of using neurotransmitters to label neural cell cultures is to add them to the medium and allow them to be concentrated by high-affinity uptake in appropriate cells.

Cells using neurotransmitters, such as glycine, glutamate, GABA, catecholamines, and indoleamines, all have high-affinity transport mechanisms with affinities in the micromolar range. Incubation of cultures in submicromolar concentrations of radioactive neurotransmitter for as little as 5 min can give labeling that is easily detected by autoradiography. Because all the compounds contain a primary amine group, they can be fixed within cells with glutaraldehyde or paraformaldehyde fixatives. For the catecholamines and indoleamines, an additional method of detection is available. By using aldehyde fixatives containing glyoxylic acid, fluorescent derivatives of the compounds are created and can be viewed microscopically (Lindvall et al., 1973).

For the visualization of cholinergic neurons, a different method has to be used. These cells have a high-affinity uptake system for the precursor choline that is readily available in radioactive form. Neither choline nor acetylcholine can be fixed with standard aldehyde fixatives. Two approaches have been devised to overcome this. One is to use long postlabeling periods to allow some of the radioactive choline to become incorporated into phospholipids. The second is to freeze-dry the cultures and carry out autoradiography with dry films for emulsion (Baughman and Bader, 1977). Both methods are difficult and have been superseded by immunocytochemical methods, such as those described later in this chapter.

The specificity of cell labeling can often be enhanced pharmacologically. For example, glia as well as GABAergic neurons have transport systems for GABA. Accumulation of GABA by glia in cultures can be selectively inhibited by β-alanine.

One advantage of neurotransmitter uptake and autoradiography as a cell labeling method is that the cell processes, as well as the cell body, can be detected. We have previously used this labeling method to identify glycinergic and gabaergic subclasses of amacrine cells in cultures of neonatal rat retina (Akagawa and Barnstable, 1987a,b). The cells that accumulated glycine had short, highly branched processes, whereas those that accumulated GABA had longer, less-branched processes. Interestingly, these are similar to the respective morphologies found in the intact retina. This suggests that under appropriate culture conditions, cell types can maintain complex structural features, although the nature of the features determining morphology is unknown.

The third class of pharmacological markers uses transmitter receptor ligands. Although these have been used extensively to show regional and laminar differences in receptor density in tissue sections, their use in tissue culture is rare. The major reason for this is that the density of receptors on a single cell is often too low to give sufficient levels of bound radioactive ligand that can be detected autoradiographically. An increasing number of antibodies are becoming available that detect subclasses of receptors and have sufficient sensitivity to be useful in tissue culture.

2.4. Immunological Markers

Antibodies provide by far the most useful way of characterizing cells in culture. Not only is a large battery of antibodies readily available, but the methods for producing antibodies against any cell type of interest have now become routine. The specificity and sensitivity of antibodies are ideally suited for studies in culture. Perhaps even more important, antibodies can be used in combinations that allow even more precise definition of cell types and their physiological or developmental status. A further advantage of antibodies that extends beyond studies in culture is that the same reagent can be used to identify cells in intact tissue, and can be used for biochemical studies.

As will be clear from the rest of this chapter, antibodies have become available against a wide variety of neurotransmitters, transmitter synthesizing enzymes, and receptors. Thus, in many ways, antibodies have replaced many of the other categories of cell markers.

3. The Range of Cell-Type Specific Antibodies Available

In 1982, I reported that most of the known monoclonal antibodies reacting with neural antigens had been discussed at a recent meeting (Barnstable, 1982). Such a statement cannot be made now because the numbers have multiplied so much. From numerous studies in many species and in many areas of the nervous system, it has become clear that there is extensive molecular diversity between cell types in the nervous system that can be detected immunologically, and thus, it is possible to generate antibodies of sufficient specificity to allow precise marking of cell types.

At the simplest level, good antibodies are available to distinguish glia from neurons. Astrocytes can be labeled by antisera or monoclonal antibodies against the intermediate filament protein GFAP. It is possible to be more precise by using antibodies that distinguish mature astrocytes from either embryonic radial glia (rat 402; Hockfield and McKay, 1985) or from retinal Müller glia (R4 and R5; Dräger et al., 1984). Oligodendrocytes can be labeled by antibodies against galactosylcerebroside, or can be followed through several steps of maturation using antibodies against a number of cell surface antigens (Ranscht et al., 1982; Sommer and Schachner, 1981). Microglia share a number of properties with macrophages. Thus, they can be labeled by antibodies originally generated against macrophages (Dräger, 1983; Hume et al., 1983), or by a functional method involving phagocytosis of fluorescent latex beads (Giulian and Baker, 1986). The labeling of neurons is more complex. A number of general neuronal markers have been described, but upon close examination, few of them are able to label all or only neurons. When antibodies against Neuron Specific Enolase (NSE) are used to label tissue sections of brain, it is clear that there are differences in intensity of labeling of different cell types. This can result in some cells in culture not being labeled, and thus, misclassified. Similarly, many but not all neurons express neurofilaments. Small interneurons have fewer or no neurofilaments, as compared with larger pyramidal cells. Use of neurofilament antibodies to label cells in culture will, therefore, underestimate the number of neurons and the extent of neuronal process outgrowth.

In some species, antibodies against the neural cell adhesion molecule N-CAM can provide a good neuronal marker. This is particularly true in cultures of the chick nervous system (Edelman et al., 1983). In mammals, N-CAM is also expressed on glia, and so cannot be used as definitively (Goridis et al., 1983).

The most common need for cell markers, however, is to distinguish different types of neuron or neural cells at different stages of development. There are now antibodies available against almost all known neurotransmitters and most of their biosynthetic enzymes. Even amino acid neurotransmitters, such as glutamate and aspartate, can be fixed and labeled because they are at a higher concentration in cells that use them as transmitters than other cells.

There is also an increasing number of antibodies available against cell-type specific molecules of unknown function. For example, our own work has led to the production of antibodies with which we can label each of the five major neuronal and two glial cell types of the rat retina Some markers have been identified and are well-characterized. The visual pigment protein opsin is a good marker for rod photoreceptors (Fig. 2). It is an integral membrane protein and is expressed early in the cell differentiation pathway (Hicks and Barnstable, 1987; Hicks et al., 1989). Importantly, it is also expressed well by rod photoreceptors in culture, as shown in Fig. 3 (Akagawa and Barnstable, 1986; Akagawa et al., 1987; Sparrow et al., 1990). Similarly, the membrane glycoprotein Thy-1 is an excellent marker for retinal ganglion cells. In many regions of the CNS, all types of neurons seem to express Thy-1 so that it can be used as a general neuronal marker (Fields et al., 1978). In the retina, only ganglion cells express Thy-1, and so it can be used as a cell-type specific marker (Barnstable and Dräger, 1984; Leifer et al., 1984). Figure 4 shown the labeling of neuritic outgrowth from retinal explant cultures with polyclonal and monoclonal anti-Thy-1 reagents (Sparrow et al., 1990). Horizontal cells share expression of neurofilaments with ganglion cells and a carbohydrate epitope, HNK-1, with amacrine cells. Thus, they can be uniquely identified by a combination of the two markers. Similarly, bipolar cells share a cell surface antigen, RET-B2, of unknown function with photoreceptors, but do not express opsin (Akagawa and Barnstable, 1986). Amacrine cells express a 35-kDa membrane antigen recognized by the antibody HPC-1 (Barnstable et al., 1985). Although the function of this molecule is not known, it is expressed early and

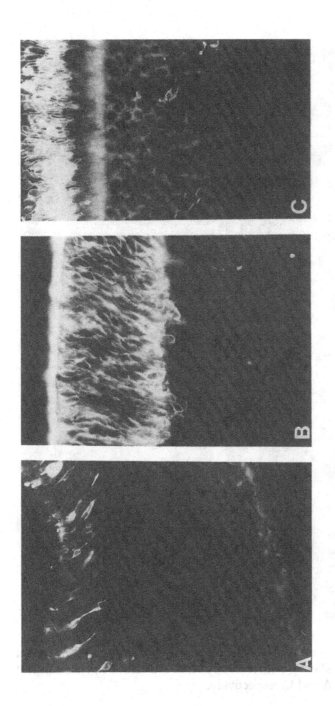

Fig. 2. Changing expression of photoreceptor antigen RET-P1, a determinant on the N-terminal portion of opsin. (A) At PN2, a few immature cells are labeled at the ventricular surface of the tissue. (B) By PN8, most of the cells in the forming photoreceptor layer are labeled, although they still have an immature morphology, and the outer segments have only just begun to extend. (C) In the adult, the whole rod photoreceptor is labeled. Taken from Barnstable (1987).

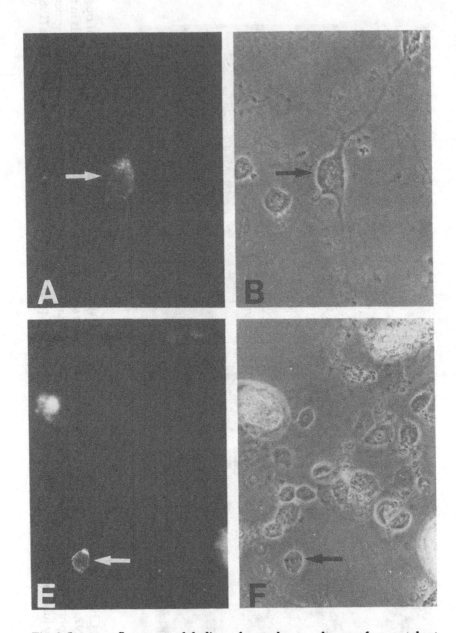

Fig. 3. Immunofluorescent labeling of monolayer cultures of neonatal rat retina. (**A and C**) Labeling with antibody HPC-1. (**B and D**) phase-contrast of same field as in **A** and **C**, respectively.

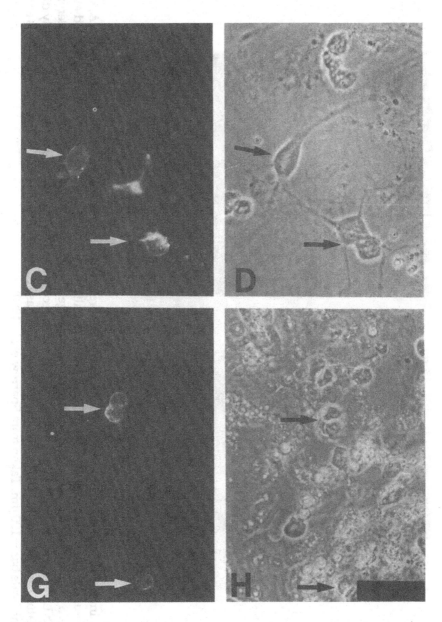

(E and G) Labeling with antibody RET-P1. (F and H) Phase-contrast of the same field as E and G, respectively. Calibration bar = 25 μm. Taken from Sparrow et al., 1990.

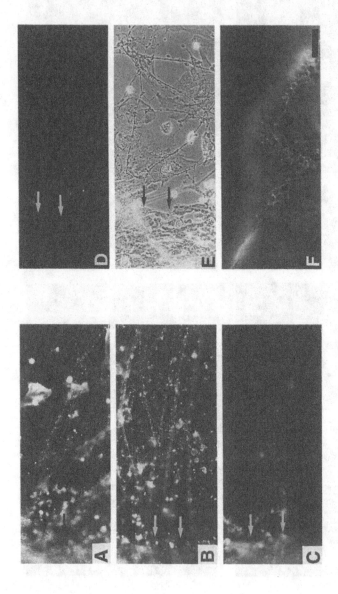

Fig. 4. Immunofluorescent labeling of neuritic outgrowth from retinal explant using antibodies to Thy-1. (A, C, D, and E) Neural retinae were explanted at E13 and grown in culture for 7 d. (B and F) Neural retinae explanted at E16 and grown in culture for 7 d. Arrows indicate approximate edge of explant. (A,B) Labeling with polyclonal antiserum to Thy-1. (C) Labeling with monoclonal antibody 2G12, which recognizes the Thy-1 in tissue sections and freshly dissociated mature ganglion cells. (D) Rabbit serum control. (E) Phase contrast of field in D. (F) Thy-1 positive cells at edge of explant. Scale bar = 100 μm. Taken from Sparrow et al. (1990).

is found in cell cultures (e.g., Akagawa and Barnstable, 1986), retinal transplants (Aramant et al., 1990), and transformed cells (Hammang et al., 1990).

Comparison of retinal cultures labeled with HPC-1 or RET-B2 clearly shows the need for such markers. The bipolar or amacrine cells labeled in Fig. 5 cannot be distinguished morphologically, and the relative numbers of the two cell types can vary according to the age or developmental stage of the tissue used to prepare the cultures.

As mentioned previously, retinal astrocytes and Müller glia can be distinguished with appropriate antibodies. In the intact retina, only astrocytes express GFAP. After injury Müller glial cells also become GFAP+ve. This result means that GFAP is not a good marker to distinguish these cell types because its pattern of expression depends on the cell physiology. Other markers reactive with either membrane or cytoplasmic antigens can distinguish the two cell types. RET-G1 is a glial cell surface marker that is common to Müller cells and astrocytes. RET-G2 and RET-G3, on the other hand, are restricted to Müller cells (Barnstable, 1980). The intermediate filament associated protein R5 is found in both astrocytes and Müller cells, but a similar protein, R4, is found only in Müller cells (Dräger et al., 1984). The labeling pattern of these antibodies in culture, together with a method of preparing Müller cell cultures, has recently been reported by two laboratories (Hicks and Courtois, 1990; Sarthy, 1985).

It is important to point out that the retina is not unique in its antigenic diversity. From our own work we have used antibodies and lectins to obtain a good characterization of subpopulations of GABAergic interneurons in several areas of mammalian cerebral cortex (Arimatsu et al., 1987; Kosaka et al., 1989; Naegele and Barnstable, 1991; Kosaka et al., 1990). In cerebellum, an antigen expressed in a subset of Purkinje cells has been described (Leclerc et al., 1988). In the leech, single neurons in each segmental ganglion can be identified with monoclonal antibodies (Zipser and McKay, 1981). The general conclusion seems to be that good cell type specific markers can be found whenever sufficient effort is put into looking for them.

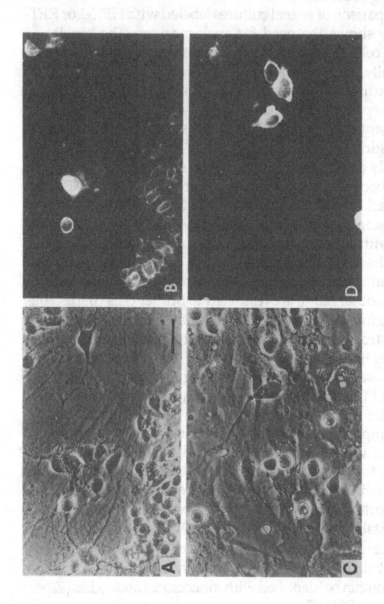

Fig. 5. Immunofluorescent labeling of monolayer cultures with antibody RET-B2 (A, B) or HPC-1 (C, D). (A, C) Phase contrast; (B, D) fluorescence. RET-B2 labels both photoreceptors and a subset of other cells of neuronal morphology. HPC-1 labels a subset of cells of neuronal morphology. Scale bar, 20 μm. Taken from Akagawa and Barnstable (1986).

4. What Antibody Should I Use?

4.1. Polyclonal or Monoclonal?

Although the differences between monoclonal and polyclonal antibody reagents are well known, it is worth examining how some of these differences might affect the use of such reagents to label cell cultures. Consider the case of antibodies against a single protein. For a monoclonal antibody, all the reacting antibody molecules will bind to the same site on the antigen and will bind with the same affinity. A polyclonal reagent may have many antibodies that can react with several epitopes and with several affinities for each epitope. When the amounts of antigen expressed by a cell are low, this can result in greater sensitivity from the polyclonal reagent because more antibody molecules can be attached. This is exemplified in Fig. 4 where the labeling of identical cultures is much stronger with a Thy-1 antiserum than with a Thy-1 monoclonal antibody. All too frequently, however, this advantage is counterbalanced by the problems of crossreactivity. In an antiserum, a subpopulation of antibodies may also react with another molecule with a different affinity. This will give rise to differences in the observed labeling patterns according to the concentration of antiserum, length of incubation, and temperature at which the reaction is carried out. These variables can be controlled, but they can cause problems as serum properties change with storage or different investigators carry out experiments under different conditions. True crossreactivity, when an epitope is shared between two different molecules, can also give rise to misleading conclusions. Several workers reported that horizontal cells in the retina contained the peptide α-MSH. It later turned out that many anti-α-MSH sera crossreacted with an epitope on neurofilament proteins that were the molecules being detected in these cells (Dräger et al., 1983).

4.2. Cell Surface or Cytoplasmic Markers?

The choice of antibodies to use or produce depends on the experiments being carried out. If the aim is to label cells intensely, even into their fine processes, then a cytoplasmic marker, such as GFAP (for astrocytes) or neurofilaments (for neurons), would be best. Cell surface markers give fainter labeling but have a variety of advantages, such as being useful for living cells, and for manipulating the composition or physiological behavior of cells. In most cases, cell surface markers are evenly distributed over the cell body and processes even though they may demonstrate specific cell membrane domains in vivo.

A number of cytoplasmic markers are restricted to dendrites, cell bodies, or axons. One of the most commonly used cytoplasmic markers has been the microtubule associated protein MAP-2. This protein is found exclusively in dendrites, and so can be used to differentiate dendrites and axons in culture (Cáceres et al., 1986). Tau proteins, another group of microtubule associated proteins, are expressed preferentially in axons. They have not, however, been used extensively as axonal markers in culture.

4.3. Should I Make My Own?

There are three methods of obtaining a panel of antibodies suitable for labeling cells in culture. Many good antibodies against standard markers, such as neurofilaments, GFAP, Thy-1, and a variety of neurotransmitters and their biosynthetic enzymes are available commercially. Alternatively, one may approach colleagues and ask for antibodies. Most, but not all, laboratories are willing to share reagents, including cell type specific antibodies. Sending out antibodies can be very time-consuming and so requests can be held for a month or so until there is some slack time in the lab. On the other side, such delays can mean several months' delay in a laboratory's ability to assemble all the reagents to carry out a series of experiments that involve cell labeling.

Faced with the possibility of such delays, it is worth considering the advantages and disadvantages of making one's own antibodies. The major advantage, of course, is that there will be an adequate supply of reagent. Detailed protocols for immunization, antibody production, and characterization have been collected together in a recent volume (Harlow and Lane, 1988).

5. The Use of Antibodies as Cell Markers in Cultures

Antibody labeling experiments can be carried out as direct or indirect procedures. Direct labeling procedures use antibodies coupled to an appropriate fluorochrome or enzyme. Indirect procedures use unlabeled primary antibody followed by a secondary antiimmunoglobulin that is coupled to the fluorochrome or enzyme. Indirect procedures have the important advantages that only one or a few reagents need to be coupled and, they are more sensitive because several secondary antibody molecules can bind per primary molecule. In this Section I will discuss the various steps of indirect immunolabeling of neuronal cultures.

5.1. Preparation of Cultures for Labeling

5.1.1. Rinsing

If living cells are to be labeled, then all that is required is that growth medium and any cell debris are removed by gently washing with either fresh medium or a serum-free salts solution. It is, however, important to use solutions containing calcium and magnesium to prevent cell detachment.

The shearing action of fluid movement across culture wells or dishes can damage fine neural structures. A gentle and effective way of washing the cultures is to set them at a shallow angle and aspirate medium from the lower edge while gently adding buffer or medium to the upper. The temperature of the wash solution is also important. Adding ice-cold buffer to a culture with many fine processes can lead to process detachment and retraction.

5.1.2. Fixation

Most labeling of cultures is carried out after fixation. Some of the many fixatives are listed below.

1. *Paraformaldehyde.* This can be used at concentrations between 1–4%. It can be made by heating solid paraformaldehyde in a phosphate buffered saline or 0.1*M* sodium phosphate

buffer at pH 7.4 to approx 80°C in a fume hood. Formation of solution is aided by adding a few drops of 1N NaOH. When the solution has cooled, the pH can be readjusted with 1N HCl. Paraformaldehyde fixatives are often used ice-cold, although it is debatable whether this is either necessary or advantageous for cells in culture.

2. *Glutaraldehyde.* This preserves morphology better than paraformaldehyde because it is much more efficient at covalently crosslinking proteins. For electron microscopic studies, glutaraldehyde is almost essential. Its efficiency at crosslinking proteins is also one of its disadvantages for immunocytochemical labeling. Many more antigenic sites can be blocked or masked by the crosslinking. In addition, in tissue explants or reaggregates, antibodies will penetrate less well through glutaraldehyde-fixed tissue. Glutaraldehyde is generally used at concentrations from 0.1–2%. It is worth remembering that a 1% solution of glutaraldehyde is approximately 0.1M, orders of magnitude higher than the concentrations of reactive lysine residues in most tissue cultures.

3. *Cold acetone.* Acetone works as a fixative by denaturing proteins and removing some lipid so that membranes are disrupted. It is not a good fixative for preserving ultrastructure, or when antigenic determinants require intact three-dimensional structures of proteins. On the other hand, it does preserve structures, such as the cytoskeleton, with many antigenic sites intact.

4. *Acetic acid-methanol.* This is usually used as a cold solution of 5% acetic acid in methanol. As with acetone, this is a denaturing fixative that can be used for some antigens, but many antigenic determinants will be lost.

5. *Bouin's fluid.* This is an excellent fixative for preserving ultrastructure, but it needs to be tested carefully because not all labeling procedures are compatible with its use. It is made from 75 mL saturated aqueous picric acid, 25 mL formalin (40% formaldehyde in water), and 5 mL glacial acetic acid. After fixation, the tissue may look yellow because of the picric acid. If later steps include washing with alcohol, this will remove much of the color.

5.1.3. Blocking

After fixation, the cultures need to be washed to remove traces of fixative. In the case of aldehyde fixatives, the cultures also need to be treated to deactivate any remaining aldehyde groups that could interact with antibodies and give nonspecific labeling. This can be done with a small molecule, such as glycine, but is more conveniently carried out with a protein solution, such as bovine serum albumin or serum We routinely use a 5% solution of serum from the animal species from which the secondary antibody was derived. In many cases this is goat, and so a 5% goat serum in PBS is used before primary antibody incubation. Blocking is almost instantaneous, and so we have found no advantage in carrying out this step for longer than 15 min. If cytoplasmic antigens are to be studied, it is useful to use detergents in the blocking step to ensure that antibodies will have good access to the antigens. Buffers containing 0.1% Triton X-100 are usually adequate. For most procedures it is not necessary to have detergent in all subsequent steps, but this should be confirmed for each antigen.

5.2. Light Microscopic Labeling Methods

After removal of the blocking solution, primary antibody is added. This usually requires dilution from a stock solution. A monoclonal antibody tissue culture supernatant can be used from no dilution to 1:50, an ascites fluid from 1:100–1:10,000 and a polyclonal serum from 1:200–1:10,000. These figures are approximate and individual reagents can work at much higher dilutions. A preparation of one of our monoclonal antibodies, recognizing β-tubulin (Gozes and Barnstable, 1982), has worked consistently at dilutions of $1:10^6$. Sera and ascites fluids cannot be used at concentrations higher than about 1:100 because nonspecific labeling from other immunoglobulins becomes severe. An alternative is to use affinity purified antibodies. Typical concentrations for such reagents would be 0.1–1.0 µg/mL. The effective concentration of an antibody is also dependent on the other variables of labeling, namely time and temperature. If incubations longer than several hours are planned, it is better to carry them out at 4°C than at room temperature. With longer incubation times it is often possible to use orders of magnitude less primary antibody. Although the advantages of this of are considerable in tissue sections and explants, they are less apparent in monolayer cultures. Unless the antibody is extremely scarce, it is probably best to choose a concentration that works well with incubation times of an hour. At the other extreme, incubation times should be at least 15 min. Using dissociated retinal cells we have obtained labeling of ganglion cells by a Thy-1 antibody in solution within 5 min, but the intensity plateaued at 15 min.

After incubation, all unbound primary antibody needs to be removed. This is straightforward but, as with all washing steps, must be done very gently so as not to dislodge cells or damage cells processes. Three washes with 5-min incubations in serum-free medium or buffered salts solution should be sufficient.

To detect bound primary antibody, it is now necessary to incubate the culture in a secondary antibody. These are commercially available from many suppliers and come in the two categories of fluorescent conjugates or enzyme-linked conjugates. Fluorescent dyes coupled to antibodies range from AAMCA, which is excited by UV light and emits a blue fluorescence, through fluorescein which is excited by blue light and emits a green fluorescence, to rhodamine or Texas red, which are excited by green light and emit a red fluorescence. The reasons for choosing one over the other can include cost, sensitivity of photographic film being used to record data, or background autofluorescence of cultures. The last can occasionally be a problem with aldehyde fixed cells, since some autofluorescence emits in a yellow range that passes through most filters used for fluorescein.

Even more choices are available for enzyme coupled reagents. The two enzymes most commonly used are horseradish peroxidase (HRP) and alkaline phosphatase (AP). The enzymatic activity is used to produce a colored insoluble product from soluble substrates. A wider variety of substrates is available for HRP, and some of the insoluble products are suitable for both light and electron microscopy. The enzymes can be coupled directly to the secondary antibodies but, because this is usually achieved with crosslinking fixatives, such as glutaraldehyde, some of the complexes formed may have low activity. A more sensitive approach is to a peroxidase-antiperoxidase (PAP) complex or a biotin-avidin linking system. In the PAP method, an antibody against HRP is produced in the same species as the primary antibody, for example, mouse for monoclonal antibodies. After washing the cultures after primary antibody, the secondary antibody, an anti-mouse immunoglobulin, is added. This is incubated for 1 h, the culture washed, and then a soluble PAP complex formed from HRP and an anti-HRP is added. This can bind to the remaining arm of the bivalent secondary antibody. In addition to the increased sensitivity of fully active enzyme molecules, more sensitivity is gained because each PAP complex can have several enzyme molecules.

A similar enhancement is obtained with avidin biotin systems. A typical sequence of steps would be primary antibody, biotinylated antiimmunoglobulin, avidin, and finally biotinylated horseradish peroxidase or alkaline phosphatase.

In choosing between these methods the added sensitivity of an HRP or avidin biotin method needs to be balanced against their extra steps. Each extra step can cause loss of cells from the culture. If a more direct method is adequate for the markers, then it should be used.

5.3. Electron Microscopic Labeling Methods

Electron microscopy (EM) is ideally suited for detailed analysis of a single cell or closely adjacent groups of cells. It is much less appropriate for general characterization of neural cell types in culture because of the effort needed to survey a whole culture dish. Electron microscopy has been used for many neural cultures, but two examples will illustrate its use. Sympathetic neurons can show plasticity in their neurochemical phenotype (*see* Patterson, 1978). The phenotype of a particular cell can be defined by extracellular recording and it has been possible to examine the synaptic morphology of such cells to correlate structure with physiology (Landis, 1980). The detailed structure of vertebrate rod photoreceptors in culture has been examined and compared with their in vivo structure to show that they remain intact, and that the electrophysiological responses to light flashes are a good indicator of in vivo responses (Townes-Anderson et al., 1985).

In treating neural cultures for EM immunocytochemistry, most of the methods are similar to some of those used for right microscopic labeling. The fixatives used need to be strong enough to preserve detailed ultrastructure. A glutaraldehyde/paraform-aldehyde mixture or Bouin's fluid are the fixatives of choice. Some forms of HRP-coupled secondary antibody with a substrate that gives an electron dense product, such as DAB, is often use. With EM immunocytochemistry two other types of secondary antibody can also be used. The first consists of secondary antibody coupled to ferritin. The iron cores of the ferritin molecule are sufficiently electron dense to be visible in the electron microscope. Even better are secondary antibodies coupled to colloidal gold particles. These are very clear in the electron microscope and can be made to almost any diameter. Making colloidal gold particles of specific sizes is something of an art, but good particles with adsorbed secondary antibodies are now widely available commercially.

Most EM immunocytochemistry is performed before embedding in epon or araldite. Postembedding labeling is rarely used on cultures because it has little advantage of increased access to antigens and has lower sensitivity. Procedures for preparing the cultures for embedding and the embedding itself are beyond the scope of this chapter, but details can be found in any EM textbook.

If cultures have been grown on glass or hard plastic, it is necessary to detach the embedded culture from the coverslip. The easiest way to do this is to throw the coverslips and embedded culture into liquid nitrogen. Because the coverslip and the resin contract at different rates, the two become separated. The embedded culture can then be sectioned as normal. An alternative method is to grow cultures on one of the flexible and porous supports now available. Depending on which of these is chosen, the support can be dissolved (if nitrocellulose), or left on the block and sectioned with the culture.

5.4. Double-Labeling Procedures

A single antibody can show how many of a particular cell type are present in a given culture. Two different antibodies can show not only the numbers of two cell types, but also their interrelationships.

Two-color immunofluorescence is perhaps the simplest form of double labeling. The requirement for this is that the two primary antibodies can be separately recognized by the secondary antibodies. Usually, this means that the primary antibodies come from different species, although monoclonal antibodies of different subclasses can also be used. Unless the primary antibodies interact with each other in any way, they can be added to the culture as a mixture. After incubation and washing, the secondary antibodies can also be added as a mixture. An important control is to show that each secondary antibody only reacts with the appropriate primary antibody. After washing the culture to remove unbound secondary antibody, the culture can be viewed with the fluorescence microscope. Each secondary antibody can be viewed with its own excitation and emission filters to show labeled cells.

An alternative to fluorescence double labeling is to use enzyme-coupled secondary antibodies and appropriate substrates to give different colored reaction products. We have used this approach in tissue to identify the neurotransmitter phenotype of cells expressing a membrane antigen (Naegele et al., 1988). The membrane antigen was detected by an IgM monoclonal antibody followed by a peroxidase-conjugated secondary antibody with a DAB substrate. The neurotransmitter GABA was detected by a rabbit anti-GABA serum, followed by a biotin conjugated secondary antibody, and then an alkaline phosphatase-avidin system with a Vector red substrate. Double-labeled cells showed a dark brown membrane labeling and a purple cytoplasmic label.

It is also possible to combine immunocytochemistry with other types of labeling procedure. For example, we have confirmed that GABAergic neurons in rat retinal cultures are amacrine cells by combining radioactive GABA uptake and autoradiography with immunocytochemical labeling with an amacrine cell specific antibody HPC-1 (Akagawa and Barnstable, 1987a). Although the technique is not as easy as labeling with two antibodies, double-labeled cells can clearly be seen, as shown in Fig. 6.

6. The Need to Correlate Biochemical and Immunocytochemical Findings

6.1. Some Artifacts of Tissue Culture

In this Section I want to consider a number of examples where labeling patterns in culture do not correspond to what might be expected from the in vivo patterns. The simplest example of this is where cells are in culture, but lose the expression of a particular marker. We have observed this phenomenon with retinal pigment epithelial cells (Neill and Barnstable, 1990). In vivo and in freshly dissociated preparations, these cells express a number of specific antigens. With time in culture, however, many of these antigens are no longer expressed. In such a case, using these antigens as markers would give a false estimate of the number of retinal pigment epithelial cells in the culture.

We have also documented the converse effect, namely new expression of antigens in culture. In the embryonic development of the optic cup, the neuroepithelial cells express the adhesion molecule N-CAM. The cell layer that forms the retina continues to express this molecule, but in the other layer that forms retinal pigment epithelium, expression is lost. When the pigment epithelial cells are dissociated and placed in tissue culture, however, they reexpress the N-CAM polypeptides, as illustrated in Fig. 7 (Barnstable, 1990; Neill and Barnstable, 1990).

Both of the previous examples suggest that expression of a particular marker may be under dynamic control, and that expression in culture may differ from expression in vivo. A further example of this can be seen in the developmental expression of antigens in culture. In the rat retina, rod photoreceptor development can be divided into at least three phases, defined by the expression of different membrane antigens (Barnstable,

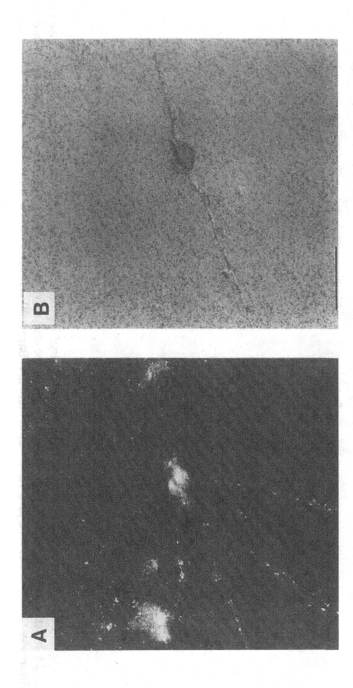

Fig. 6. Double-labeling of cells in retinal monolayer cultures with (^{3}H) GABA and monoclonal antibody HPC-1. (A) Immunofluorescence with antibody HPC-1 showing labeled amacrine cell bodies and processes. (B) (^{3}H) GABA uptake showing a labeled cell body and process that were also labeled in A. Other HPC-1-positive cell bodies and processes were not labeled by (^{3}H) GABA. Bar in B = 25 μm. Taken from Akagawa and Barnstable, (1987a).

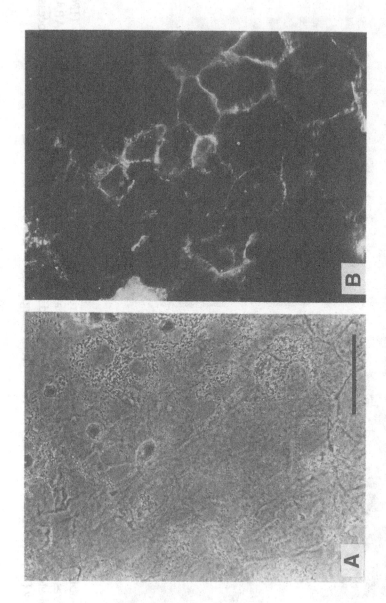

Fig. 7. Indirect immunofluorescence of cultured Retinal Pigment Epithelial cells using a monoclonal mouse anti-N-CAM. Most of the RPE cells are labeled even though these cells do not express N-CAM in the intact eye. (A) Phase and (B) fluorescent images. Bar = 63 μm. Taken from Neill and Barnstable (1990).

1987). The visual pigment protein is expressed shortly after the photoreceptors become postmitotic. About 1 wk later, other markers are expressed as the cells begin to elaborate outer segments. Toward the end of the second postnatal week, as the retina becomes competent to process and transmit light signals, additional markers are detectable. In monolayer cultures, rod photoreceptors can be identified by the expression of opsin (Akagawa and Barnstable, 1986). Later markers are not detected. If the cultures are grown at high density, or as reaggregates, then some of the these markers can be detected (Akagawa et al., 1987). Such an example has been useful in alerting us to possible cell interactions important for rod development, but it also serves as an additional example that culture conditions can affect marker expression. Cells can also take up abnormal positions in three-dimensional cultures, such as reaggregates. Using an amacrine cell-specific antibody, HPC-1, we found that the outer layer of cells in retinal reaggregate cultures consisted of amacrine cells rather than ganglion cells, as would be expected from the normal lamination of the retina (Fig. 8). Without good cell markers we would not have detected this culture abnormality.

One final example can illustrate a different problem. A number of gangliosides are synthesized by glial cells. Expression of these sometimes cannot be detected immunocytochemically, even though their presence can be detected biochemically (Goldman et al., 1984). This phenomenon of masking of antigenic determinants can give false negatives. Since the exact mechanism of the masking is not understood, it is not yet possible to adjust culture conditions to overcome it.

6.2. Commonly Used Biochemical Procedures

It is often important to determine whether an antibody is recognizing the same antigen in tissue culture as in vivo or in two different culture systems. The simplest method for this is immunoblotting, although immunoprecipitation has some advantages and may sometimes be essential. Detailed descriptions of immunoprecipitation can be found elsewhere (Harlow and Lane, 1988), and this chapter will focus on immunoblotting (Towbin et al., 1979).

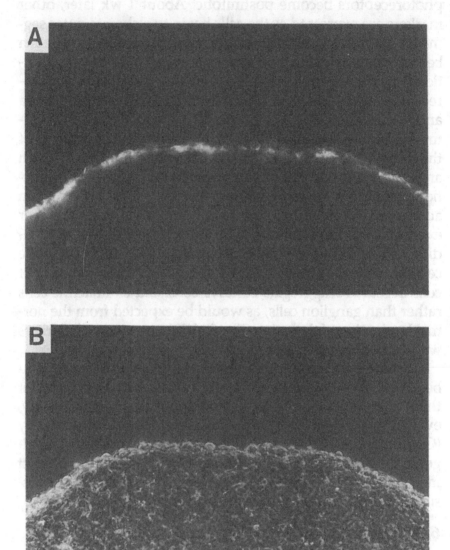

Fig. 8. Cryostat section of reaggregate labeled with amacrine cell-specific antibody HPC-1. (A) Fluorescence. (B) Phase-contrast. Only the outer layer of cell bodies and the processes immediately underneath are labeled. In the intact retina, amacrine cells have a wider distribution, and many are found in the middle layers of the tissue. Bar in B, 100 μm. Taken from Akagawa et al. (1987).

Two methods for the preparation of cultures are available. In the first, cultures are rinsed with serum-free medium and then dissolved in gel sample buffer (0.625M Tris-HCl, pH 6.8, 2% SDS, 10% glycerol, 5% 2-mercaptoethanol, 0.01% bromophenol blue) at a ratio of 0.15–0.25 mL/35-mm dish. The sample is then boiled and loaded on a gel. If nuclear DNA makes the sample too viscous, an alternative method is possible. The cultures can first be dissolved in 0.5% Nonidet-P-40. This lyses the cells but keeps the nuclei intact so that they can be removed by centrifugation. The nuclei-free supernatant can then be brought to gel sample buffer concentrations by the addition of an appropriate vol of 4X concentrated sample buffer.

One of the most convenient gel systems was originally described by Laemmli (1970). The concentration of acrylamide in the resolving gel should be chosen to give good resolution of the molecules being analyzed; the lower the concentration, the larger the molecules that are well-resolved. Table 1 gives the mixtures necessary to prepare resolving gels of 5–15% acrylamide. The gel mixture is cooled on ice and degassed. TEMED (20 µL) is added, mixed, and the gel poured to the appropriate height. After polymerization, a 3% stacking gel is poured using the mixture given in Table 1.

Samples are loaded on the gel, which is then run in 0.025M Tris-HCl, pH 8.3, 0.192M glycine, 0.1% SDS at either 40 V constant voltage overnight, or at 25–40 mA constant current for about 4 h (for a 15-cm gel, minigels take much less time). When the bromophenol blue marker is about 1 cm from the bottom of the gel, the current is switched off and the gel removed.

A sandwich of filter paper, gel, nitrocellulose filter, and filter paper is then made and placed in a holder in a blotting tank with the nitrocellulose toward the anode. It is vital at this stage to ensure that no air bubbles are trapped between any layers of the sandwich. The tank is filled with 0.025M Tris-HCl, pH 8.3, 0.192M glycine, to which 20% by vol of methanol is added. The blotting can be carried out at 0.5–1 A for 1–2 h, or at lower currents for longer times.

After blotting, the nitrocellulose filter is removed. Proteins bound to the filter can be visualized by incubating in Ponceau S stain for 5 min. The stain can be bought as a concentrated

Table 1
Mixtures for SDS Polyacrylamide Gel

(A) Resolving gel[a]	Percentage acrylamide, final vol, mL				
	5	7.5	10	12.5	15
29.2% Acrylamide/0.8% bis-acryl	5.00	7.50	10.00	12.50	15.00
0.75M Tris-HCl (pH 8.8)	15.00	15.00	15.00	15.00	15.00
10% SDS	0.30	0.30	0.30	0.30	0.30
H_2O	9.55	7.05	4.55	2.05	0.00
10% $(NH_4)_2S_2O_7$	0.15	0.15	0.15	0.15	0.15
Total (mL)	30.00	30.00	30.00	30.00	~30.00

(B) Stacking gel[b]	
29.2% Acrylamide/0.8% bis-acryl	0.8
1M Tris-HCl (pH 6.8)	1.0
10% SDS	0.08
H_2O	6.08
10% $(NH_4)S_2O_7$	0.04

[a]Cool on ice, degas for 5 min, add 20 µL TEMED, mix, and pour to appropriate height.
[b]Cool, degas, add 12.5 µL TEMED, mix, and pour.

solution that is diluted in water. Excess stain can be rinsed off with water. The filter is then blocked by incubation in 0.1% Tween-40 in PBS containing 1% nonfat dried milk. Many other blocking solutions are equally effective and can be used if necessary.

After blocking, the filters are treated with primary and secondary antibodies with washing steps just as for immunohistochemistry. Alkaline phosphatase-coupled secondary antibodies give more sensitive detection than peroxidase coupled antibodies when colored reaction products are used. An even more sensitive method uses a peroxidase-conjugated secondary antibody and a chemiluminescence detection. Kits for this reaction are available from many suppliers. In our hands, it has given approx tenfold greater sensitivity than colored reaction products.

Immunoblot analysis can be used to ensure that the same molecules can be detected in culture as can be detected in tissue. If the molecules are not the same, the use of the antibody as a defined marker becomes suspect. Changes in migration of particular bands can also be informative, since they may indicate changes in patterns of glycosylation or phosphorylation as the cells grow in culture.

An almost identical blotting procedure is available for the detection of glycolipids and gangliosides (Blum and Barnstable, 1987; Towbin et al., 1984). An appropriate extract is resolved by thin-layer chromatography (TLC) using standard chromatographic procedures. The chromatograms are then dried. They are wet uniformly with isopropanol/water (3:1, v/v), and then blotted by firmly compressing the moistened chromatogram onto a dry nitrocellulose sheet. Solvent appears to wick onto the nitrocellulose and carry the lipids with it. The nitrocellulose is dried and then processed as described earlier. Immunostaining of TLC plates directly has also been described. We prefer the blotting procedure because we have been able to carry out additional reactions, such as base treatment to cleave O-acetyl groups (Blum and Barnstable, 1987), with the blots that are not as easy with TLC plates.

7. Other Uses of Cell Type-Specific Antibodies

So far in this chapter, antibodies have been described as passive reagents that can document the types of cells, and their state of differentiation, in a culture. They can, however, be used more actively to manipulate the composition of cultures. In this section, three such uses of antibodies will be given, namely, negative selection by antibody-dependent cell killing, positive selection by "panning," and using antibodies to create adhesive substrates for neural growth.

7.1. Antibody-Dependent Cell Killing

This method has been used extensively in the field of immunology, but has received little attention in neurobiology. The theory behind the procedure is very straightforward. In its simplest form, we have a culture composed of cell types X and Y, and an antibody against a cell surface antigen expressed by X but not Y. The culture is treated with the antibody and then with a source of complement, such as normal rabbit serum. Cell type X will be lysed and the culture will now be composed only of Y. There are three problem areas that can influence the effectiveness of this procedure. The first is the specificity and type of the antibody. If the antibody crossreacts with any other cell type, more than one cell type may be killed. With a monoclonal antibody, the procedure will also only be effective with subclasses of antibodies that interact with complement. A single IgM molecule can interact, as can multiple molecules of IgG_2 and IgG_3 antibodies. Most, but not all IgG_1 antibodies do not interact with complement.

The second potential problem is the sensitivity of cells to complement-dependent lysis. Binding of antibody to fine processes may not be an effective way of causing cell lysis because these processes may be isolated from the cell bodies. Some cells may also have a sufficient density of appropriate pumps and ion channels to overcome the effects of the membrane pores created by complement.

The third problem is of nonspecific effects on other cells. Bystander lysis can occur when terminal complement complexes formed in solution attach to adjacent cells rather than the cell to which the antibody is found. A more serious problem can arise if the cell killed is important for the survival of other cells. The simplest example of this would be in a mixed culture of glia and neurons. Most of the neurons are likely to be growing on the glia, so killing these cells could lead to detachment of many of the neurons. Other, more complicated interactions, such as secretion of trophic factors, can also be disrupted, but that is an experimental observation that could be very useful.

7.2. Antibody-Mediated Positive Selection

The use of antibodies to purify cell types also has a long history in many fields of biology. The antibodies can be coupled to a fluorescent dye, a magnetic particle, or an insoluble matrix for separations by fluorescence-activated cell sorting, magnetic separations, and panning, respectively. Although fluorescence-activated cell sorting is extremely accurate, it is relatively slow, and many neural cells do not survive the long procedure or the shearing forces encountered in the machine. The other two methods have the advantage that they work on bulk cell preparations. Magnetic spheres have found limited use, but new styles of magnets are becoming commercially available that may make this a more routine procedure.

Panning has the advantage of being very simple. A bacterial culture dish is coated with the antibody of interest. The mixed cell suspension is added and incubated at room temperature for 45–60 min with occasional swirling to ensure access of all cells to the surface of the plate. After incubation, the nonadherent cells are removed by agitating the plate in up to ten washes of an isotonic salts solution. The major problem with this procedure is the removal of specifically found cells. If a high concentration of antibody is used to coat the panning plate, the cells adhere so tightly that they cannot be removed without damaging many of them.

If the amount of antibody is reduced to make removal easier, the yield of cells decreases. Cells can be eluted from the panning plates by proteolytic digestion with papain or trypsin. It has even been found that competition with a high concentration of antibody is effective (Barres et al., 1988).

In many areas of the nervous system, it is necessary first to remove cells that adhere nonspecifically to immunoglobulins. The most common of these are microglia and other macrophage. These can be removed most effectively in a panning step involving an antimacrophage antibody.

Using the procedures outlined above, retinal ganglion cells were isolated with a Thy-1 antibody in good yield and almost 100% purity (Barres et al., 1988). Ganglion cells comprised 0.57% of the starting retinal suspensions, so the procedure can obtain at least a 200-fold purification. The specifically bound cells that were then eluted from the panning plates showed high viability and were suitable for electrophysiological recording.

7.3. Antibodies as Growth Substrates

The previous section showed that antibodies bound to plastic dishes could bind neural cells. Some years ago, we took this observation one step further and tested whether or not such antibody-coated dishes could provide a long-term substrate for neural attachment and growth (MacLeish et al., 1983). The initial stimulus for these experiments was a need to grow retinal cells from Ambystoma tigrinum, the tiger salamander. Unlike most mammalian neural cells, cells from lower vertebrates adhere poorly to most readily available culture substrates. We were able to develop an antibody against cell surface antigens of every salamander cell type we examined, including red blood cells. Glass coverslips were acid washed, rinsed with ethanol, then distilled water, and incubated with antibody (Sal-1) overnight at room temperature. The coverslips were then rinsed with salamander salts solution and used for tissue culture.

When dissociated salamander retinal cells were plated on control, untreated, coverslips few cells attached. Although the cells survived, they showed abnormal morphology with few processes and halos of microspikes. In contrast, when the retinal cell suspensions were plated onto Sal-1 coated coverslips, virtually all the cells adhered within seconds. The attached cells appeared healthy, retained processes present at the time of dissociation, and elaborated more extensive networks of processes over time.

Electrophysiological recording and electron microscopic examination were used to verify that the cells were normal. The two simplest reasons for the success of this procedure are that either the antibody is simply increasing adhesion, or the antibody binds to and triggers a specific growth receptor. Sal-1 antibody appears to recognize many glycoproteins on the cell surface, and thus, it most probably acts simply by increasing the overall level of adhesion. We have been working with a different antibody where a more specific effect may be involved.

From general effects upon the cell suspension plated, we moved to a more selective application (Leiter et al., 1984). As described earlier, Thy-1 antibodies react with most neurons, but in retina react with only the ganglion cells. Coverslips were coated with monoclonal anti-Thy-1 (antibody 2 G12; Barnstable and Dräger, 1984), and dissociated rat retinal cells were plated. As compared with other adhesive substrates, such as polylysine, there was no increase in the number of ganglion cells attached. What was striking, however, was that ganglion cell process outgrowth was greatly enhanced on the antibody substrate, whether measured as percentage of cells with processes or as the length of processed per cell. Other antibodies that bound to ganglion cell surface antigens did not have this stimulatory effect. Further evidence for the specificity of this effect has come from the finding that not all Thy-1 antibodies are effective, even though all appear to label the cells to the same extent. It seems likely that there is an "active site" on the Thy-1 molecule for this process. It seems clear that the effect is not peculiar to retinal ganglion cells, because anti-Thy-1 substrates have also been used for the successful culture of Purkinje cells from cerebellum (Messer et al., 1984).

If the Thy-1 antibodies were mimicking a natural process, there might be a Thy-1 receptor on cells that interacts with neurons, in particular astrocytes. Antiidiotype antibodies that react with the active Thy-1 antibodies have been used as probes for this receptor and recognize a single polypeptide on cultured astrocytes. This recognition is blocked by purified Thy-1, lending further support to the designation of this polypeptide as a Thy-1 receptor (Dreyer et al., 1990).

The previous discussion has been presented in some detail not only because it represents an interesting extension of the use of cell type-specific antibodies, but also because it serves to point out that antibodies can be more than passive labeling reagents. On fixed cell cultures this is not important, but if antibodies are used to identify living cells, it may be very important to show that the antibody itself does not perturb the physiology of the culture.

8. Conclusions

From the beginning of this chapter, I have presented my view that antibodies provide an excellent tool for precisely defining the composition of neural cultures. The ways in which these reagents can be used are more extensive than it has been possible to describe here. The procedures outlined will, however, provide reliable data given suitable antibodies. In providing examples, I have drawn on the work of my own laboratory using retina, although there are now many examples from other regions of the nervous system.

Antibodies can be more than passive labeling tools. I have tried to show that they can be used to monitor the development of cultures and to manipulate their composition. It can also be emphasized that antibodies can also be used to study the intact nervous system, and are thus, excellent tools to verify that what we can see in a culture dish bears some relation to what happens in vivo.

Acknowledgments

I thank H. Stopka and R. Brown for help with the manuscript. Work from my laboratory described in this article has been supported by NIH grants EY 00785, EY 05206, and EY 07119, as well as by the Darien Lions and Research to Prevent Blindness, Inc. C. J. B. is a Jules and Doris Stein Research to Prevent Blindness, Inc., professor.

References

Akagawa K. and Barnstable C. J. (1986) Identification and characterization of cell types in monolayer cultures of rat retina using monoclonal antibodies. *Brain Res.* **383,** 110–120.

Akagawa K. and Barnstable C. J. (1987a) Identification and characterization of cell types accumulating GABA in rat retinal cultures using cell type. specific monoclonal antibodies. *Brain Res.* **408,** 154–162.

Akagawa F. and Barnstable C. J. (1987b) Selective sorting out of glycine-accumulating cells in reaggregate cultures of rat retina *Dev. Brain Res.* **31,** 124–128.

Akagawa K., Hicks D., and Barnstable C. J. (1987) Histiotypic organization and cellular differentiation of rat retinal neurons maintained in reaggregate culture. *Brain Res.* **437,** 298–308.

Aramant R., Seiler M., Ehinger B., Bergstrom A., Adolph A. R., and Turner J. E. (1990) Neuronal markers in rat retinal grafts. *Dev. Brain Res.* **53,** 47–61.

Arimatsu Y., Naegele J. R., and Barnstable C. J. (1987) Molecular markers of neuronal subpopulations in layers 4, 5 and 6 of cat primary visual cortex. *J. Neurosci.* **7,** 1250–1263.

Bader C. R., MacLeish P.R., and Schwartz E. A. (1978) Responses to light of solitary rod photoreceptors isolated from tiger salamander retina. *Proc. Natl. Acad. Sci.* **75,** 3507–3511.

Barnstable C. J. (1980) Monoclonal antibodies which recognize different cell types in the rat retina *Nature* **286**, 231–235.

Barnstable C. J. (1982) Monoclonal antibodies—tools to dissect the nervous system. *Immunol.Today* **3**, 157–159, 167,168.

Barnstable C. J. (1987) A molecular view of mammalian retinal development. *Mol. Neurobiol.* **1**, 9–46.

Barnstable C. J. (1991) Molecular aspects of development of mammalian optic cup and formation of retinal cell types. *Prog. Retina Res.* **10**, 69–88.

Barnstable C. J. and Dräger U. C. (1984) Thy-1 antigen: A ganglion cell specific marker in rodent retina. *Neuroscience* **11**, 847–855.

Barnstable C. J., Hofstein R., and Akagawa K. (1985) A marker of early amacrine cell development in rat retina. *Dev.Brain Res.* **20**, 286–290.

Barres B. A., Silverstein B. E., Corey D. P., and Chun L. L. Y. (1988) Immunological, morphological, and electrophysiological variation among retinal ganglion cells purified by panning. *Neuron* **1**, 791–803.

Bartlett W. P. and Banker G. A. (1984) An electron microscopic study of the development of axons and dendrites by hippocampal neurons in culture. *J. Neurosci.* **4**, 1944–1953.

Baughman R. W. and Bader C. R. (1977) Biochemical characterization and cellular localization of the cholinergic system in the chicken retina. *Brain Res.* **138**, 469–485.

Blum A. S. and Barnstable C. J. (1987) O-acetylation of a cell surface carbohydrate creates iscrete molecular patterns during neural development. *Proc. Natl. Acad. Sci.* **84**, 8716–8720.

Burden S. J., Sargent P. B., and McMahan U. J. (1979) Acetylcholine receptors in regenerating muscle accumulate at original synaptic sites in the absence of the nerve *J. Cell Biol.* **82**, 412–425.

Cáceres A., Banker G. A., and Binder, L. (1986) Immunocytochemical localization of tubulin and microtubule-associated protein 2 during the development of hippocampal neurons in culture. *J. Neurosci.* **6**, 714–722.

Dimpfel W., Huang R. T. C., and Habermann E. (1977) Gangliosides in nervous tissue cultures and binding of ^{125}I-labelled tetanus toxin-neuronal marker. *J. Neurochem.* **29**, 329–334.

Dräger U. C. (1983) Coexistence of neurofilaments and vimentin in a neurone of adult mouse retina. *Nature* **303**, 169–172.

Dräger U. C., Edwards D. L. , and Kleinschmidt J. (1983) Neurofilaments contain a-melanocyte-stimulating hormone (a-MSH)-like immunoreactivity. *Proc. Natl. Acad. Sci.* **80**, 6408–6412.

Dräger U. C., Edwards D. L., and Barnstable C. J. (1984) Antibodies against filamentous components in discrete cell types of the mouse retina *J. Neurosci.* **4**, 2025–2042.

Dreyer E. B., Leifer D., Levin L. A., and Lipton S. A. (1990) An endogenous glial receptor for Thy-1 and its role in retinal ganglion cell neurite outgrowth. *Invest. Ophthalmol. Vis. Sci.* (Suppl.) **31**, 286.

Edelman G. M., Hoffman S., Chuong C.-M. Thiery J.-P., Brackenbury R., Gallin W. J., Grumet M., Greenberg M. E., Hemperly J. J., Cohen C., and Cunningham B. A. (1983) Structure and modulation of neural cell adhesion molecules and early and late embryogenesis. *Cold Spring Harbor Symp. Quant. Biol.* **48**, 515–526.

Farb D. H., Berg D. K., and Fischbach G. D. (1979) Uptake and release of (3H) γ-aminobutyric acid by embryonic spinal cord neurons in dissociated cell culture. *J. Cell Biol.* **80**, 651–661.

Fields K. L., Brockes J. P., Mirsky R., and Wendon L. M. B. (1978) Cell surface markers for distinguishing different types of rat dorsal root ganglion cells in culture. *Cell* **14**, 43–51.

Furshpan E. J., MacLeish P. R., O'Lague P. H., and Potter D. D. (1976) Chemical transmission between rat sympathetic neurons and cardiac myocytes developing in microcultures: Evidence for cholinergic, adrenergic, and dual-function neurons. *Proc. Natl. Acad. Sci.* **73**, 4225–4229.

Giulian D. and Baker T. J. (1986) Characterization of ameboid microglia isolated from developing mammalian brain. *J. Neurosci.* **6**, 2163–2178.

Goldman J. E., Hirano M., Yu R. K., and Seyfried T. N. (1984) G_{D3} ganglioside is a glycolipid characteristic of immature neuroectodemal cells *J. Neuroimmunol.* **7**, 179–192.

Goridis C., Deagostini-Bazin H., Hirn M., Hirsch M. - R., Rougon G., Sadoul R. Langley O. K., Gombos G., and Finne J. (1983) Neural surface antigens during nervous system development. *Cold Spring Harbor Symp. Quant. Biol.* **48**, 527–537.

Gozes I. and Barnstable C. J. (1982) Monoclonal antibodies that recognize discrete forms of tubulin. *Proc. Natl. Acad. Sci. USA* **79**, 2579–2583.

Hammang J. P., Baetge E. E., Behringer R. R., Brinster R. L., Palmiter R. D., and Messing A. (1990) Immortalized Retinal Neurons Derived from SV40 T-Antigen-Induced Tumors in Transgenic Mice. *Neuron* **4**, 775–782.

Harlow E. and Lane D. (1988) *Antibodies: A Laboratory Manual.* Cold Spring Harbor Laboratory, Cold Spring Harbor, NY.

Hicks D. and Courtois Y. (1990) The growth and behaviour of rat retinal Müller cells In Vitro 1. An improved method for isolation and culture. *Exp. Eye Res.* **51**, 119–129.

Hicks D. and Barnstable C. J. (1987) Different monoclonal antibodies reveal different binding patterns on developing and adult retina. *J. Histochem Cytochem.* **35**, 1317–1328.

Hicks D., Sparrow J., and Barnstable, C. J. (1989) Immunoelectron microscopic examination of the surface distribution of opsin in the rat retinal photoreceptor cells. *Exp. Eye Res.* **49**, 13–29.

Hockfield S. and McKay R. D. G. (1985) Identification of major cell classes in the developing mammalian nervous system. *J. Neurosci.* **5**, 3310–3328.

Hume D. A., Perry V. H., and Gordon S. (1983) Immunohistochemical localization of a macrophage-specific antigen in developing mouse retina: Phagocytosis of dying neurons and differentiation of microglial cells to form a regular array in the plexiform layers. *J. Cell Biol.* **97**, 253–257.

Jessell T. M., Siegel R. E., and Fischbach G. D. (1979) Induction of acetylcholine receptors on cultured skeletal muscle by a factor extracted from brain and spinal cord. *Proc. Natl. Acad. Sci.* **76,** 5397–5401.

Kosaka, T., Heizman C. W., and Barnstable, C. J. (1989) Monoclonal antdbody VC1.1 selectively stains a population GABAergic neurons containing a calcium binding protein parvalbumin in the rat cerebral cortex. *Exp. Brain Res.* **78,** 43–50.

Kosaka T., Isogai K., Barnstable C. J., and Heizmann C. W. (1990) Monoclonal antibody HNK-1 selectively stains a population of GABAergic neurons containing a calcium-binding protein parvalbumin in the rat cerebral cortex. *Exp. Brain Res.* **82,** 566–574.

Laemmli U. K. (1970) Cleavage of structural proteins during the assembly of the head of bacteriophase T4. *Nature* **227,** 680–685.

Landis S. C. (1980) Developmental changes in the neurotransmitter properties of dissociated sympathetic neurons: A cytochemical study of the effects of medium. *Dev. Biol.* **77,** 349–361.

Leclerc N., Gravel C., and Hawkes R. (1988) Development of parasagittal zonation in the rat cerebellar cortex: MabQ113 antigenic bands are created postnatally by the suppression of antigen expression in a subset of purkinje cells. *J. of Comp. Neuro.* **273,** 399–420.

Leifer D., Lipton S. A., Barnstable C. J., and Masland R. H. (1984) Monoclonal antibody to Thy-1 enhances regeneration of processes by rat retinal ganglion cells in culture. *Science* **224,** 303–306.

Lindvall O., Bjorklund A., Hokfelt T., and Ljungdahl, A. (1973) Application of the glyoxylic acid method to Vibratome sections for improved visualization of central catecholamine neurons. *Histochemie* **35,** 31–38.

MacLeish P. M., Barnstable C. J., and Townes-Anderson, E. (1983) Use of a monoclonal antibody as a substrate for mature neurons in vitro. *Proc. Natl. Acad. Sci. USA* **80,** 7014–7018.

Messer A., Snodgrass G. L., and Maskin, P. (1984) Enhanced survival of cultured cerebellar Purkinje cells by plating on antibody to Thy-1. *Cell. Molec. Neurobiol.* **4,** 285–290.

Naegele J. R. and Barnstable C. J. (1991) A carbohydrate epitope defined by monoclonal antibody VC1.1 is found on N-CAM and other cell adhesion molecules. *Brain Res.* **559,** 118–129.

Naegele J. R., Arimatsu Y., Schwartz P., and Barnstable, C. J. (1988) Selective staining of a subset of GABAergic neurons in cat visual cortex by monoclonal antibody VC1.1. *J. Neurosci.* **8,** 79–89.

Neill J. M. and Barnstable C. J. (1990) Differentiation and dedifferentiation of rat retinal pigment epithelial cells. Expression of RET-RE2, and RPE antigen, and N-CAM durng development and in tissue culture. *Exp. Eye Res.* **51,** 573–583.

Patterson P. H. (1978) Environmental determination of autonomic neurotransmitter functions. *Ann. Rev. Neurosci.* **1,** 1–17.

Patterson P. H. and Chun L. L. Y. (1977) The induction of acetylcholine synthesis in primary cultures of dissociated rat sympthetic neurons. I. Effects of conditioned medium. *Dev. Biol.* **56,** 263–280.

Rauscht B., Clapshaw P. A., Price J., Nobel M., and Siefert, W. (1982) Development of oligodendrocytes and Schwann cells studied with a monoclonal antibody against galactocerebroside. *Proc. Natl. Acad. Sci.* **79,** 2709–2713.

Sarthy P. V. (1985) Establishment of Müller cell cultures from adult rat retina. *Brain Res.* **337,** 138–141.

Sommer I. and Schachner M. (1981) Monoclonal antibodies (01 to 04) for oligodendrocyte cell surfaces; an immunological study in the central nervous system. *Dev. Biol.* **83,** 311–327.

Sparrow J. R., Hicks D., and Barnstable C. J. (1990) Cell commitment and differentiation in explants of embryonic rat retina. Comparison to the developmental potential of dissociated retina. *Dev. Brain Res.* **51,** 69–84.

Towbin H., Schoenenberger C., Ball R., Braun D. G., and Rosenfelder G. (1984) Glycosphingolipid-blotting: An immunological detection procedure after separation by thin layer chromatography. *J. Immun. Methods.* **72,** 471–479.

Towbin H., Staehelin T., and Gordon J. (979) Electrophoretic transfer of proteins from polyacrylamide gels to nitrocellulose sheets: Procedure and some applications. *Proc. Natl. Acad. Sci.* **76,** 1318–1322.

Townes-Anderson E., MacLeish P.R, and Raviola E. (1985) Rod cells dissociated from mature Salamander retina: Ultrastructure and uptake of horseradish peroxidase. *J. Cell Biol.* **100,** 175–188.

Willinger M. and Schachner M. (1980) GM1 Ganglioside as a marker for neuronal differentiation in mouse cerebellum. *Dev. Biol.* **74,** 101–117.

Zipser B. and McKay R. (1981) Monoclonal antibodies distinguish identifiable neurones in the leech. *Nature* **289,** 549–554.

Environmental Influences
on Cells in Culture

Jane E. Bottenstein

1. Introduction

The technique of culturing cells derived from the nervous system has been in use for over 80 years, with the objective of simplifying the experimental system to provide readily manipulatable models of neural function. These preparations are useful for testing hypotheses relevant to cell adhesion, motility, survival, proliferation, longevity, and expression of cell type-specific properties. Many significant modifications of these methods have resulted over the years, as new data have emerged from cell biology and neurobiology studies (*see* reviews by Bottenstein, 1983a,1985,1988).

Once the selection is made to use in vitro methods, several important decisions must be made at the onset of the study. First, what will constitute the best model for the phenomenon to be analyzed? Should explant, dissociated, or continuous cell line cultures be employed? Should the cultures be of mixed cell types or enriched for a particular cell type? Each has its advantages and disadvantages. Second, it is important to identify the types of cells present in the preparation. Only cell lines that have been well-characterized should be employed in experiments. When primary cultures are used, the cellular profile can change with time in vitro, since different cell types may proliferate at different rates, or some not at all, depending on the culture

From: *Neuromethods, Vol. 23: Practical Cell Culture Techniques*
Eds: A. Boulton, G. Baker, and W. Walz ©1992 The Humana Press Inc.

conditions and stage of development. Morphology is sugges-
tive, but can be misleading. There are many markers available
for this identification, and generally single or double immunostaining for cell type-specific antigens is definitive. In addition,
neurons can be identified electrophysiologically by their action
potential capability. Finally, how can one best approximate in
vivo conditions in the culture vessel? What aspects of the culture environment are important for cell attachment to the
substratum, migration of cells on the culture surface, extension
of processes, proliferation of cells by mitosis, survival beyond
days or a few weeks in vitro, and expression of the myriad differentiated properties typical of individual cell types? There is
extensive literature on the ionic, nutrient, and physicochemical
requirements of cultured cells that includes ion balance, amino
acid, carbon source and energy, vitamin, trace element, pH, osmolarity, oxygen, carbon dioxide, and temperature optima (*see* review
by Bottenstein, 1983a). More recently, a growing number of hor-
monal, growth factor, and extracellular matrix influences have
been described. Less understood at present are the mechanisms
underlying cell-cell interactions that involve both diffusible
factors and contact-mediated events. Future studies and improved
technology will no doubt lead to further refinements of the
methods and of our modeling capabilities.

This chapter will briefly discuss four critical environmental
parameters of a cell culture system: culture surface, basal culture medium, medium supplements, and physicochemical
variables. Specific examples of five rat CNS serum-free culture
preparations will be provided: B104 neuronal cell line, C6 glial
cell line, embryonic brain neurons, neonatal cerebral cortical
type 1 astrocytes, and neonatal brain O-2A lineage cells.

2. Critical Environmental Parameters

2.1. Culture Surface

Cells are most often cultured on polystyrene surfaces treated
for cell culture, and sometimes on glass surfaces if fluorescence
or Nomarski optical analysis will be used. Cells are generally
not directly attached to this surface, but rather to molecules

adsorbed to the negatively charged plastic or glass surfaces. Adding culture medium with a high protein content, e.g., serum-containing medium, results in adsorption of many proteins that form a 50 Å layer, and are the sites of attachment of the cells. To better mimic in vivo conditions, a variety of extracellular matrix (ECM) components have been used to coat culture surfaces, e.g., collagens, fibronectin, laminin, glycosaminoglycans, or less-defined complex mixtures of ECM molecules. In addition, because cells have a net negative charge, basic polymers like polylysine have been used to modify these surfaces to enhance adhesion. Combinations of these two types of modifications have also been employed with excellent results. This choice is not trivial, since subsequent cellular behaviors, including responses to growth factors, can be greatly influenced by the surface cells contact. For example, rat CNS B104 neuroblastoma cells exhibit a threefold increase in cell number 4 d after treatment, when the substratum is modified with fibronectin or polylysine individually, or a synergistic tenfold increase with both treatments compared to controls (Bottenstein and Sato, 1980). Moreover, cells in culture produce ECM molecules that they deposit on the surface of the culture vessel, and their synthesis can be influenced by both soluble factors in the medium and the surface the cells contact.

2.2. Basal Culture Medium

Not fully appreciated is the importance of the choice of basal culture medium, since the growth and differentiation of many cells can be significantly different depending on this variable. As one example, U251 MGsp human glioma cells exhibit a tenfold increase in cell number after 4 d, when the basal culture medium is Dulbecco's modified Eagle's medium (DMEM) vs Ham's F12 medium (F12; Michler-Stuke and Bottenstein, 1982b). Table 1 shows a comparison of the commonly used Eagle's minimum essential medium (MEM), DMEM, and F12 medium. Each contains amino acids, vitamins, inorganic salts, an energy source, and other components. These and other basal media can be purchased from vendors, and they vary appreciably in their ratio and types of amino acids and in the content of other molecules.

Table 1
Commonly Used Basal Media for Neural Cells*

Component, mg/L	MEM	DMEM	F12
Amino acids			
L-Alanine			8.9
L-Arginine	126.0	84.0	211.0
L-Asparagine			15.0
L-Aspartic acid			13.3
L-Cysteine	31.3	62.6	35.1
L-Glutamic acid			14.7
L-Glutamine	292.0	584.0	146.0
Glycine		30.0	7.5
L-Histidine	42.0	42.0	21.0
L-Isoleucine	52.0	105.0	3.9
L-Leucine	52.0	105.0	13.1
L-Lysine	72.5	146.0	36.5
L-Methionine	15.0	30.0	4.5
L-Phenylalanine	32.0	66.0	5.0
L-Proline			34.5
L-Serine		42.0	10.5
L-Threonine	48.0	95.0	11.9
L-Tryptophan	10.0	16.0	2.0
L-Tyrosine	51.9	103.8	7.8
L-Valine	46.0	94.0	11.7
Vitamins			
B-12			1.4
Biotin			0.007
Choline chloride	1.0	4.0	14.0
Folic acid	1.0	4.0	1.3

	MEM	DMEM	F12
i-Inositol	2.0	7.2	18.0
Niacinamide	1.0	4.0	0.04
Nicotinamide		4.0	0.5
D-Pantothenic acid	1.0	4.0	
Pyridoxal	1.0	4.0	0.06
Pyridoxine			0.04
Riboflavin	0.1	0.4	0.3
Thiamine	1.0	4.0	
Inorganic salts			
Calcium chloride	200.0	200.0	33.2
Copper sulfate			0.0025
Iron nitrate		0.1	
Iron sulfate			0.83
Magnesium chloride			57.2
Magnesium sulfate	97.7	97.7	
Potassium chloride	400.0	400.0	223.6
Sodium chloride	6800.0	6400.0	7599.0
Sodium phosphate	140.0	125.0	142.0
Zinc sulfate			0.9
Other			
D-Glucose	1000.0	4500.0	1802.0
Hypoxanthine			4.8
Linoleic acid			0.084
Lipoic acid			0.2
Putrescine			0.16
Sodium pyruvate			110.0
Thymidine			0.7

*MEM, Eagle's minimum essential medium; DMEM, Dulbecco's modified Eagle's medium; F12, Ham's F12 medium. Formulations are for powdered media and are from the Gibco Laboratories' technical brochure.

It is advisable to test several different basal media or combinations of them to optimize results. For all cell culture reagents, including basal media made from powder, triple distilled or cartridge-processed water (prefilter, carbon, two ion exchange, and pyrogen removal cartridges in series) of 10–18 megohm-cm is suggested. Water of lesser purity may contain contaminants that are toxic to cells. In the absence of serum molecules that bind and inactivate many of these contaminants, cells in vitro are more vulnerable to their deleterious effects. Culture media should be stored in the dark at 4°C.

2.3. Medium Supplements

Most cultured cells, however, even on a permissive surface, will not survive or exhibit optimal phenotypic properties for long in basal culture medium alone, and require supplementation. This reflects additional requirements for growth and survival factors, hormones, transport proteins, trace elements (e.g., iron, copper, zinc, selenium, manganese, molybdenum, vanadium), and/or ECM factors. Serum supplementation has been the traditional choice, although it is not representative of the biological fluids that normally contact CNS cells, such as interstitial or cerebrospinal fluid. Moreover, it contains a multiplicity of proteins, lipids, and other known and unknown factors, whose concentrations can vary greatly depending on the age of the donor and the particular batch used. In addition to its high cost, this additive can introduce variability in experimental results. Since it is a complex mixture of positive and negative growth and differentiation regulators, serum supplementation can mask responses to exogenous factors introduced in experiments, and can also obscure the interpretation of the results obtained. Finally, serum can be a source of mycoplasma, endotoxins, and other deleterious contaminants. To avoid these problems, addition of purified components is generally preferable. Serum-free formulations that substitute purified factors have been found suitable for many in vitro growth and differentiation studies (Bottenstein, 1983b,1985). Factors common to most chemically defined media are the polypeptide hormone insulin, the iron-transport protein transferrin, and the trace element selenium. The other components vary depending on the cell type.

2.4. Physicochemical and Other Variables

The gaseous environment, temperature, osmolarity, and pH of the culture system also play critical roles in cell survival, adhesion, and growth (*see* review by Bottenstein, 1983a). The necessity for physiological osmolarity and pH is well known, and even small changes in these parameters can have disastrous effects on experiments. Most vertebrate cultures are maintained at 37°C in an atmosphere of 5% carbon dioxide/95% air, and contain sodium bicarbonate and/or synthetic buffers like N-2-hydroxyethylpiperazine-N'-2-ethanesulfonic acid (HEPES; H-3375) to maintain a pH of about 7.3. Other factors influencing the behavior of cultured cells are the medium volume-to-surface area and cell density ratio, as well as the frequency and percentage of medium changes. It is best to maintain the medium volume-to-surface area ratio at 0.12–0.25 (mL/cm^2), the cell density as high as possible, and the medium changes to the minimum necessary (usually every 3–5 d) to replenish depleted components without removal of vital conditioning factors made by the cells themselves. Careful attention to all these variables is an important prerequisite for consistent experimental results.

3. Serum-Free Culture Methods for CNS-Derived Neural Cells

We have developed a series of serum-free chemically defined media for culturing the major neural cell types found in the CNS (*see* Table 2). N1 and N2 media are optimized for cells of neuronal phenotype (Bottenstein and Sato, 1979; Bottenstein, 1983a,1985), G2 and G5 media for cells of an astrocytic phenotype (Michler-Stuke and Bottenstein, 1982,a,b; Michler-Stuke et al., 1984), and O1 and O3 media for cells of an oligodendrocytic or O-2A glial progenitor phenotype (Bottenstein, 1986; Bottenstein et al., 1988; Hunter and Bottenstein, 1989,1990,1991).

Table 2
Serum-Free Chemically Defined Media for Neural Cells

Supplement	N1	N2	G2	G5	O1	O3
Insulin	5 µg/mL	5 µg/mL		5 µg/mL	5 µg/mL[a]	15 µg/mL
Transferrin	5 µg/mL	100 µg/mL	50 µg/mL	50 µg/mL	50 µg/mL	1 µg/mL
Progesterone	20 nM	20 nM				
Putrescine	100 µM	100 µM				
Selenium	30 nM	30 nM	30 nM	30 nM	30 nM	30 nM
Hydrocortisone			10 nM	10 nM		
Biotin			10 ng/mL	10 ng/mL	10 ng/mL	10 ng/mL
Fibroblast growth factor			5 ng/mL	5 ng/mL		
Epidermal growth factor				10 ng/mL		

[a]100 ng/mL IGF 1 can replace insulin. N1 and N2 media supplements are added to a 1:1 mixture of Dulbecco's modified Eagle's medium (DMEM) and Ham's F12 medium (F12). G1, G5, O1, and O3 media supplements are added to DMEM only. Values shown are the final concentrations. Supplements are made 50–100-fold concentrated and stored at –20°C if the solvent is Hank's balanced salt solution (HBSS), or at 4°C if it is ethanol or 0.01N HCl. Solvents: HBSS except for insulin and primary stocks of progesterone and hydrocortisone; insulin is solubilized in 0.01N HCl and progesterone and hydrocortisone in ethanol for primary stocks and HBSS for secondary stocks. Reagent sources are: bovine insulin (Sigma I-5500, St. Louis, MO), human transferrin (Sigma T-2252), progesterone (Sigma P-0130), putrescine (Sigma P-7505), sodium selenite (spectrographically pure; Johnson Matthey, London), hydrocortisone (Sigma H-4001), biotin (Calbiochem-Behring 2031, San Diego, CA), fibroblast growth factor (Bethesda Research Labs 6111LA), and epidermal growth factor (Bethesda Research Labs 6100LA, Gaithersburg, MD).

Substratum modification with polylysine and fibronectin is recommended when using any of these media. Substitution of laminin for fibronectin has little or no effect other than increasing neuronal process length. Medium is changed when the pH becomes acidic because of increased lactate production as cells increase in number, or at 4-5 d intervals in less dense cultures. Buffering agents are added to the culture medium (sodium bicarbonate and HEPES at 1.2 g/L and 15 mM, respectively), and cells are maintained at 37°C in a humidified atmosphere of 95% air/5% carbon dioxide. Serum-free culture media are made fresh or stored 1–2 d at 4°C, since we have seen some loss of activity after a week of storage at 4°C. Storage at −20°C results in some inactivation of insulin after thawing. Adsorptive losses of low concentrations of protein components are minimized by the use of plastic (polypropylene, preferably, or polystyrene) vessels, pipets, and micropipet tips when storing and handling serum-free medium supplemented with defined factors. All solutions, pipets, and culture vessels need to be sterile, and aseptic technique is used in all the procedures discussed. Antibiotics are not necessary when culturing cell lines. However, it is recommended that antibiotics be present in primary cultures at the time of plating but be removed at the first medium change. Dissection instruments are from Roboz Surgical (Washington, DC). All reagents are from Sigma (St. Louis, MO) unless otherwise noted.

3.1. Neuronal Cell Cell Lines

Neuroblastoma and somatic cell hybrid cell lines derived from rodent and human neurons of either the central or peripheral nervous system can be maintained in a proliferative stage with expression of their neuronal properties in N2 medium (Table 2; Bottenstein and Sato, 1979; Bottenstein, 1984) on culture surfaces modified with polylysine and fibronectin (Bottenstein and Sato, 1980). These neuronal cell lines include B104, NIE-115, NS20, N2A, N18, NB41A3, LA-N-l, CHP 134, SK-N-SH-SY5Y, NX31, and NG108-15 (Bottenstein, 1983a,1984). In contrast, C62B glial, 3T3 fibroblast, or L6 skeletal muscle cell lines do not fare well in this medium (Bottenstein, 1983a). Figure 1 shows a culture of rat CNS B104 neuroblastoma cells maintained

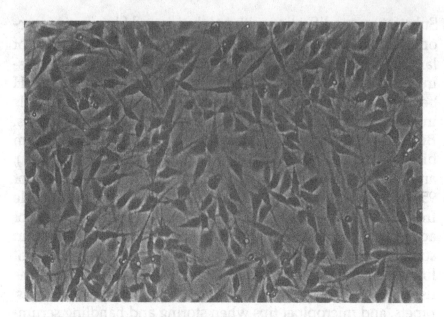

Fig. 1. Phase-contrast micrograph of rat CNS B104 cells of neuronal phenotype on day two after subculturing. Cells were maintained in N2 medium on polylysine and fibronectin-modified surfaces. Most of the cells are bipolar or tripolar and have long processes.

in N2 medium. This medium supports the division of these cells, enhanced process length, action potential generation, and similar levels of choline acetyltransferase activity compared to serum-containing medium. Adrenal medullary PC12 pheochromocytoma cells, used as sympathetic neuronal cell models, also do well in a variant of N2 medium using DMEM as the basal medium (Bottenstein, 1983a,1984).

1. For the rapid amplification of B104 cells, they can be grown in serum-containing medium, e.g., a 1:1 mixture of DMEM (GIBCO 430-2100, Grand Island, NY) and F12 (GIBCO 430-1700) containing 10% fetal bovine serum, before culturing in N2 medium on polylysine and fibronectin-modified surfaces.
2. Substratum modification: Culture surfaces are first coated with 0.05 mg/mL solution of poly-D-lysine (P-7886; M_r 30,000–70,000 in water) for 5 min at room temperature. After aspiration and a water wash, surfaces can be used

immediately or stored hydrated for a few hours. The final concentration of polylysine is 2 µg/cm². Fibronectin is added to the culture medium prior to addition of the cell inoculum and rapidly adsorbs to the culture surface. It is not necessary to precoat and wash the surface as for polylysine. This ECM glycoprotein is affinity-purified from human plasma using a gelatin-Sepharose column and elution with 0.05M Tris-HCl buffer (pH 7.4) containing 4M urea (ICN 821527, Costa Mesa, CA) at room temperature, followed by dialysis of the peak against 1M urea in phosphate buffered saline con-taining 0.01M sodium citrate (pH 7.2; MCB SX0445-1, Cincinnati, OH), and storage at 4°C only (Bottenstein and Sato, 1980). The final concentration of fibronectin is usually about 1 µg/cm², depending on the batch. It is added directly to the medium in the culture vessels and they are tilted back and forth immediately to ensure even distribution of this highly adhesive molecule.

3. Cells are harvested from subconfluent flask cultures with 0.05% trypsin/0.5 mM EDTA (Irvine Scientific 9336 and 9314, respectively, Santa Ana, CA), after washing with Hank's balanced salt solution (HBSS) lacking calcium and magnesium (Irvine Scientific 9230) to facilitate detachment. After about 1–2 min at room temperature, a sharp tap or shake releases the cells. An equal vol of 0.05% soybean trypsin inhibitor (Type 1-S; T-9003) is added, and the cells are transferred to a centrifuge tube. After centrifugation in a tabletop clinical centrifuge at the lowest speed for 2 min, the cell pellet is resuspended in N2 medium (Table 2). The cell inoculum is then added to culture vessels containing N2 medium that has been equilibrated with the incubator atmosphere and temperature (≥30 min).

4. B104 cells are capable of clonal growth (about 25 cells/cm²) and can be serially subcultured using this method. Moreover, they can be stored indefinitely under liquid nitrogen in N2 medium containing 10% dimethylsulfoxide (DMSO; Grade I, D-5879). After thawing, cells are centrifuged at the lowest speed in a tabletop centrifuge to remove the DMSO and are resuspended in plating medium.

3.2. Glial Cell Lines

Both rodent and human glial cells derived from tumors that express an astrocytic phenotype can be maintained in serum-free G2 medium (Table 2) on polylysine and fibronectin-modified surfaces (Michler-Stuke and Bottenstein, 1982a,b). Division of U251 MGsp, C62B, and RN-22 cells is supported by G2 medium, whereas LA-*N*-1 neuronal and WI38 fibroblast cells die after 2–3 d (Michler-Stuke and Bottenstein, 1982a). RN-22 schwannoma cells of PNS origin have an increased growth rate if FGF and hydrocortisone are deleted from G2 medium. All of these glial cell lines can be serially subcultured several times using this serum-free medium. Figure 2 shows rat CNS C62B glioma cells maintained in G2 medium. Cell division is supported by this medium, and processes are predominantly multipolar and highly ramified, rather than bipolar or tripolar, as seen in serum-supplemented medium (Bottenstein and Michler-Stuke, 1983). These cells can be amplified rapidly using DMEM supplemented with 10% fetal bovine serum. The procedure for growing C62B cells is the same as for B104 cells (*see* Section 3.1.) except that G2 medium is substituted for N2 medium.

Somatic cell hybrid cell lines expressing an oligodendrocytic phenotype can be successfully cultured in O1 medium (Table 2) on polylysine and fibronectin-modified surfaces. CO-13-7 and ROC-1 cells can be serially subcultured using this medium, whereas C62B glioma, 3T3 fibroblast, and B104 neuroblastoma cells cannot (Bottenstein, 1986). CO-13-7 cells maintained in O1 medium continue to express the oligodendrocyte marker galactocerebroside and exhibit a twofold increase in 2',3'-cyclic nucleotide 3'-phosphodiesterase specific activity after 4 d of treatment. The procedure for growing these two cell lines is identical to that for B104 cells (*see* Section 3.1.), except that O1 medium is substituted for N2 medium.

3.3. Embryonic Neurons

Dissociated rat brain cells obtained from midterm embryos and cultured in N2 medium on surfaces modified by polylysine and fibronectin give rise to predominantly neuronal cultures. Neurons can be readily identified in these cultures by immuno-

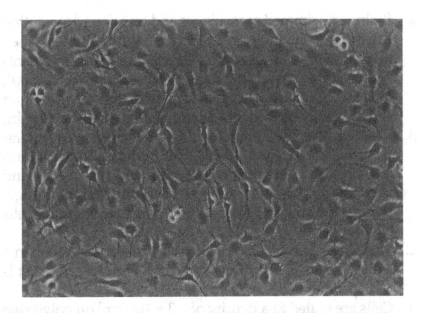

Fig. 2. Phase-contrast micrograph of rat CNS C62B cells of glial pheno-type on day three after subculturing. Cells were maintained in G2 medium on polylysine and fibronectin-modified surfaces. Most of the cells are multi-polar with branched processes.

staining for neuron-specific enolase (Marangos and Schmechel, 1987). Similar methods have been used to successfully culture neurons from specific regions of the nervous system, e.g., embry-onic rat cerebellum, locus coeruleus, substantia nigra, caudate nucleus, or hypothalamus and embryonic chick dorsal root gan-glia (Bottenstein, 1985; Bottenstein et al., 1980). Kingsbury et al. (1985) have maintained 8 d-old cerebellum-derived granule cell neurons in N2 medium with >90% purity of this cell type. Other examples are described by Bottenstein (1983b).

1. A pregnant rat containing ED13 or 14 embryos is anesthe-tized with methoxyflurane (Pitman-Moore 55685, Washing-ton Crossing, NJ), and the uterus is removed and placed in trituration medium (DMEM supplemented with 10% fetal bovine serum). The embryos are aseptically removed and placed in a Petri dish containing trituration medium. Whole brains are then dissected free using a microscope set up in a laminar flow hood.

2. Brains are trimmed of meninges and transferred to another Petri dish containing trituration medium. They are then minced with small scissors for 4 min. These tissue fragments are transferred to a 140-µm metal sieve (Bellco 1985-85000, #100 mesh, Vineland, NJ) prewetted with trituration medium and are gently pressed through using a glass pestle for 2 min.

3. This filtrate is then poured onto a prewetted filter apparatus (Millipore 25 mm, XX1002530, Bedford, MA) containing a 20-µm nylon mesh (Tetko 460, Elmsford, NY), and is stirred gently with a glass rod for 10–15 min. Single cells pass through the mesh by gravity and clumps of cells are retained.

4. This second filtrate is centrifuged at the lowest speed in a tabletop clinical centrifuge for 5 min, and the cell pellet is resuspended in N2 medium.

5. Cells are plated at a density of $\geq 3 \times 10^4 / cm^2$ on polylysine and fibronectin-modified surfaces (*see* Section 3.1.). For neuron-specific enolase immunostaining, it is convenient to plate the cells onto glass coverslips in 24-well tissue culture plates.

3.4. Neonatal Type 1 Astrocytes

Type 1 and 2 astrocytes appear to arise from different progenitor cells during development (Raff et al., 1984). They can be distinguished from each other by immunostaining criteria, and also exhibit some functional differences (Miller and Raff, 1984; Wilkin et al., 1983). Type 1 astrocytes are best identified by the presence of glial fibrillary acidic protein (GFAP) and the absence of other neural cell type-specific antigens (Raff et al., 1983). G5 medium (Table 2) was developed for optimal growth and differentiation of this cell type. When epidermal growth factor (EGF) is present in the culture medium, it is not necessary to modify the substratum with fibronectin. This is probably attributable to stimulation by EGF of ECM production by the astrocytes. Figure 3 illustrates the expansive process formation of neonatal rat cortical astrocytes in G5 medium, which is significantly greater than in serum-containing medium where they have a fibroblast-like morphology. Cells were immunostained for GFAP after 5 d

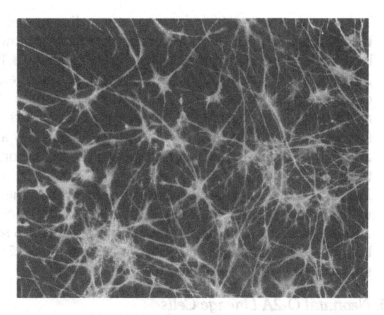

Fig. 3. Fluorescence micrograph of neonatal rat cortical astrocytes after 5 d in vitro. Cells were maintained in G5 medium on polylysine and fibronectin-modified surfaces. Cells were immunostained for glial fibrillary acidic protein (GFAP), and about 90% of the cells are GFAP-positive in these primary cultures.

in vitro. After the first subculture, the cells are ≥98% GFAP-positive. Type 1 astrocytes obtained with this method can be subcultured several times and have been maintained as long as 2 mo. The following procedure is a brief version of that described by Bottenstein and Michler-Stuke (1989).

1. After decapitation, whole brains are removed from 1–3 d old neonatal rat pups. The cerebral hemispheres are dissected free, minced with small scissors in trituration medium (DMEM supplemented with 10% fetal bovine serum), and the tissue fragments are transferred to a centrifuge tube for incubation with 0.25% trypsin in phosphate buffered saline for 8 min at 37°C. An equal vol of trituration medium is then added to inactivate the trypsin and inhibit clumping, followed by centrifugation in a tabletop clinical centrifuge for 2 min. The cell pellet is resuspended in serum-free DMEM.

2. The trypsin-treated fragments are then dissociated by trituration through a 10-mL pipet about 15 times, followed by filtration by gravity through a 230-μm metal sieve (Bellco 1985-85000, #60 mesh). Cells are then counted in a hemocytometer and plated at $\geq 1.25 \times 10^5$ cells/cm^2 in G5 medium on polylysine and fibronectin-modified surfaces (*see* Section 3.1.). The medium is changed the next day and at 4–5 d intervals thereafter, or sooner if the pH becomes too acidic.

3. Primary cultures can be subcultured just before confluency using 0.1% trypsin/0.9 mM EDTA in HBSS lacking calcium and magnesium with inactivation of trypsin by 0.1% soybean trypsin inhibitor in HBSS (Irvine Scientific 9222). A split ratio of 1:2 or 1:4 is suggested.

3.5. Neonatal O-2A Lineage Cells

O-2A Lineage cells include the bipotential O-2A glial progenitor and its progeny: oligodendrocytes and type 2 astrocytes. These cells can be identifed by the following immunostaining criteria: O-2A cells express A2B5 gangliosides but not galactocerebroside (GalC) or glial fibrillary acidic protein (GFAP); oligodendrocytes express GalC but not GFAP, and when immature, also express A2B5 antigens; type 2 astrocytes express A2B5 antigens and GFAP but not GalC.

The number of O-2A progenitors can be significantly amplified in culture by exposure to growth factor-containing conditioned medium produced by B104 neuroblastoma cells. Figure 4 shows a field of bipolar, A2B5-positive O-2A progenitors after 4 d in vitro. Differentiation of some progenitors into primarily oligodendrocytes will occur in the presence of the growth factor, resulting in large numbers of them 7–10 d after plating. Figure 5 shows cells of typical oligodendrocyte morphology that are prominent in 10 d-old cultures. They exhibit highly branched processes and express the oligodendrocyte-specific marker GalC (Bottenstein et al., 1988). However, if the growth factor is removed earlier or cultures have never been exposed to it, all of the progenitors rapidly differentiate into oligodendrocytes. Since the progenitors are fewer in number in the latter condition, the final

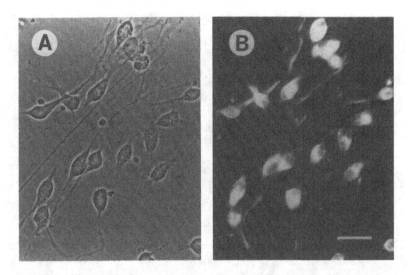

Fig. 4. Neonatal rat brain glial progenitors after 4 d in vitro. Cells were maintained in O3 medium containing 10 μg protein/mL of B104 cell conditioned medium on polylysine and fibronectin-modified surfaces. Cells were mechanically dissociated and applied to a Percoll density gradient to remove cellular debris. Cultures were fixed with 3.7% p-formaldehyde in phosphate buffered saline and immunostained with A2B5 antibody. **(A)** Phase-contrast micrograph shows a field of bipolar glial progenitors; **(B)** fluorescence micrograph of the same field shows that these progenitors are A2B5-positive. Double immunostaining for galactocerebroside or glial fibrillary acidic protein indicated that the bipolar cells did not express these antigens. Bar, 20 μm.

number of oligodendrocytes is lower. In contrast, if serum-containing medium replaces the growth factor, differentiation of progenitors into type 2 astrocytes is induced and fewer oligodendrocytes are produced. The active factor in serum is thought to be ciliary neurotrophic factor (Hughes et al., 1988). We have also described methods for (1) further enrichment of O-2A progenitors that involve differential cell adhesion only (Bottenstein and Hunter, 1990) or include addition of a heparin treatment step (Hunter and Bottenstein, 1990); and (2) culture of mature O-2A progenitor cells from 30 or 60 d-old rat brain (Hunter and Bottenstein, 1991). The following methods for neonatal O-2A lineage cells are described in greater detail by Bottenstein and Hunter (1990).

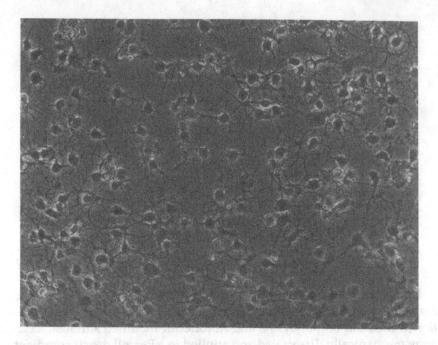

Fig. 5. Phase-contrast micrograph of neonatal rat brain cells after 10 d in vitro. Cells were maintained in O3 medium containing 10 μg protein/mL of B104 conditioned medium on polylysine and fibronectin-modified surfaces. Cells of characteristic oligodendrocyte morphology are prominent; they exhibit multiple highly branched processes. Galactocerebroside immunostaining of these cultures confirmed the abundance of oligodendrocytes.

1. Growth factor-containing conditioned medium from B104 cell cultures is prepared as follows: B104 cells are plated into N4 medium (identical to N2 medium except the transferrin concentration is reduced to 1 μg/mL; *see* Table 2) on polylysine and fibronectin-modified surfaces as described earlier at a density of 10^4 cells/cm². Four days later the conditioned medium is removed, phenylmethylsulfonyl fluoride (protease inhibitor; P-7626) is added to give a final concentration of 1 μg/mL, and it is centrifuged at 4°C for 15 min at 3500 rpm to remove nonadherent cells/debris. The supernatant is aliquoted and stored at −70°C. Freeze-thawing in excess of 3× results in significant loss of activity. The protein concentration of this conditioned medium should be about 20–35 μg/mL.

2. Newborn rat pups (1–3 d old) are decapitated and whole brains are dissected free. The olfactory bulb and cranial nerves are severed with a spatula and the brains are placed in a Petri dish containing trituration medium (DMEM supplemented with 10% calf serum and antibiotics: penicillin/streptomycin/neomycin/gentamycin at 50, 50, 10, and 50 µg/mL, respectively; GIBCO 600-5640 and 600-5710). Meninges are trimmed from the brain, and they are minced with small scissors for 4 min. Brain fragments are then transferred to a 380 µm metal sieve (Bellco 1985-85000, #40 mesh) prewetted with trituration medium, and pressed through gently with a glass pestle (Bellco 1985-11516) for 2 min. This filtrate is then applied to a prewetted 140 µm metal sieve (Bellco #100 mesh) and pressed through gently with the plunger of a 5 mL disposable syringe for 2 min. Finally, the second filtrate is added to a filter apparatus (Millipore 25 mm, XX1002530) with a 20 µm nylon mesh (Tetko 460) and stirred gently with a glass rod. Single cells pass through the mesh by gravity and are collected in a 50 mL centrifuge tube.

3. The final filtrate is centrifuged at the lowest speed in a tabletop clinical centrifuge for 3 min, and the cell pellet is resuspended in O3 medium with penicillin/streptomycin/neomycin without gentamicin at the above concentrations, and counted in a hemocytometer. About $25–30 \times 10^6$ cells/brain are usually obtained. Cells are plated at 5×10^5 cells/cm^2 in O3 medium containing 10 µg/mL of B104 conditioned medium protein (about 33% v/v) on polylysine and fibronectin-modified surfaces.

4. The above procedure is simple and useful for most studies. However, if cultures substantially free of cellular debris and erythrocytes are desired, Percoll gradient centrifugation can be performed after dissociation and prior to plating. First, about 10^8 brain cells are added to a 10 mL Oak Ridge-style polycarbonate tube (Thomas Scientific 2611-B18, Swedesboro, NJ), followed by 2.8 mL of 80% Percoll/20% 1.25M sucrose (P-1644, 5-9378) and then trituration medium to bring the vol to 9.5 mL. The tube is centrifuged at 30,000g

for 45 min at 4°C in a Ti50 rotor in a Beckman L8M ultra-centrifuge, or its equivalent. The upper layers of debris are aspirated, and the clear middle cell layer (2–3 mL) is transferred by pipet to a 15 mL centrifuge tube. Remaining in the original tube are an erythrocyte layer and an amber bottom layer of Percoll. Cells are washed by adding 12 mL of trituration medium, followed by centrifugation at 190g in a tabletop clinical centrifuge for 10 min. A second wash is performed with HBSS (Irvine Scientific 9222, Santa Ana, CA), and then the cells are resuspended in O3 medium and counted in a hemocytometer. Cells are plated at a density of $1.3–2.5 \times 10^5/cm^2$ in O3 medium containing 10 µg/mL (33% v/v) B104 conditioned medium protein on polylysine and fibronectin-modified surfaces. Although there is substantial loss of cells during this procedure (10% yield) the plating efficiency is 2.5–5-fold greater compared to the standard method above. Subsequent immunostaining of Percoll gradient-processed cells shows significantly lower background.

With either method after 4 or 8 d in vitro, O-2A progenitors are increased by 12–15-fold over the control unconditioned N4 medium. After 8 d in vitro, the number of oligodendrocytes are increased fourfold over the control. After 3 wk in vitro, the activity of oligodendrocyte-specific 2',3'-cyclic nucleotide 3'-phosphodiesterase is increased 12-fold over the control (Bottenstein et al., 1988). To hasten differentiation into oligodendrocytes, conditioned medium can be omitted at the onset or at a later time. Fewer O-2A progenitors will be generated, however, resulting in decreased numbers of oligodendrocytes. For example, if conditioned medium is removed after 4 d in vitro, only a fivefold increase in the number of oligodendrocytes occurs, compared to the control, but the percentage of oligodendrocytes is about 50% greater. To induce differentiation of O-2A progenitors into type 2 astrocytes, conditioned medium is replaced with O3 medium containing 10% fetal bovine serum. If this is done at 4 d in vitro, after three more days type 2 astrocytes are increased 32-fold compared to the control. However, type 1 astrocytes also proliferate vigorously under these conditions.

4. Conclusion

The use of serum-free chemically defined media for culture of neural cells has many advantages over the use of serum-supplemented medium (*see* Section 2.3.). The range, reproducibility, and ease of interpretation of experiments are increased if chemically defined media are used. Suppression of fibroblast overgrowth is another clear benefit, as well as enrichment for specific cell types. The serum-free culture methods described here have proven beneficial in many studies of the proliferation and differentiation of neural cells. In addition, they have been invaluable in the identification and bioassay of cell-produced factors that mediate cell-cell interactions.

References

Bottenstein J. (1983a) Growth requirements of neural cells in vitro. *Adv. Cell. Neurobiol.* **4**, 333–379.

Bottenstein J. (1983b) Defined media for dissociated neural cultures, in *Current Methods in Cellular Neurobiology* (Barker J. and McKelvy J, eds.), John Wiley, New York, pp. 107–130.

Bottenstein J. (1984) Culture methods for growth of neuronal cell lines in defined media, in *Methods in Molecular and Cell Biology, vol. 4: Methods for Serum-free Culture of Neuronal and Lymphoid Cells* (Barnes D., Sirbasku D., and Sato G., eds.), Alan Liss, New York, pp. 3–13.

Bottenstein J. (1985) Growth and differentiation of neural cells in defined media, in *Cell Culture in the Neurosciences* (Bottenstein J. and Sato G., eds.), Plenum, New York, pp. 3–43.

Bottenstein J. (1986) Growth requirements in vitro of oligodendrocyte cell lines and neonatal rat brain oligodendrocytes. *Proc. Natl. Acad. Sci. USA* **83**, 1955–1959.

Bottenstein J. (1988) Advances in vertebrate cell culture methods. *Science* **239**, G42,G48.

Bottenstein J. and Hunter S. (1990) Culture methods for oligodendrocyte cell lines and oligodendrocyte-type 2 astrocyte lineage cells, in *Methods in Neurosciences, vol. 2: Cell Culture* (Conn P. M., ed.), Academic, San Diego, pp. 56–75.

Bottenstein J. and Michler-Stuke A. (1983) Proliferation of glial-derived cell lines in serum-free defined medium, in *Developing and Regenerating Vertebrate Nervous Systems, vol 6: Neurology and Neurobiology* (Coates P., Markwald R., and Kenny A., eds.), Alan Liss, New York, pp. 185–189.

Bottenstein J. and Michler-Stuke A. (1989) Serum-free culture of dissociated neonatal rat cortical astrocytes, in *A Dissection and Tissue Culture Manual of the Nervous System* (Shahar A., de Vellis J., Vernadakis A., and Haber B., eds.), Alan Liss, New York, pp. 109–111.

Bottenstein J. and Sato G. (1979) Growth of a rat neuroblastoma cell line in serum-free supplemented medium. *Proc. Natl. Acad. Sci. USA* **76,** 514–517.

Bottenstein J. and Sato G. (1980) Fibronectin and polylysine requirement for proliferation of neuroblastoma cells in defined medium. *Exp. Cell Res.* **129,** 361–366.

Bottenstein J., Hunter S., and Seidel M. (1988) CNS neuronal cell line-derived factors regulate gliogenesis in neonatal rat brain cultures. *J. Neurosci. Res.* **20,** 291–303.

Bottenstein J., Skaper S., Varon S., and Sato G. (1980) Selective survival of neurons from chick embryo sensory ganglionic dissociates utilizing serum-free supplemented medium. *Exp. Cell Res.* **125,** 183–190.

Hughes S.M., Lillien L., Raff M., Rohrer H., and Sendtner M. (1988) Ciliary neurotrophic factor induces type-2 astrocyte differentiation in culture. *Nature* **335,** 70–72.

Hunter S. and Bottenstein J. (1989) Bipotential glial progenitors are targets of neuronal cell line-derived growth factors. *Develop. Brain Res.* **49,** 33–49.

Hunter S. and Bottenstein J. (1990) Growth factor responses of enriched bipotential glial progenitors. *Develop. Brain Res.* **54,** 235–248.

Hunter S. and Bottenstein J. (1991) O-2A glial progenitors from mature brain respond to CNS neuronal cell line-derived growth factors. *J. Neurosci. Res.* **28,** 574–582.

Kingsbury A., Gallo V., Woodhams P., and Balazs R. (1985) Survival, morphology and adhesion properties of cerebellar interneurons cultured in chemically defined and serum-supplemented medium. *Develop. Brain Res.* **17,** 17–25.

Marangos P. and Schmechel D. (1987) Neuron specific enolase, a clinically useful marker for neurons and neuroendocrine cells. *Ann. Rev. Neurosci.* **10,** 269–295.

Michler-Stuke A. and Bottenstein, J. (1982a) Proliferation of glial-derived cells in defined media. *J. Neurosci. Res.* **7,** 215–228.

Michler-Stuke A. and Bottenstein J. (1982b) Defined media for growth of human and rat glial-derived cell lines, in *Cold Spring Harbor Conferences on Cell Proliferation, vol. 9* (Sirbasku D., Sato G., and Pardee A., eds.), Cold Spring Harbor Laboratory, Cold Spring Harbor, New York, pp. 959–971.

Michler-Stuke A., Wolff J.R., and Bottenstein, J. (1984) Factors influencing astrocyte growth and development in defined media. *Int. J. Develop. Neurosci.* **2,** 575–584.

Miller R. and Raff M. (1984) Fibrous and protoplasmic astrocytes are biochemically and developmentally distinct. *J. Neurosci.* **4,** 585–592.

Raff M., Miller R.H., and Noble, M. (1983) A glial progenitor cell that develops in vitro into an astrocyte or an oligodendrocyte depending on culture medium. *Nature* **303**, 390–396.

Raff M., Abney E.R., and Miller R.H. (1984) Two glial cell lineages diverge prenatally in rat optic nerve. *Develop. Biol.* **106**, 53–60.

Wilkin G., Levi G., Johnstone S., and Riddle P. (1983) Cerebellar astroglial cells in primary culture: expression of different morphological appearances and different ability to take up [^{3}H]D-aspartate and [^{3}H]GABA. *Develop. Brain Res.* **10**, 265–277.

Detection and Analysis
of Growth Factors
Affecting Neural Cells

Marston Manthorpe, David Muir,
Brigitte Pettmann, and Silvio Varon

1. Introduction

1.1. Cells as Probes for "Growth Factors"

Individual cellular behaviors are controlled by extrinsic signals within the environmental milieu. These signals may be ions, nutrients, hormones, or, pertinent to this review, a collection of proteins generally termed "growth factors." Growth factors, which may have autocrine, paracrine, or endocrine origins, can be presented to cells in a soluble form from the extracellular fluid, or in an insoluble form from adjacent cells or extracellular matrices. Growth factors operate on cells through specific high-affinity cell-surface receptors that when occupied, produce a series of biochemical reactions that collectively comprise alterations of cell behavior. An important field of research aims to understand how, within tissues, individual and combinations of growth factors control cell behaviors. Most of the present information on growth factors derives from the direct examination of factor influences on cells in culture (Varon et al., 1983; Manthorpe et al., 1989). This chapter will focus on nervous system cells (i.e., neurons and glial cells) and will summarize the ways that in vitro cellular behaviors can be affected by growth

From: *Neuromethods, Vol. 23: Practical Cell Culture Techniques*
Eds: A. Boulton, G. Baker, and W. Walz ©1992 The Humana Press Inc.

regulating factors. Since our context is "Neuromethods," particular attention will be given to detail different methods for detecting and quantifying the action of growth-regulating factors on neural cells.

1.2. Survey of Factor Influences on Neurons and Glia

A myriad of individual proteins have been reported to influence specific behaviors of cultured neurons and glial cells. Table 1 presents a list of the better-characterized factors categorized according to the type of influences they have on neurons or glial cells. Neuronotrophic factors (NTFs) are those proteins that support the survival or general growth of neurons. The first factor to be discovered, Nerve Growth Factor (NGF), remains by far the best characterized NTF, and in some ways, it is almost unfair to place this factor in the same list as the remaining ones that are newly discovered and much less well-characterized than NGF. Two factors, Brain-derived Growth Factor (BDNF) and Neurotrophin-3 (NT-3), are related to NGF and to one another and belong to what is being recognized as an NGF family of NTFs. Neurite-promoting factors (NPFs) do not support neuronal survival or general growth by themselves but are required in addition to an NTF for the neuron to extend an axonal or dendritic (collectively called "neuritic") process. The best characterized NPFs are the extracellular matrix molecules, such as laminin and fibronectin, and the cell-surface adhesion molecules, such as the cadherins. There are also factors that inhibit neurite outgrowth, such as certain extracellular matrix proteoglycans. Factors that act on glial cells include mitogens, such as the fibroblast growth factors (FGFs) and platelet-derived growth factors (PDGFs). The cell adhesion molecules promote and sustain glial cell attachment to extracellular matrices and to other cells, and the antiproliferative factors inhibit mitogen-stimulated glial cell proliferation.

Several points should be made regarding the list of factors in Table 1. Most of the factors can act on either neuronal or glial cells. A notable example is NGF, which has been studied for almost 40 years (Levi-Montalcini, 1952,1987). NGF supports

Table 1
Growth Factors Affecting Neural Cells

Factors affecting neurons	Selected references
Neuronotrophic factors	
NGF	Levi-Montalcini, 1987; Thoenen et al., 1988; Varon et al., 1990
FGFs	Walicke et al., 1986; 1990; Gospodarowicz et al., 1986; Morrison et al., 1988
Epidermal Growth Factor	Morrison et al., 1987
Brain-derived Growth Factor	Barde et al., 1987
Neurotrophin-3	Maisonpierre et al., 1987
Ciliary Neuronotrophic Factor	Lam et al., 1991; Manthorpe and Varon, 1985; Manthorpe et al., 1989
Activin	Schubert et al., 1990
Purpurin	Schubert et al., 1986
Neurite Promoting Factors	
Laminin	Manthorpe et al., 1983b; Rogers et al., 1983; Wewer et al., 1983; Baron van Evercooren et al., 1982; Faivre-Bauman et al., 1984
Merosin	Engvall et al., 1992; Ehrig et al., 1990
Fibronectin	Manthorpe et al., 1983b; Rogers et al., 1983; Akers et al., 1981
Collagens	Adler et al., 1979
Cell Adhesion Molecules	Seilheimer and Schachner, 1988; Letourneau et al., 1988; Drazba and Lemmon, 1990; Bixby et al., 1988; Neugebaur et al., 1988; Tomaselli et al., 1988
S100	Kligman, 1982
NGF	Greene et al., 1987; Rudkin et al., 1989; Rydel and Greene, 1987
Glia Maturation Factor	Lim et al., 1990
Proteoglycans	Hantaz-Ambroise et al., 1987; Gurwitz and Cunningham, 1987

Neurite-inhibiting factors

Proteoglycans	Muir et al., 1989a
Myelin-derived proteins	Caroni and Schwab, 1988; Schwab and Caroni, 1988
Proteases	Gurwitz and Cunningham, 1988

Mitogens for neuron-related cells

NGF	Cattaneo and McKay, 1990; Ludecke and Unsicker, 1990
FGF	Rydel and Greene, 1987; Claude et al., 1988

Antiproliferative factors for Neuron-related cells

NGF	Lazarovici et al., 1987; Greene et al., 1987
Glia Maturation Factor	Lim et al., 1986

Factors Affecting Glial Cells

Mitogens

FGFs	Besnard et al., 1989; Perraud et al., 1988b
Platelet-derived Growth Factor	Besnard et al., 1987; Pringle et al., 1989; Hunter and Bottenstein, 1990; Davis and Stroobant, 1990
Transforming Growth Factor-β	Ridley et al., 1989; Eccleston et al., 1989a
Interleukins	Benveniste and Merrill, 1986
Proteases	Perraud et al., 1987; Puro et al., 1990
Protease inhibitors	Perraud et al., 1988a
Insulin-like growth factors	McMorris and Dubois-Dalq, 1988
Glia Maturation Factor	Lim et al., 1990; Hunter and Bottenstein, 1990; Bosch et al., 1989
Fibronectin	Goetschy et al., 1987; Muir et al., 1989b
Laminin	Muir et al., 1989b

Adhesion promoting factors

Laminin	Dillner et al., 1988; Kleinman et al., 1985
Merosin	Engvall et al., 1991
Fibronectin	Rogers et al., 1987; Cardwell and Rome, 1988a
Collagens	Adler et al., 1979
Cell Adhesion Molecules	Edelman, 1987

Proliferation inhibitors

Antiproliferative Proteins	Muir et al., 1990b
Astrostatin	Rogister et al., 1990
Glial Growth Inhibitory Factor	Kato et al., 1987
Interleukin-2	Saneto et al., 1986; Hunter and Bottenstein, 1990
Type-I, IV Collagens	Goetschy et al., 1987; Eccelstein et al., 1989b
Glia Maturation Factor	Lim et al., 1986
Ciliary Neurotrophic Factor	Lillien et al., 1988; Hughes et al., 1989
Colony Stimulating Factor	Sawada et al., 1990
Transforming Growth Factor-β	Hunter and Bottenstein, 1990; Davis and Stroobant, 1990
Epidermal Growth Factor	Hunter and Bottenstein, 1990

Differentiation factors

FGF	Perraud et al., 1988b; Pettman et al., 1991
NGF	Yavin et al., 1992
Fibronectin	Cardwell and Rome, 1988b

neuronal survival and can stimulate neurite outgrowth from peripheral and central neurons in vitro and in vivo (Varon et al., 1990). NGF can also promote the proliferation of brain-derived neuroblasts (Cattaneo and McKay, 1990) and neuroblastoma cells (Ludecke and Unsicker, 1990) and inhibit the proliferation and promote several specific differentiative properties of pheochromocytoma (PC12) cells (Greene et al., 1987; Rydel et al., 1987). NGF also inhibits the proliferation and promotes the differentiation of C-6 glioma cells (Yavin et al., 1992) and can even act on cells outside the nervous system by stimulating the proliferation of lymphocytic (Otten et al., 1989), hemopoietic (Matsuda et al., 1988), carcinoma (Kahan and Kramp, 1987; Goretzki et al., 1987; Rakowicz and Koprowski, 1989), and spleen (Dean et al., 1987; Thorpe and Perez-Polo, 1987) cells. Another example of factor multiplicity is the Ciliary Neuronotrophic (CNTF) and Fibroblast Growth (FGF) Factors that can support the survival and general growth of neurons (Manthorpe et al., 1986b; Walicke et al., 1986a). Both CNTF and FGF can also act on glial cells. For example, FGFs are mitogens for astroglial cells (Perraud et al., 1988a) and CNTF inhibits the proliferation and causes the conversion of 02A glial precursor cells into astroglial cells (Lillien et al., 1988; Hughes et al., 1989). Other examples include the extracellular matrix glycoproteins, such as laminin and fibronectin, which promote axonal outgrowth from various neurons (Manthorpe et al., 1983b; Rogers et al., 1983) and also promote the adhesion and differentiation of glial cells (Dillner et al., 1988).

In addition to the the various promoting factors, there exist factors that inhibit cellular responses to other factors. For example, a 55 kDa neural antiproliferative protein (NAP) inhibits the proliferation of mitogen-stimulated Schwann cells (Muir et al., 1990b). Other proteins such as Glial Maturation Factor, Interleukins, and astrostatin inhibit astroglial cell proliferation (Lim et al., 1986; Saneto et al., 1986; Rogister et al., 1990). There are also substances that inhibit neurite outgrowth from neurite-promoting factor-stimulated neurons (Caroni and Schwab, 1988; Schwab and Caroni, 1988; Muir et al., 1989a; Manthorpe et al., 1990).

At least three concepts have emerged from the repertoire of factor actions:

1. A given factor can act on either neuronal or glial cells, as well as outside of the nervous system;
2. A given factor may elicit similar or dissimilar cellular responses in different cells;
3. Several factors can act on one cell type in agonistic, antagonistic, or independent manners.

Such concepts raise interesting questions regarding how factor specificity is regulated in vivo. For example, what determines whether a cell will be exposed to a given factor? Even if the cell is exposed, what determines whether the cell will have factor receptors and if so, how will the cell respond to the accessible factor? When a cell responds to multiple factors, what is the hierarchy of responses? Can one factor act on the cell to cause it to produce another factor and can one factor cause a cell to respond more or less efficiently to another factor? Answers to such basic questions will require the extensive use and further development of quantitative in vitro methods for the detection and isolation of factors and for the determination of relative factor potencies and response hierarchies.

1.3. Principles of Quantification

The importance of cell culture methods for quantifying the biological activities of growth factors was recognized following the discovery of NGF (Levi-Montalcini, 1987). In this early case (Cohen et al., 1954), intact embryonic day 10 (E10) chick sympathetic ganglia were cultured as an explant in a plasma clot containing 33% each of rooster plasma, embryonic day 10 chick extract, and the serially diluted NGF-containing test sample. Within 18 h, the neurons inside ganglia not receiving NGF would degenerate and die. In the presence of NGF, ganglionic neurons survived and, possibly aided by other factors present in the culture system, would extend neurites into the surroundings to produce a neuritic halo. The diameter and density of this halo was related to the NGF dose and the ganglionic neuritic response was scored as 0–4+, with 0 being the response without NGF and 4+ being the maximum possible NGF response (*see* Fig. 1 A–D). The 4+ response was assigned as that elicited by one Biological Unit (BU) per mL of NGF. The scoring system was such that a 2+ or 1+ response could be elicited by, respectively, a 1/2 or 1/4 dilution of the sample that barely gave a 4+ response. Figure 1E shows a dose-response curve for an NGF-containing sample. This particular test sample contains 0.3 mg/mL of NGF that after dilution by 30,000-fold elicits a 4+ response. Therefore, its titer is defined as 30,000 BU/mL and its specific activity as 100,000 BU/ mg or 10 ng/BU.

Such an assay was used for 30 yr to monitor NGF biological activity during biochemical purification and treatments and to show that purified NGF is able to elicit its half-maximal effects at very low concentrations ($10^{-9}M$). The main limitations of the explant halo assay are that it: (1) is time-consuming to set up individual ganglionic explant cultures; (2) only allows a limited number of individual samples to be assayed, with about 120 ganglia (from four chick embryos) one could assay in triplicate about ten samples using four successive tenfold dilutions of NGF; and (3) does not readily allow one to measure potential NGF influences on the cells within the explant, such as neuronal cell survival and hypertrophy, neurite outgrowth, biochemical changes, and glial cell responses.

Ideally, quantitative growth factor bioassays using neural cells as probes should be simple, well-controlled, and reproducible. If possible, bioassays should:

1. Use defined populations of purified, dissociated cells;
2. Use well-defined culture media;
3. Be of the shortest possible duration, preferably 24 h or less;
4. Be convenient and inexpensive enough to allow thousands of assays to be performed daily;
5. Allow assessments of individual responses or, if desired, multiple cellular responses using the same culture; and
6. Be amenable to rapid and error-free data analyses.

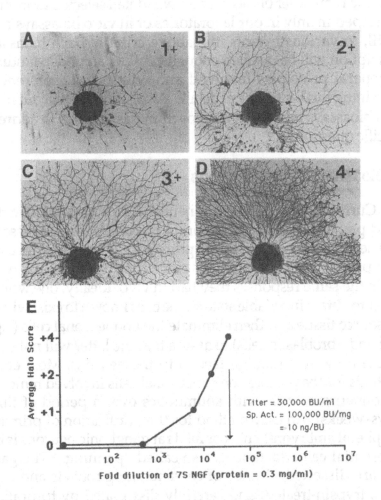

Fig. 1. Explant bioassay for NGF. Embryonic day 10 (E10) chick paravertebral sympathetic ganglionic explants were cultured in a rooster plasma clot containing different doses of purified β NGF according to Cohen et al. (1954). After 18 h, the explants were stained with silver and scored for the extent of neurite outgrowth. No neurite outgrowth occurs in the absence of NGF (not shown). Depending on NGF dose, neurite halos were assigned scores of 1+ (A), 2+ (B), 3+ (C), or maximally 4+ (D). A dose-response curve of halo score and NGF dose is shown in (E). The titer of this particular NGF sample is based on the dose just giving a maximal response and is 30,000 BU/mL of NGF sample which, based on the 0.3 mg/mL NGF sample protein content, represents a specific activity of 10 ng/BU.

The remainder of this chapter will detail selected examples developed mainly in our laboratories of in vitro bioassays that fulfill, to various extents, the above criteria. Each bioassay description will be detailed enough to allow an understanding of general principles involved and will include references containing further technical details. Also, we will point out how such bioassays have been employed in past studies to address specific questions.

2. Bioassays Using Neurons

Currently popular in vitro neuronal bioassays involve the rapid preparation of purified primary populations of dissociated neurons. Rapidly prepared primary cells are preferred over post-primary cells or cell lines since they are more likely to retain the same responses they had in vivo. Ideally, one would prefer to obtain in a viable state all the cells known to exist within the source tissue and then eliminate the nonneuronal cells (e.g., glial and fibroblastic cells) to assure that the latter will not contribute growth-regulating factors to the neurons. Most early methods for the purification of neuronal cells involved removal of nonneuronal cells with antimitotics over a period of time (days–weeks). An easy method for the purification of primary peripheral embryonic chick or fetal rat ganglionic neurons is by differential cell attachment, also called "panning." The ganglia are dissected, cleaned of contaminating rootlets and capsules, trypsin-treated, and carefully dissociated by trituration through a constricted Pasteur pipet in a protein-containing medium (e.g., Dulbecco's Modified Eagle's Medium plus 10% fetal calf serum). During dissociation, the pipet diameter, force of trituration, and number of up-and-down passages should be optimized to assure a maximal neuronal cell yield per ganglion, preferably close to the number of neurons previously established by histology to be present in the ganglion. One tissue from which nearly *all* the component neurons can be dissociated is the chick embryo ciliary ganglion (Manthorpe et al., 1985,1989). The cell suspension, usually 10 mL/40–100 ganglia equivalent of cells, is then seeded with fetal calf serum-containing medium into

100 mm diameter plastic tissue culture plates. After a defined period, usually 2.5 h, most of the ganglionic nonneuronal cells will attach to the culture plate. At that time, the unattached neurons are collected and used for bioassays.

Embryonic chick or fetal rat brain and spinal cord neurons can be purified from most nonneuronal cells by simply culturing them at low density in serum-free defined medium for 24 h, after which time practically only neuronal cells will live (Barbin et al., 1984b; Varon et al., 1984). Other methods, such as retrograde labeling in vivo with fluorescent dyes followed by tissue dissociation and fluorescence-activated cell sorting have been used to purify spinal motor neurons (Calof and Reichardt, 1984; Martinou et al., 1989). Velocity sedimentation centrifugation of the cell suspension helps to purify larger, more dense motoneurons from a mixed cell population (Juurlink et al., 1990).

2.1. Neuronal Survival and Growth Assays

The current view is that neurons in vivo require an adequate NTF supply from their postsynaptic target cell and/or their associated cells (Varon et al., 1988,1989). Most purified perinatal ganglionic neurons will not survive for even 24 h in vitro without the addition of NTFs. Thus, the presence of living neurons in vitro would indicate and possibly reflect quantitatively the presence of an NTF supply. The number of surviving neurons can be determined directly under a microscope after their fixation or indirectly by measuring the amount of a vital dye that they can metabolize.

2.1.1. Neuronal Counting Assays

The following method, or its variations, have been used to monitor the biological activity of NTFs for peripheral (PNS) and central nervous system (CNS) neurons (Manthorpe et al., 1981a; Longo et al., 1982a). Each well of 96-well Costar A/2 microplates is coated for 1 h with 25 μL of polyornithine solution (0.1 mg/mL in borate buffer, pH 8.0) followed by water washing (2 × 50 μL) and a 2 h coating with 25 μL of laminin solution (1 μg/mL in buffered balanced salt solution, e.g., Eagle's salts, BSS). The laminin-containing wells can be used immediately or stored for months at –20°C individually wrapped in plastic sandwich bags. The wells are washed with 50 µL of the culture medium and supplied with 25 µL of fresh medium. It is critical not to allow the laminin coating to dry out (and therefore to denature and become inactivated) in the aspirated wells. An effective technique is to aspirate and immediately add medium to one row at a time. Sterile multiple channel (8 or 12 tipped) micropipets and vacuum-connected metal aspirators are particularly useful for these fluid addition/removal steps. The test sample is diluted appropriately in tubes, and 25 µL of the diluted sample is added to the first well in a given row (for 2^{11} twofold dilutions) or column (for 2^7 twofold dilutions) and the samples serially diluted twofold directly in successive wells. Other dilution schemes (e.g., three-, five-, or tenfold) can be used as long as care is taken to assure complete mixing of the well contents. Finally, the test neuron suspension is added in 25 µL of medium containing 500–2000 neurons/well. A typical daily dissection of 100 ciliary or dorsal root ganglia yields about 10^6 purified neurons (i.e., 10^4 neurons/ganglia) for setting up 10–20 microplates (i.e., 960–1920 individual cultures at 500–1000 neurons/well).

The culture plates are then incubated for 24 h followed by 20 min fixation with 100 µL 4% glutaraldehyde in medium. This fixative will firmly fix all attached cells to the bottom of each well. The fixative is removed by flicking, tap water is added to fill each well, and a glass plate is used to cover the microplate without trapping air bubbles. This procedure eliminates the fluid meniscus in the well and allows the well contents to be examined by phase contrast microscopy free of Schlieren distortion. Viable neurons can easily be recognized by their phase-bright appearance and possession of neurites terminating with growth cones (*see* Fig. 2A,B).

The number of neurons/well at 24 h is determined within 4.5×0.1 mm diametral strips (or other area depending on the available magnification and eyepiece grid) and multiplied by 35 [= $\pi(2.25$ mm well radius$)^2/4.5$ mm $\times 0.1$ mm] to yield the total number of neurons/well. The number of neurons/well at 2 h represents the number of input neurons and is useful for comparison to the number surviving at subsequent times. To determine

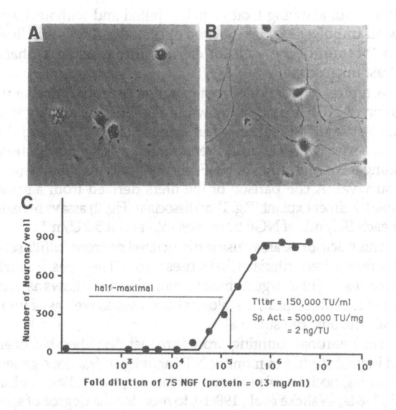

Fig. 2. Dissociate bioassay for NGF. Embryonic day 8 (E8) chick dorsal root ganglionic neurons were purified by differential attachment (panning) and cultured by seeding 1000 neurons into each microwell containing different doses of purified β NGF according to Manthorpe et al. (1981a). After 24 h, the cultures were fixed and the number of neurons/well determined from diametral counts. No survival occurs in the absence of NGF (A), but most of the 1000 input neurons survived and extended processes with a high dose of NGF (B). A dose response curve of neuronal number/well and NGF dose is shown in (C). This NGF sample is the same as that assayed with explants in Fig. 1. The titer here is based on half-maximal activity and is 150,000 Trophic Units (TU)/mL of NGF sample or 2 ng/TU.

the *exact* number of input cells, it is important to set up a small number of wells (e.g., six) in a separate plate for fixation and counting after 2 h in vitro. Since not all the neurons may attach strongly by 2 h, the well contents can be preserved by filling the well with fixative (4% glutaraldehyde in BSS) just to the rim of the well and then placing a coverslip or glass plate over the well without allowing fixative to be spilled and without trapping air bubbles. Any unattached or dislodged cells will float to the bottom of the well for easy counting using a phase contrast microscope.

A typical plot of dorsal root ganglion neuronal number vs titration of NGF activity is shown in Fig. 2C. The activity in a given test sample is expressed in Trophic Units (TUs) and the titer of a sample in TU/mL is represented as the fold-dilution, yielding a concentration of NGF, eliciting half-maximal neuronal survival. A comparison of the titers derived from a same sample by either explant (Fig. 1) or dissociate (Fig. 2) assays reveals that each BU/mL of NGF translates into about 5 TU/mL.

This microplate assay using dissociated neurons fulfills several of the desired criteria in that it uses purified neurons, is short, permits testing of a large number of samples, and allows assessment of morphological/histological analyses as well as of individual neuronal cell survival.

The neuronal counting procedure just described has been used in past studies to monitor NTF activity in fractions generated during biochemical purification (Manthorpe et al., 1980; Barbin et al., 1984a; Walicke et al., 1986b), to monitor the degree of specific inactivation of NTF by anti-NTF antibodies (Manthorpe et al., 1981a), and to measure NTF activities in fluids from rat brain lesions (Manthorpe et al., 1983a), peripheral nerve regeneration chambers (Longo et al., 1982b,1983), human cerebrospinal fluid from head trauma patients (Longo et al., 1984b), or in conditioned media from cultured neurons (Manthorpe et al., 1982b) or glial cells (Varon et al., 1981). The method has been modified to monitor NTF activities on different types of peripheral (Manthorpe et al., 1982a; Longo et al., 1983) and central (Varon et al., 1984; Walicke et al., 1986b; Barbin et al., 1984b) neurons.

2.1.2. Vital Dye Neuronal Assays

The neuronal cell counting microassay, although advantageous, remains inconvenient in that direct neuronal counting is tedious and time-consuming, requiring about 2 h/96-well plate of direct microscopic counting plus data compilation time (i.e., for a total of about 20–40 h/10–20 microplates). A modification of the neuronal survival assay has been developed that allows a faster and equivalent measurement of trophic factor activities (Manthorpe et al., 1986a,1989). This colorimetric assay relies on the ability of viable, but not dying or dead, neurons to convert a soluble yellow tetrazolium dye, 3-(4,5-diaminothiazol-2yl)-2,5-diphenyl tetrazolium bromide or "MTT," into an insoluble blue formazan product that is easily visible within living cells (*see* Fig. 3A–D). The blue product can be solubilized with acidic alcohol and the absorbance of the resulting solution is measured using a microplate spectrophotometer (ELISA reader). This method was adapted from a similar assay used to measure lymphocyte cell number increases under the influence of lymphokines (Mosmann, 1983).

Essentially the same type of microcultures are set up as previously described for the neuronal counting assay except that after 16 h in vitro, the cultures receive 5 µL of MTT solution (1.5 mg/mL in culture medium and stored at –20°C in light-shielded aliquots). After 24 h in vitro (i.e., 8 h after dye addition), 50 µL of acid alcohol (8 mL concentrated HCl/L isopropanol) is added and the well contents are mixed thoroughly by 15 up-and-down triturations using a multichannel pipet. The alcohol-trituration treatment will lyse all cell bodies and release the solubilized blue product. The acid component disallows phenol red interference by converting the phenol red to yellow color. The absorbance difference at 570–630 nm of each well is measured with an ELISA reader and the numbers are fed directly into a computer for immediate data analyses. The optical density is related to the number of neurons surviving in each well. As with the cell counting assay, the relative activity of the NTF sample in TU/mL is expressed based on half maximal effects and is equivalent to the activity values obtained with the counting assay (*see* Fig. 3E). Standard curves for purified NTFs, such as β-NGF or rat nerve CNTF should always be run in parallel for comparison of plateau, background, and slope values with those of the test samples.

This spectrophotometric method for assessing neuronal survival has the advantages over the direct counting method because it sums viability (i.e., optical density [O.D.]) of all

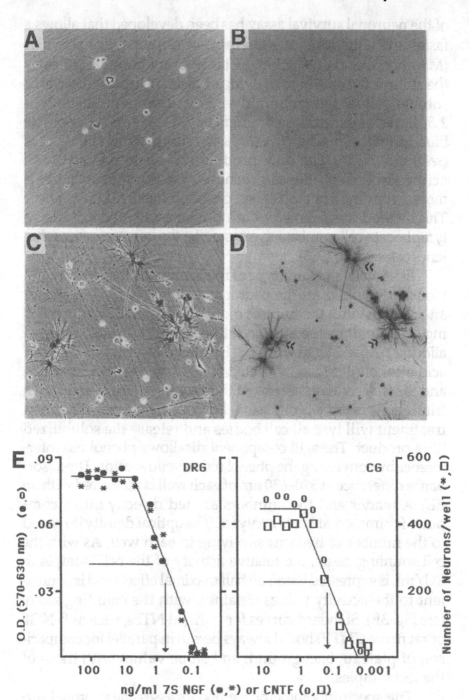

neurons within a well (rather than determining cell number within a restricted diametral area) and the ODs of ten 96-well plates can be read in as little as 10 min (rather than 20 h). In addition, the OD values can be fed from the spectrophotometer directly into a computer spreadsheet/graphics program for averaging replicates and constructing graphs (rather than manual transcription of numbers, calculations, and plotting of graphs). The spectrophotometric method has the distinct disadvantage that the cells are necessarily destroyed, although sister plates can be prepared for direct observation if desirable.

This spectrophotometric procedure for NTF measurement has been used in past studies to monitor NTF activities in fractions generated during factor purification and biochemical treatments (e.g., Manthorpe et al., 1986b), to measure NTF activities in conditioned media and extracts from cultured glial cells (e.g., Rudge et al., 1985; Muir et al., 1989b), and to quantify the NTF activity of tumor-promoting phorbol diesters (e.g., Montz et al., 1985). The method also has been used to monitor residual NGF activity in spent osmotic pumps after their use for extended delivery of NGF into brain ventricles (Williams et al., 1987; Vahlsing et al., 1989).

Fig. 3 *(opposite page)*. Colorimetric bioassay for NGF and Ciliary Neuronotrophic Factor (CNTF). Embryonic day 8 chick dorsal root (DRG) or ciliary (CG) ganglionic neurons were purified by differential attachment (panning) and cultured by seeding 500 neurons into each microwell containing different doses of either purified NGF or CNTF according to Manthorpe et al. (1986a). After 16 h, a vital dye, MTT, was added to the cultures that were maintained for an additional 8 h. Wells were either fixed and the number of neurons/well determined, or the metabolized MTT was solubilized and its optical density/well determined using a microplate spectrophotometer. Corresponding phase contrast (A,C) or light (B,D) microscopic photos were taken of MTT-stained ciliary ganglion neurons cultured in the absence (A,B) or presence (C,D) of CNTF. Note the presence of MTT-derived crystalline product (arrowheads in D) in neurons surviving in response to CNTF. A dose-response curve of neuronal number/well or OD/well vs NGF or CNTF dose is shown in (E). Note that both the cell counting (*,□) and colorimetric (•,o) methods yield the same ED_{50} values of about 2 ng/mL for NGF and 0.08 ng/mL of CNTF.

2.2. Neurite Promotion Assays

Even in the presence of an appropriate NTF, neurons will not extend neuritic processes (axons and dendrites) unless encouraged to do so by other sets of factors, collectively termed NPFs. Extracellular matrix protein NPFs can operate on neurons through receptors, such as the "integrins" (Buck and Horwitz, 1987; Hynes, 1987; Ruoslahti and Pierschbacher, 1987). Cell-surface adhesion molecules (CAMs) operate homotypically by binding to one another from different cells (Edelman,1987) and, like the extracellular matrix protein NPFs, can also operate on cultured cells after being bound to substrata (Bixby and Zhang, 1990). Using modifications of the cell counting NTF bioassay one can readily set up microcultures to detect and quantify the activities of NPFs.

2.2.1. Assays for the Presence of Neurites

Purified neuronal microcultures are established in A/2 microplates as with the NTF assays except for the following modifications. For assays of substratum-binding NPFs (i.e., those that act after being bound to the culture substratum), the polyornithine-coating step is followed by a 2 h coating with serial dilutions of the test NPF sample. The test sample is aspirated, the well is washed with culture medium, and 25 µL medium is added/well. The same precautions used with the laminin-coating for an NTF assay (see Section 2.1.1.) are taken to prevent substratum drying. The neurons are then added with 25 µL of an appropriate medium. The assay can be carried out in the absence of an NTF if the culture time is limited to a few hours, but an adequate NTF concentration (e.g., 100 TU/mL) will be required for longer term NPF assays. At the selected time, the assay is terminated by adding 100 µL of 4% glutaraldehyde fixative, the culture is washed to remove dead cells and debris, and the culture is set up for observation by phase contrast microscopy as described earlier (see Section 2.1.1.). Neurons with or without neurites are easily distinguishable, as shown in Fig. 4A–D.

The proportion of neurons bearing a neurite of defined minimal length is determined. A plot of the percentage of neurons with neurites vs concentration of several NPFs (laminin, merosin, fibronectin, and collagen) is shown in Fig. 4E. As with NTF assays, NPF activity is assigned based on half-maximal effects.

This type of neurite-promoting bioassay has been used to monitor NPF activity in fractions generated during biochemical purification (Manthorpe et al., 1981b; Engvall et al., 1986; Davis et al., 1985b,1986,1987). The assay was used to quantify the relative neurite-promoting potencies of laminin, fibronectin, and collagen (Manthorpe et al., 1983b) and to detect NPF activity in fluids released by injured peripheral nerve (Longo et al., 1984a).

Another utility of the above neuronal survival and neurite-promoting bioassays is to measure the amount of growth factor activity *produced* by cells in microculture (Rudge et al., 1985; Manthorpe et al., 1986c,1987; Muir et al., 1989b). Microcultures are established to contain glial cells (14,000/well) as described later in Section 3. After 1, 2, 3, and 4 d in microculture, the microplates are put on ice and the wells are individually extracted by a brief pulse from a probe sonicator. Alternatively, conditioned media can be collected separately, fresh medium added to the adherent cells, and sonication carried out on the cells. The total culture extract or separated conditioned media and cell extract from each well can be transferred to microplates and stored at –70°C. The plates are thawed and serial dilutions of the well contents assayed for their ability to:

1. Support the 24 h survival of embryonic day 8 dorsal root ganglionic neurons (NGF-like trophic activity);
2. Support the 24 h survival of embryonic day 8 ciliary ganglionic neurons (CNTF-like trophic activity); and
3. Promote the 6 h neuritic elongation by embryonic ciliary neurons (laminin-like neurite-promoting activity), all using microculture assays as described above.

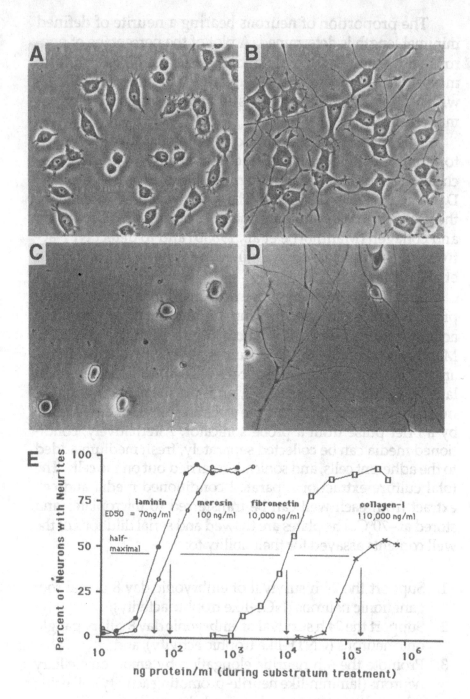

The activities of microculture extracts are expressed in tro-phic(TU/mL) or neurite-promoting (NPU/mL) U/mL as described earlier. Figure 5 shows the temporal profiles of accumulation of CNTF, NGF and laminin-mediated activities extracted from rat nerve Schwann and rat brain astrocytic cells.

2.2.2. Assays for Selected Features of Neurite Outgrowth

The neurite promotion bioassay can be modified to quan-tify the potencies of substances affecting different characteris-tics of the neurite outgrowth response. Serial titrations of test sample can be presented to neurons for various times in the presence of a constant, maximally active NTF, and the cultures fixed and the neurons scored for the time of onset of neurites (neurite initiation), the number of neurites/cell body (neuronal polarity), the number of growth cones/neurite (which, divided by 2 yields the number of neurite branch points), the length of each neurite (individual neurite length), or the sum of lengths of all neurites from each neuron (total neuritic output). One can determine the length of neurites at various times and from these data calculate the average rate of neurite growth. One can also stain the cultures to determine which processes are axons (neurofilament M- and F-positive) or dendrites (MAP-2-positive) (Bruckenstein et al., 1989) and evaluate the selective effects of

Fig. 4 (opposite page). Dissociate bioassay for NPFs. Rat phaeochromo-cytoma cells (A,B) or purified chick embryo day 8 ciliary ganglionic neu-rons (C,D) were cultured by seeding 1000 cells into microwells individually pretreated with different amounts of purified extracellular matrix proteins according to Manthorpe et al. (1983b). After 24 h, the cultures were fixed and the proportion of neurons bearing neurites was determined from diametral counts. Relatively little neurite outgrowth occurs in the absence of an NPF (A,C), but most of the input cells extended neurites in the presence of an NPF (B,D). A dose-response curve of neurite outgrowth vs NPF dose for ciliary neurons is shown in (E). Note that laminin and merosin are more potent (ED_{50} = 70–100 ng/mL) neurite-promoting agents in this assay than are fibronectin (ED_{50} = 10,000 ng/mL) or collagen Type I (ED_{50} = 110,000 ng/mL).

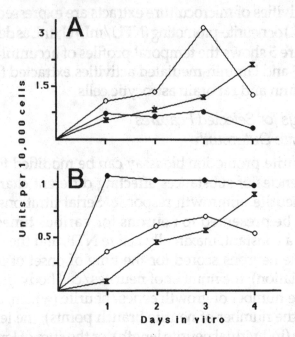

Fig. 5. Bioassays of trophic and neurite promoting activity outputs from cultured glial cells. Microcultures containing about 14,000 purified rat nerve Schwann cells **(A)** or rat brain astroglial cells **(B)** were sonicated after the specified days in vitro and the extracts assayed for trophic activity toward embryo day 8 chick ciliary (•, CNTF) or dorsal root (O, NGF) ganglionic neurons or neurite-promoting activity toward ciliary neurons ($\overline{x}$,Laminin). Titers are expressed in Trophic or Neurite Promoting Units per 10,000 cells as described by Muir et al. (1989b) and Rudge et al. (1985). Note that most activities tend to accumulate in these glial cultures.

different factors on axonal or dendritic production. Neurite outgrowth assays were used to show that neurite initiation time and neuronal polarity were increased with increasing doses of laminin protein, but that laminin doses did not affect neurite branching or the rate of elongation (Davis et al., 1985a).

2.3. Neurite Inhibition Assays

The presence of maximally active doses of NTF and NPF activities does not necessarily assure that neurite outgrowth will occur, even in longer term cultures. There are naturally occurring growth-regulating substances that *inhibit* neurite outgrowth and this negative activity can be detected and quantified by a two-dimensional NPF assay. In such an assay, maximal neurite outgrowth is assured when neurons are cultured on a substratum that stimulates maximal neurite outgrowth (e.g., polyornithine coating followed by 25 ng/well; 1 µg/mL × 25 µL of purified laminin). The substratum is then washed and exposed to 25 µL of serially diluted putative inhibitory test sample and washed again. Neurons are then added and fixed after culturing for a period of time, and the proportion of neurite-bearing neurons determined by direct count as described previously (*see* Section 2.1.1.). The titer of the test sample in Neurite Inhibitory Units (NIU)/mL is determined as the fold-dilution of the sample that suppresses 50% of the laminin-mediated neurite response (Fig. 6).

Antibodies generated against the NPF can act as neurite inhibitors. For example, the neurite outgrowth response to purified human placental laminin will be inhibited if the laminin substratum is previously exposed to antilaminin antibodies (Davis et al., 1985b; Engvall et al., 1986; Dillner et al., 1988) or the cells are exposed to antilaminin receptor (integrin) antibodies (Engvall et al., 1992; Tomaselli et al., 1988). Antilaminin antibodies, however, will not inhibit the neurite-promoting activity of a laminin-proteoglycan complex derived from rat glial cells (Davis et al., 1985b,1986), a finding that suggests the two laminin forms may be different. Antilaminin plus anti-β_1 integrin antibodies do not inhibit all the neurite outgrowth on Schwann cell surfaces (Bixby et al., 1988), suggesting that Schwann cell-attached NPF activity is not all owing to laminin and the β_1-containing laminin receptors (i.e., integrins). Figure 6 illustrates the effects of an antilaminin monoclonal antibody on purified rat yolk sac tumor laminin. Other inhibitors of neurite outgrowth are naturally occurring substances. One such substance is a laminin-binding proteoglycan released from RN22 Schwannoma cells (Muir et al., 1989a). Inhibitory potency can be similarly assayed by following the laminin coatings with the proteoglycan-containing sample (*see* Fig. 6).

The neurite inhibitory bioassay was used to monitor activity fractions during the biochemical purification of the RN22 Schwannoma cell-derived proteoglycan (Muir et al., 1989a).

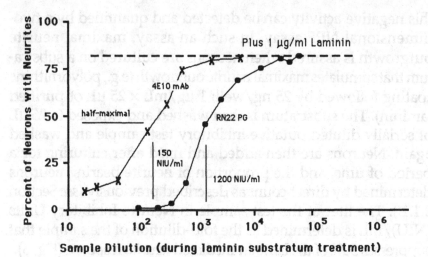

Sample Dilution (during laminin substratum treatment)

Fig. 6. Bioassay for inhibitors of laminin-induced neurite promoting activity. Microwells were coated with 50 µL of 1 µg/mL purified human laminin followed by serial dilutions of phosphate-buffered saline containing either a monoclonal anti-laminin antibody (4E10) or a heparan sulfate proteoglycan purified from rat RN22 Schwannoma cells (RN22 PG). After washing off the unbound materials, purified embryo day eight chick ciliary ganglionic neurons were presented for 4 h and the cultures were fixed. The number of neurons with or without at least one neurite was determined by microscopy according to Muir et al. (1989a). The percent of neurons possessing a neurite is plotted vs the dilution of sample used in substratum coating. Note that the inhibitory titers of each sample are expressed in Neurite Inhibitory Units (NIU)/mL of sample medium.

Quantification of the inhibitory activity in fractions from molecular sieve columns or after various treatments allowed the determination of the size of the proteoglycan complex (>1000 kDa), its potency (400 NPU/mg protein), and its inactivation by glycosaminoglycan lyases (which cleave off the carbohydrate side-chains).

The NPF assay has been used in combination with rotary shadowing electron microscopy to investigate antilaminin monoclonal antibodies and correlate their potency in reducing laminin NPF activity with the position of their binding on the laminin cruciate structure (Dillner et al., 1988). The results from these studies suggest that the neurite promoting activity of laminin resides with the end of the long arm of the laminin cross. A summary of such results is shown in Fig. 7. Incubation of

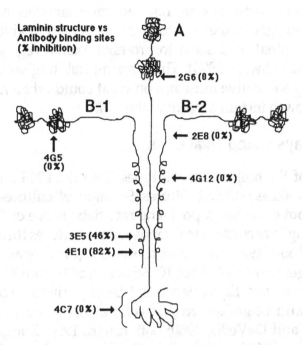

Fig. 7. Localization of antilaminin antibody binding sites and their influences on laminin neurite promoting activity. The laminin molecule contains three chains (B1, A, B2) in a cross-shaped configuration. The binding sites of several monoclonal antibodies are presented, along with a value (in parenthesis) representing the percent of maximal inhibition that the antibody elicits in inhibitory bioassays performed as shown in Fig. 6. Note that these results suggest that the domain within laminin that elicits neurite-promoting activity resides in the long arm of the molecule. The antibody 4C7 binds at an undetermined site at the end of the "A" chain. *See* Dillner et al. (1988).

laminin with one specific antilaminin antibody, called 2E8, prevented the binding and laminin-directed inhibitory activity of the proteoglycan. Since the 2E8 antibody is known from ultrastructural mapping studies (*see* Fig. 7) to bind to laminin at a region where the long and short arms of laminin intersect, this result suggested that this cross region of the laminin molecule is somehow involved in conveying the inhibitory activity of the proteoglycan.

Soluble neurite inhibitory substances have been recently described in association with cultured rat oligodendrocytes and myelin (Caroni and Schwab, 1988; Schwab and Caroni, 1988).

Although these materials are not yet well-characterized, antibodies raised against these factors and presented in vivo to the injured rat spinal cord seem to promote axonal regeneration (Schnell and Schwab, 1990). These axon-inhibiting substances may convey a negative influence on what could otherwise be a positive growth terrain present in the CNS.

3. Bioassays Using Glial Cells

Each of the major glial cell types from the PNS and CNS can be obtained as purified cultures. *Schwann cell* cultures can be purified from dissociated postnatal rat sciatic nerve or fetal rat sensory ganglia according to established procedures (Brockes et al., 1979; Wood, 1976). A recent procedure (Muir et al., 1989b) allows the generation of about 10^9 Schwann cells from 24 neonatal rat sciatic nerves. Equivalent numbers of purified rat cerebral astroglial and oligodendroglial cells also can be prepared (McCarthy and DeVellis, 1980; Bottenstein, 1990; Rudge et al., 1985; Louis et al., 1992). Such glial cell cultures can be established in 96-well plates for subsequent assays of factors affecting their morphological, proliferative, or other performances.

3.1. Morphological Assays

As with purified neuronal microculture assays, purified glial cells can be transferred to microwells and presented with selected growth factor test samples. Individual cells can then be examined for factor influences on cell morphology and proliferation. Glial cells (Schwann, astroglial, or oligodendroglial) are seeded on uncoated or fibronectin-coated ($10 \, \mu g/mL \times 25 \, \mu L/well$) substrata at 14,000/6 mm Ø well in 100 µL of DMEM + 1% calf serum. The amount of serum is to be chosen empirically to assure cell survival (e.g., using the MTT assay described in Section 2.1.2.) without having a significant mitogenic effect. After 2 h, the medium is aspirated to remove unattached cells and cell debris and replaced with 100 µL fresh medium. At 24 h, by which time the cells have formed an evenly distributed cell monolayer, the medium is replaced with test medium and the culture maintained for a selected time. The cultures are then fixed by the addition of 100 μL 4% glutaraldehyde in medium, washed and prepared for examination by phase-contrast microscopy (*see* Section 2.1.1.). Figure 8A,B shows the effects of FGF on the morphology of rat cerebral astroglial cells in microculture. Figure 8C presents a typical dose-response curve for FGF-induced transformation of astrocytes from a flattened to a fibrous morphology. Similar assays have been used to examine the effects of dibutyryl cyclic AMP and gangliosides on astroglial cell morphology (Skaper et al., 1986).

3.2. Proliferation Assays

3.2.1. Purified Glial Cell Proliferation Assays

Glial cell microplate cultures can be set up to measure the effects of growth factors on cell proliferation. One such method relies on the use of an antibody against bromodeoxyuridine (BrdU) to measure BrdU incorporation into cell nuclei (Muir et al., 1990a). In this assay, BrdU is added at a final concentration of 10 μM to the mitogen-stimulated test microcultures. After 24 h, BrdU incorporation is terminated by carefully aspirating the treatment medium and washing with 200 μL/well of BSS. The cells are fixed by replacing the BSS with 200 μL/well of 70% ethanol for 20 min at room temperature. Next, the wells are washed with water and the DNA denatured by incubation with 100 μL/well of 2M HCl for 10 min at 37°C. Following aspiration, the wells are filled with borate buffer (0.1M, pH 9) to neutralize the residual acid. After a wash with PBS, the cells are treated with 50 μL of blocking buffer (PBS containing 0.1% Triton X-100 and 2% normal goat serum) for 15 min at 37°C. Monoclonal mouse anti-BrdU antibody (50 μL/well; 1 μg/mL) diluted in blocking buffer is then added for 60 min at 37°C. Unbound antibody is removed by three washes with PBS containing 0.1% Triton. Bound antibody is detected by adding peroxidase-conjugated goat antimouse IgG (50 μL/well; 2 μg/mL) diluted in blocking buffer and incubating for 30 min at 37°C. After several washes with PBS/Triton and a final rinse with PBS only, the wells receive 100 μL/well of 50 mM phosphate/citrate buffer, pH 5, containing 0.05% of the *soluble* chromagen o-phenylenediamine

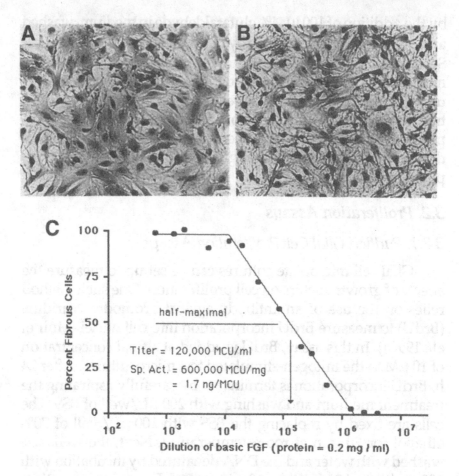

Fig. 8. Bioassay of morphology-altering activity of FGF toward glial cells. Purified rat brain astroglial cells were left unexposed or exposed for 24 h to an FGF-containing sample. In the absence of FGF, the cells express a flattened morphology **(A)**, whereas in the presence of FGF, the cells express a fibrous or stellate morphology **(B)**. **(C)** Dose-response curve of FGF vs the percent of astroglial cells expressing the fibrous morphology. Note that this FGF sample has a titer of 120,000 Morphology Conversion Units (MCU)/mL and a specific activity of 1.7 ng/MCU. *See* Pettmann et al. (1991).

(OPDA) and 0.02% H_2O_2. The reaction is terminated after 5–20 min by pipeting 80 µL from each well into a clean microtiter plate containing 40 µL/well of 2N sulfuric acid. The optical density in the new plate is measured at 490 nm by a microplate reader interfaced with a computer. The assay plate containing the cell monolayer is washed twice with PBS and immunocytochemically stained for BrdU-DNA by adding 100 µL/well of PBS containing 0.05% of the *insoluble* chromagen diaminobenzidine (DAB) and 0.02% H_2O_2. The staining reaction is monitored by light microscopy and stopped by washing the cells with water after 10–20 min. Unlabeled nuclei can be visualized after counterstaining with 0.1% toluidine blue in borate buffer at pH 9.

With this method, a single microculture can be examined for cell morphology, cell number, DNA synthesis (with the same sensitivity as that given by the radiothymidine technique; *see* Muir et al., 1990a); and expression of cell antigens. This latter staining can be carried out by adding with the mouse anti-BrdU antibody another antibody against a cell antigen. The BrdU technique is ideally suited to measure in parallel the effects of hundreds of different culture conditions on cell proliferation and phenotypic expression. Figure 9 shows the appearance of BrdU-stained Schwann cells cultured in the absence (Fig. 9A) or presence (Fig. 9B) of neuronal cells that contain a Schwann cell mitogen on their axonal surfaces. Figure 9C shows a typical dose response curve of OD by ELISA for BrdU-incorporation for rat nerve Schwann cells that have proliferated in response to increasing numbers of living neurons.

3.2.2. Mixed Cell Culture Proliferation Assays

To demonstrate the utility of the BrdU assay in mixed cultures, one can set up cocultures in microwells with a mixture of rat nerve Schwann and rat brain astroglial cells. The cultures are maintained in medium containing 10% fetal calf serum, conditions known to stimulate the proliferation of astroglial but not Schwann cells. Immunostaining for astroglial cell glial fibrillary acid protein (GFAP) is achieved by applying to the microcultures polyclonal rabbit anti-GFAP antiserum (1:500). The culture is incubated simultaneously with the monoclonal anti-BrdU antibody as described in Section 3.2.1. The bound anti-GFAP antibodies are detected by biotinylated goat antirabbit IgG (2 µg/mL) supplied along with the peroxidase-conjugated goat anti-mouse IgG also previously described. Following the colorimetric assay for BrdU-DNA immunoreactivity using OPDA/ H_2O_2, the wells are washed with PBS and the biotinylated-IgG is detected by adding 50 µL/well of a 1:300 dilution of an

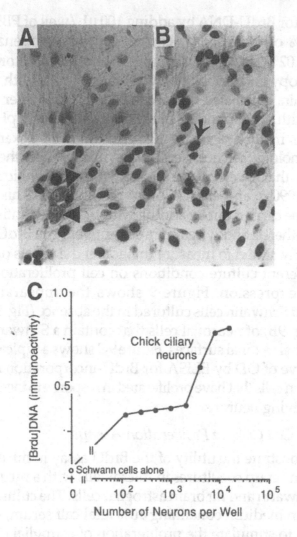

Fig. 9. Bioassay for proliferation of glial cells. About 14,000 purified and quiescent rat sciatic nerve Schwann cells were cultured in microwells in the absence or presence of different numbers of purified chick embryo ciliary neurons (arrowheads), which are mitogenic for Schwann cells. After 24 h in the presence of bromodeoxyuridine (BrdU), the cultures were fixed and the incorporation of BrdU determined as described (Muir et al., 1990a). Many more BrdU-positive Schwann cells (arrows) appeared in the presence **(B)** than in the absence **(A)** of neuronal cells. A plot of BrdU immunoactivity vs neuronal number **(C)** suggests a dose response relationship between Schwann cell proliferation and a mitogen contributed by the neuronal cells.

ABC-peroxidase reagent. Immunostaining for BrdU and GFAP protein are carried out by the addition of DAB/H_2O_2 as described in Section 3.2.1.). GFAP is an intermediate filament protein found in astroglial cells and is not expressed by sciatic nerve Schwann cells. Figure 10 shows the appearance of mixed astroglia-Schwann cell cultures labeled by BrdU and GFAP staining.

3.3. Antiproliferative Assays

Cultured rat sciatic nerve Schwann cells hardly proliferate in 10% calf serum as shown in Fig. 10. This quiescent behavior also occurs in purified Schwann cell cultures and is unexpected, since Schwann cells are known to produce and release several autocrine mitogens, including laminin (a Schwann cell product; Bunge et al., 1986; Muir et al., 1989b), transforming growth factor-β (Eccelston et al., 1989a; Ridley et al., 1989) and Glial Maturation Factor (Bosch et al., 1989). In addition to these autocrine Schwann cell mitogens, the culture medium contains fibronectin (a serum component) which has mitogenic activity for Schwann cells (Muir et al., 1989b). Using the BrdU assay, we noted that mitogen-stimulated Schwann cells undergo a rapid burst of proliferation immediately following culture medium changes (Muir et al., 1990b). In addition, we noted that nonmitogen-stimulated Schwann cells increase their proliferation if the culture medium is changed frequently (e.g., every 12 h). The possibility that Schwann cells produce an autocrine proliferation inhibitor was examined by first developing an antiproliferative assay. Schwann cells were set up in microcultures just described and presented with a constant, maximally active amount of mitogen. The microcultures were then presented with serial dilutions of conditioned medium (CM) from dense, but subconfluent Schwann cell cultures. Figure 11 (*see* page 119) shows that Schwann cell CM could completely inhibit proliferation by mitogen-stimulated Schwann cells in a dose-dependent and reversible fashion, regardless of the types of mitogen present. Analogous to other bioassays, activity titers are defined here in antiproliferative Units (APUs)/mL representing the fold dilution eliciting half-maximal responses.

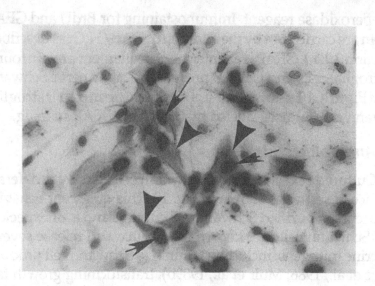

Fig. 10. Appearance of BrdU and GFAP-labelled cells in culture. Purified rat brain astroglial and rat sciatic nerve Schwann cells were cocultured in the presence of fetal calf serum and BrdU. After 48 h, the cultures were fixed and immunostained for BrdU and for GFAP (which stained the astroglial but not the Schwann cells). Note that only GFAP-positive cells (arrowheads) are also BrdU-positive (arrows), indicating that under these general culture conditions, only astroglial cells will proliferate. *See* Muir et al. (1990a).

4. Cell Blot Bioassays

All the in vitro neurocellular microassays described in the preceding sections measure the *amount* of a factor activity in a test sample. These assays have been used successfully to determine the activity levels remaining in test samples submitted to

1. Long-term storage under various conditions;
2. Chromatographic fractionation; and
3. Activation or inactivation after exposure to extremes of temperature, pH, and ionic strength, and to antibodies or degradative enzymes.

A completely different type of assay, the cell blot assay, relies on the ability of growth factor proteins to retain their biological activities after SDS-PAGE and immobilization by electroblotting to nitrocellulose paper. When responsive neurons or glial cells are cultured on the blotted lanes, the attached cells respond

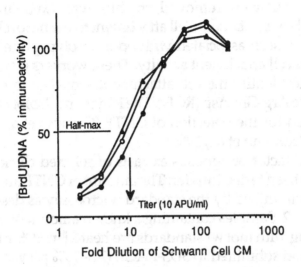

Fig. 11. Bioassay for inhibitors of glial cell proliferation. Purified rat sciatic nerve Schwann cells were cultured in microwells in the presence of cholera toxin (●), laminin (○), or chick ciliary neurons (▲) as mitogens, and each in the copresence of serial dilutions of conditioned medium from dense Schwann cell cultures as described (Muir et al., 1990b). High doses (e.g., 50%) of Schwann cell medium nearly completely inhibited BrdU incorporation into the test Schwann cells regardless of the mitogen used. The titer of the test medium was about 10 Antiproliferative Units (APU)/mL.

locally to the factor bands. The main advantage of this technique is that it allows a precise definition of the apparent mol mass of the active factor in crude samples such as tissue extracts. Two examples will be given in the next section using neuronal or glial cells to detect neuronotrophic factors or glial cell mitogens, respectively.

4.1. Cell Blot Assays for Neuronotrophic Factors

The cell blot technique was originally inspired by the work of Hayman et al. (1982). These investigators were interested in identifying an amino acid sequence within the fibronectin molecule that interacts with cell surface receptors. They submitted protease digests of fibronectin to SDS-PAGE and electroblotting and then incubated the blotted lanes with dissociated kidney cells. Conditions were adjusted such that the kidney cells would not normally attach to the nitrocellulose. After a few hours, the unattached cells were removed, the blot was fixed, and any attached cells were stained. Cell attachment to the nitrocellulose lane was only seen associated with bound polypeptide bands that retained cell attachment activity. These workers went on in later work to identify the cell attachment sequence to contain the tripeptide Arg-Gly-Asp (RGD). The Hayman cell blot method was modified for the detection of CNTF (Carnow et al., 1985) and NGF (Pettmann et al., 1989).

Purified factor or aqueous extracts of selected chick or rat tissues have been tested thus far. The amount of CNTF and NGF activity is determined by the standard microassay as described in Section 2.1.2. Samples containing a known amount of trophic activity along with mol wt standards are heated in SDS ± reducing agents and submitted to SDS-PAGE (7.5–20% polyacrylamide gradient gel) and electroblotting. The blot is rinsed in saline, stained for 30 s with 0.1% amido black in 22.5% methanol/2.5% acetic acid, and washed thoroughly with distilled water until the background is destained. The stained mol wt standards are marked with pinholes on each side and the blot is blocked with 1% ovalbumin and antibiotic solution (100 µg/mL penicillin, 100 µg/mL streptomycin, 0.25 µg/mL fungizone). The blotted lanes or groups of contiguous lanes are cut out and placed in an appropriately fitting culture chamber such that the nitrocellulose paper lies down evenly on the substratum. Culture medium (Dulbecco's Modified Eagle's plus 10% fetal calf serum) is added followed by embryonic chick ciliary or dorsal root ganglionic neurons purified by differential attachment as previously described. Neuronal cell seeding densities should be about 10,000 neurons/ cm^2 substratum. After 24–48 h culture, MTT is added to a final concentration of 0.5 mg/mL for an additional 4 h. The cell blot is then fixed by addition of 4% glutaraldehyde. Although the seeded neurons originally attach with a uniform distribution over the blotted lane, the neurons (which stain blue) will only survive over the 1–2 d culture period if they have attached on the trophic factor bands. Figure 12 (*see* page 122) shows the appearance of MTT-stained blots seeded with neurons and supported by either CNTF or NGF bands.

By counting the number of surviving neurons in incremental areas along the lane length, it is possible to determine the precise location of the the trophic factor band. A plot of neuronal number vs migration distance can be compared with a plot of mol wt markers and distance to yield the mol mass of the activity (*see* Fig. 13, page 124). The method is very sensitive; for example, even 15 TU (about 1 ng) of CNTF originally loaded on the SDS gel can be detected.

This cell blot method for detecting neuronotrophic factors has been used to demonstrate CNTF bands in different tissue extracts and to identify multiple mol wt forms of CNTF (Carnow et al., 1985; Rudge et al., 1987) and NGF (Pettmann et al., 1989). Both CNTF and NGF bands were shown to support the survival of ganglionic neurons, although the NGF, but not CNTF, also elicited neurite outgrowth on the nitrocellulose paper (*see* Fig. 12F–I). The monomeric 13 kDa form of NGF, generated by boiling NGF in SDS under nonreducing conditions, was shown to exhibit biological activity (Pettmann et al., 1989). Implicit in the success of the cell blot method is the fact that both CNTF and NGF can act on neurons locally from an immobilized state, suggesting that these two trophic factors may act on cells in vivo when bound to the extracellular matrix or to other cells. These results also raise the possibility that these NTFs do not have to be internalized by the cell in order to support neuronal survival and neurite outgrowth.

4.2. Cell Blot Assays for Glial Cell Mitogens

Fibroblast growth factor (FGF) has mitogenic activity toward cultured astroglial cells and causes the glial cells to convert from a flattened to a more spindle-shaped morphology (Perraud et al., 1988b; *see also* Fig. 8A,B). The cell blot technique was used recently to identify mitogenic and morphology-altering activities of FGF (Pettmann et al., 1991). This study was promoted by an initial observation that the mitogenic and morphology-transforming activities of purified bovine brain FGF occurred on cell blots *only* in association with a 27 kDa band. This was considered puzzling since the mol mass of FGF is known to be

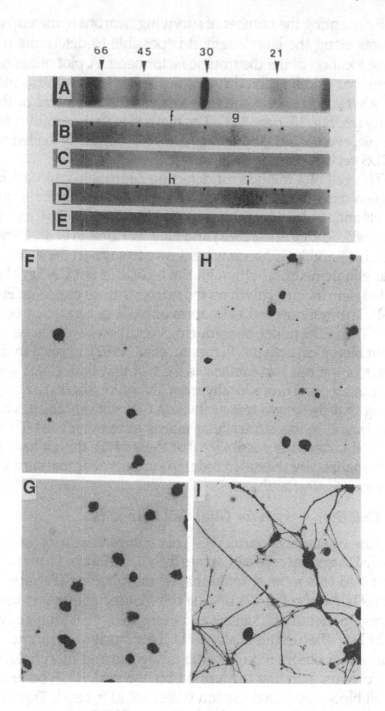

about 18 kDa and the only visible band in replicate protein-stained lanes is usually the 18 kDa band. We considered that the bovine preparation either contained a trace amount of very active non-FGF material, or that the 27 kDa protein is, in fact, one or even *the* biologically active form of FGF. Thus, we compared tissue-isolated and genetically engineered basic FGFs by cell blot analysis toward purified rat astroglial cells or NIH 3T3 mouse fibroblasts (Pettmann et al., 1991).

Bovine brain basic FGF was purified as described (Perraud et al., 1988b) and recombinant human basic FGF was provided by California Biotechnology (Mountain View, CA), and solubilized in 1% SDS. Phast precasted 0.45 mm thick 20% polyacrylamide gels were purchased from Pharmacia (Uppsala, Sweden), and electrophoresis performed using SDS buffer agarose strips. The gel was transferred to nitrocellulose by diffusion blotting after equilibration in a transfer buffer (25 mM Tris, 0.2M glycine, 20% methanol, and 0.01% SDS, pH 8.0). The blot was washed with saline/antibiotics and blocked for 30 min with medium (DMEM, 2% FCS, antibiotics). The blot was then anchored to a similarly sized sterile culture vessel and seeded with purified rat brain astroglial cells or mouse 3T3 fibroblasts at 40,000 cells/cm^2 of substratum. After 16 h, BrdU was added to 10 µM for an additional 24 h. The blot was rinsed with saline and fixed for 20 min with 4% paraformaldehyde in

Fig. 12 (*opposite page*). Cell blotting of purified neuronotrophic factors. Samples of purified CNTF or NGF were submitted to SDS-PAGE and western blotting and neurons cultured directly on the blotted lanes for 2 d and stained for viability using the vital dye, MTT as described by Pettmann et al. (1989). Mol wt markers (A); Chick embryo ciliary (B) or dorsal root (C) ganglionic neurons were cultured on blots of CNTF. Chick embryo dorsal root (D) or ciliary (E) ganglionic neurons were cultured on blots of NGF. Note that more neurons survived on the trophic factor bands. Replicate blots were stained with antineurofilament, and regions off (F,H) or on (G,I) the trophic factor bands were photographed. Neurite extension from dorsal root ganglion neurons cultured on the NGF band (I), but not from ciliary neurons cultured on the CNTF one (G).

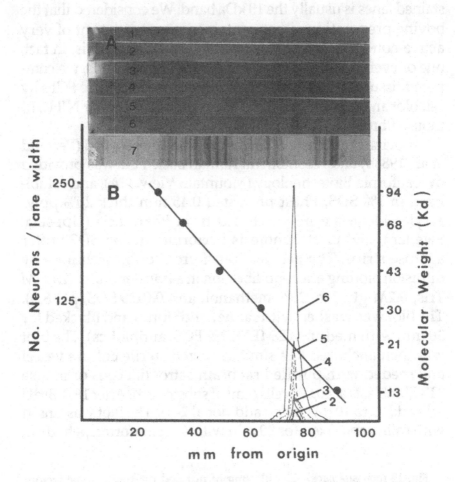

Fig. 13. Cell blotting of chick eye extract: CNTF dose response analysis. Serial dilutions of chick eye extract were submitted to SDS-PAGE, and western blotting and ciliary neurons cultured on the blotted lanes stained with the vital dye, MTT, and fixed. Viable neuronal cells are expressed as darkly stained spots forming a band in (A). The amount of loaded CNTF activity in the extract in TU/lane (i.e., TUs were determined according to Fig. 3E) was 7 (1); 15 (2); 30 (3); 60 (4); 120 (5); 240 (6). A Coomassie Blue-stained lane is shown in (7). Individually stained neurons were counted along the lane length and plotted vs the distance in mm from the lane origin (B). Note that stained neurons were only found in the 20 kDa region compared to mol wt markers (•), and sample loads containing as little as 15 TU (2) could be detected. See Carnow et al. (1985).

buffered saline and treated for 5 min with 20% methanol/3% hydrogen peroxide. The cells were then permeabilized for 5 min with 1% Triton X-100/saline, washed with saline, the DNA was denatured for 30 min with 2N HCl, the blot was neutralized for 5 min with borate buffer, pH 8.5, and finally, the blot was rinsed with saline. Peroxidase-antiperoxidase staining for BrdU was carried out by sequential treatments with:

1. 2% anti-BrdU/10% normal sheep serum/saline overnight at 4°C;
2. 5% goat antimouse IgG for 1 h at 37°C;
3. 0.1% Tween-20 3X for 5 min each; and
4. 1% mouse peroxidase-antiperoxidase complex for 2 h. Color was developed with the insoluble chromagen, 4-chloro-1-naphthol/0.04% NiCl/0.1% hydrogen peroxide/0.15 NaCl.

Light counterstaining was carried out using 0.01–0.1% toluidine blue or GFAP immunostaining. The number of anti-BrdU-labeled or morphologically altered (e.g., fibrous vs flat) cells was determined by microscopic counting along the lane length and the values plotted vs distance from the lane origin.

Figure 14 (C,D) shows the appearance of GFAP and BrdU immunostaining of astroblasts under the influence of the 18 and 27 kDa forms of recombinant human basic FGF. Normally, glial cells assume a flattened morphology, but after attachment to the 27 kDa band of FGF, the cells become fibrous to assume a more angular process-bearing or stellate shape (Fig. 14C). The 27 kDa form, but not the 18 kDa form (Fig. 14B), also stimulates BrdU incorporation. Plots of the percentage of fibrous cells and number of BrdU-positive nuclei/mm^2 confirm that both FGF activities are restricted to the 27 kDa region of the lane (Fig. 14A).

5. Conclusions and Perspectives

We have summarized some state-of-the-art cell biological methodologies currently available for detection and analyses of protein factors affecting neural cells. It is obvious that many

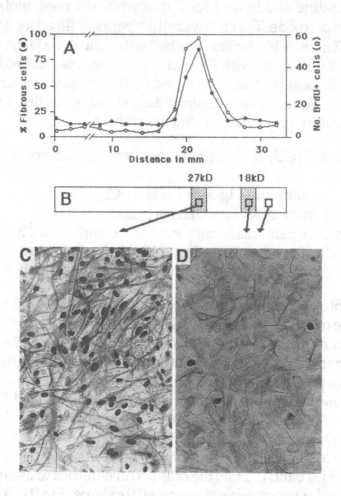

Fig. 14. Cell blotting of purified FGF. Purified human recombinant FGF was submitted to SDS-PAGE and western blotting, and purified rat brain astroglial cells cultured directly on the blotted lanes for 2 d and the cells stained for both BrdU-incorporation and for GFAP, as described by Pettmann et al. (1991). The number of BrdU-positive or GFAP-positive fibrous cells (*see* Fig. 8A,B) was determined over the lane length and plotted vs their distances from the lane origin (A). The greatest number of BrdU-positive and fibrous cells were found in the 27 kDa region of the blot. Light microscopic analysis of the lane indicated that most BrdU-positive cells were also fibrous (C, D). Such results indicate that the mitogenic and morphology-coverting activity of FGF resides in a 27 kDa molecular species.

improvements in growth factor bioassays have been made since the original ganglionic explant assay (cf. Fig. 1) was used for measuring the trophic activity of NGF. Today's assays are more convenient, rapid, and sensitive, and can be applied to dissociated and mixed or purified neural cell types. Current assays allow simultaneous assessments of the effects of a large number of test samples on a single cell type or, with appropriate cell markers, simultaneous assessments of the effects of a single factor on individual or multiple cell types in mixed cell cultures.

There is an increasing realization that cell behaviors in vivo represent the prevailing responses to the action of several concurrent growth factors. The methods outlined in this chapter will certainly continue to be modified as new factors and their responses are discovered. As more knowledge is gained concerning growth factors and their influences on cells, and as the demand for factor bioassays increases, methodological improvements are likely to be aimed at:

1. Decreasing the assay time to even minutes after factor addition;
2. Increasing sensitivity of the measurements of cell responses such that single cell assays can be used;
3. Amplification of the cell responses to measure femto- and attomolar ranges of added factors;
4. Allowing tens of thousands of simultaneous assays coupled with refinements in automated data analyses;
5. Increasing the convenience by using more readily available cell lines instead of primary cells for the most routine work;
6. Increasing the number of simultaneously measurable cellular responses per culture; and
7. Extending the range of cellular responses to include changes in the expression of individual proteins or mRNAs.

The cell blot method can be expected to extend to electroblotted fractions generated after native PAGE, isoelectric focusing, agarose gel chromatography, and perhaps even thin-layer chromatography for the determination of, respectively, factor native mol mass, isoelectric point, aggregate size, and association with lipid moieties. Other modifications may be aimed at detecting antagonistic factors by including a stimulatory factor in the blot culture and measuring the suppression of a cellular response by the blotted opposing activity.

Regardless of the pace of future improvements, one can expect that in vitro bioassays will continue to be required for the detection and analyses of neural growth factors. A thorough understanding and utilization of such bioassay methods will allow the development of a meaningful categorization of growth factor responses and target cells. Such a categorization will increase our appreciation of the degree of extrinsic regulation imposed on cells and will lay the foundation for therapeutic application of growth factors toward human nervous system disease and trauma.

References

Adler R., Manthorpe M., and Varon S. (1979) Separation of neurons and nonneuronal cells in monolayer cultures from chick embryo optic lobe. *Dev. Biol.* **69**, 424–435.

Akers R. M. , Mosher D. F., and Lilien J. E. (1981) Promotion of retinal neurite outgrowth by substratum-bound fibronectin. *Dev. Biol.* **86**, 179–188.

Barbin G., Manthorpe M., and Varon, S. (1984a) Purification of the chick eye ciliary neuronotrophic factor. *J. Neurochem.* **43**, 1468–1478.

Barbin G., Selak I., Manthorpe M., and Varon, S. (1984b) Use of central neuronal cultures for the detection of neuronotrophic agents. *Neuroscience* **12**, 33–43.

Barde Y.A., Davies A. M., Johnson J. E., Lindsay R. M., and Thoenen H. (1987) Brain-derived neurotrophic factor. *Prog. Brain Res.* **71**, 185–189.

Baron van Evercooren A., Kleinman H. K., Ohno S., Marangos P., Schwartz M., Dobois-Dalc M. E. (1982) Nerve growth factor, laminin and fibronectin promote neurite growth in human fetal sensory ganglia cultures. *J. Neurosci.* **8**, 179–194.

Benveniste E. N. and Merrill J. E. (1986) Stimulation of oligodendroglial proliferation and maturation by interleukin-2. *Nature* (Lond.) **321**, 610–613.

Besnard F., Perraud F., Sensenbrenner M., and Labourdette G. (1987) Platelet-derived growth factor is a mitogen for glial but not for neuronal rat brain cells in vitro. *Neurosci. Lett.* **73**, 287–292.

Besnard F., Perraud F., Sensenbrenner M., and Labourdette G. (1989) Effects of acidic and basic fibroblast growth factors on proliferation and maturation of cultured rat oligodendrocytes. *Int. J. Dev. Neurosci.* **7**, 401–409.

Bixby J. L. and Zhang R. (1990) Purified N-cadherin is a potent substrate for the rapid induction of neurite outgrowth. *J. Cell Biol.* 110, 1253–1260.

Bixby J. L., Lillien J., and Reichardt L. F. (1988) Identification of the major proteins that promote neuronal process outgrowth on Schwann cells in vitro. *J. Cell Biol.* 107, 353–361.

Bosch E. P., Zhong W., and Lim R. (1989) Axonal signals regulate expression of glia maturation factor-beta in Schwann cells: an immunohistochemical study of injured sciatic nerves and cultured Schwann cells. *J. Neurosci.* 9, 3690–3698.

Bottenstein J. E. (1990) Growth requirements in vitro of oligodendrocyte cell lines and neonatal rat brain oligodendrocytes. *Proc. Natl. Acad. Sci. USA* 83, 1955–1959.

Brockes J. P., Fields K. L., and Raff M. C. (1979) Studies on cultured rat Schwann cells: I. Establishment of purified populations from cultures of peripheral nerve. *Brain Res.* 165, 105–118.

Bruckenstein D., Johnson M. I., and Higgins D. (1989) Age-dependent changes in the capacity of rat sympathetic neurons to form dendrites in tissue culture. *Brain Res.* 46, 21–32.

Buck C. A. and Horwitz A. F. (1987) Integrin, a transmembrane glycoprotein complex mediating cell-substratum adhesion. *J. Cell Sci.* 8, 231–250.

Bunge R. P., Bunge M. B., and Eldridge C. F. (1986) Linkage between axonal ensheathment and basal lamina production by Schwann cells. *Ann. Rev. Neurosci.* 9, 305–328.

Calof A. and Reichardt L. F. (1984) Motoneurons purified by cell sorting respond to two distinct activities in myotube conditioned medium. *Dev. Biol.* 106, 1541–1549.

Cardwell M. C. and Rome L. H. (1988a) Evidence that an RGD-dependent receptor mediates the binding of oligodendrocytes to a novel ligand in a glial-derived matrix. *J. Cell Biol.* 107, 1541–1549.

Cardwell M. C. and Rome L. H. (1988b) RGD-containing peptides inhibit the synthesis of myelin-like membrane by cultured oligodendrocytes. *J. Cell Biol.* 107, 1551–1559.

Carnow T. B., Manthorpe M., Davis G. E., and Varon, S. (1985) Localized survival of ciliary ganglionic neurons identifies neuronotrophic factor bands on nitrocellulose blots. *J. Neurosci.* 5, 1965–1971.

Caroni P. and Schwab M. (1988) Two mewmbrane fractions from rat central myelin with inhibitory properties for neurite growth and fibroblast spreading. *J. Cell Biol.* 106, 1281–1288.

Cattaneo E. and McKay R. (1990) Proliferation and differentiation of neuronal stem cells regulated by nerve growth factor. *Nature (Lond.)* 347, 762–765.

Claude P., Parade I. M., Gordon K. A., D'Amore P. A., and Wagner J. A. (1988) Acidic fibroblast growth factor stimulates adrenal chromaffin cells to proliferate and to extend neurites, but is not a long-term survival factor. *Neuron* 1, 783–790.

Cohen S., Levi-Montalcini R., and Hamburger V. (1954) A nerve growth-stimulating factor isolated from sarcomas 37 and 180. *Proc. Natl. Acad. Sci. USA* **40,** 1014–1018.

Davis J. B. and Stroobant P. (1990) Platelet-derived growth factors and fibroblast growth factors are mitogens for rat Schwann cells. *J. Cell Biol.* **110,** 1353–1360.

Davis G. E., Manthorpe M., and Varon S. (1985a) Parameters of neuritic growth from ciliary ganglion neurons in vitro: Influence of laminin, Schwannoma polyornithine-binding neurite promoting factor and ciliary neuronotrophic factor. *Dev. Brain Res.* **17,** 75–84.

Davis G. E., Manthorpe M., Engvall E., and Varon S. (1985b) Isolation and characterization of rat Schwannoma neurite-promoting factor: Evidence that the factor contains laminin. *J. Neurosci.* **5,** 2662–2671.

Davis G. E., Manthorpe M., Williams L. R., and Varon S. (1986). Characterization of a laminin-containing neurite-promoting factor and a neuronotrophic factor from peripheral nerve and related sources. *Ann. NY Acad. Sci.* **486,** 194–205.

Davis G. E., Klier F. G., Engvall E., Cornbrooks C., Varon S., and Manthorpe M. (1987) Association of laminin with heparan and chondroitin sulfate-bearing proteoglycans in neurite-promoting complexes from rat Schwannoma cells. *Neurochem. Res.* **12,** 909–921.

Dean D. H., Hiramoto R. N., and Ghanta V. K. (1987) Modulation of immune response. A possible role for murine salivary epidermal and nerve growth factors. *J. Periodont.* **58,** 498–500.

Dillner L., Dickerson K., Manthorpe M., Ruoslahti E., and Engvall E. (1988) The neurite-promoting domain of human laminin promotes attachment and induces characteristic morphology in non-neuronal cells. *Exp. Cell Res.* **177,** 186–198.

Drazba J. and Lemmon V. (1990) The role of cell adhesion molecules in neurite outgrowth on Muller cells. *Dev. Biol.* **138,** 82–93.

Eccleston P. A., Jessen K. R., and Mirsky R. (1989a) Transforming growth factor-beta and gamma-interferon have dual effects on growth of peripheral glia. *J. Neurosci. Res.* **24,** 524–530.

Eccleston P. A., Mirsky R., and Jessen K. R. (1989b) Type I collagen preparations inhibit DNA synthesis in glial cells of the peripheral nervous system. *Exp. Cell Res.* **182,** 173–185.

Edelman G. M. (1987) Epigentetic rules for expression of cell adhesion molecules during morphogenesis. *CIBA Found. Symp.* **125,** 192–216.

Ehrig K., Lievo I., and Engvall E. (1990) Merosin and Laminin; Molecular relationship and role in nerve-muscle development, *Ann. NY Acad. Sci.* **580,** 276–280.

Engvall E., Davis G. E., Dickerson K., Ruoslahti E., Varon S., and Manthorpe M. (1986) Mapping of domains in human laminin using monoclonal antibodies: localization of the neurite-promoting site. *J. Cell Biol.* **103,** 2457–2465.

Engvall E., Muir D., and Manthorpe M. (1992) Merosin is possesses cell attachment and neurite promoting activities. *Cell Exp. Res.*, in press.

Faivre-Bauman A., Puymirat J., Loudes C., Barret A., and Tixier-Vidal A. (1984) Laminin promotes attachment and neurite elongation of fetal hypothalamic neurons grown in serum free medium. *Neurosci. Lett.* 44, 83–89.

Goetschy J. F., Ulrich G., Aunis D., and Ciesielski-Treska J. (1987) Fibronectin and collagens modulate the proliferation and morphology of astroglial cells in culture. *Int. J. Dev. Neurosci.* 5, 63–70.

Goretzki P. E., Wahl R. A., Becker R., Koller C., Branscheid D., Grussendorf M., and Roeher H. D. (1987) Nerve growth factor (NGF) sensitizes human medullary thyroid carcinoma (hMTC) cells for cytostatic therapy in vitro. *Surgery* 102, 1035–1042.

Gospodarowicz D., Neufeld G., and Schweigerer L. (1986) Fibroblast Growth Factor. *Mol. Cell. Endocrinol.* 46, 187–204.

Greene L. A., Aletta J. M., Rubenstein A., and Green S. H. (1987) PC12 pheochromocytoma cell: culture, nerve growth factor treatment, and experimental exploitation. *Methods Enzymol.* 147, 207–216.

Gurwitz D. and Cunningham D. D. (1987) Heparan sulfate proteoglycan and laminin mediate two different types of neurite outgrowth. *J Neurosci.* 7, 2293–2304.

Gurwitz D. and Cunningham D. D. (1988) Thrombin modulates and reverses neuroblastoma neurite outgrowth. *Proc. Natl. Acad. Sci. USA* 85, 3440–3444.

Hantaz-Ambroise D., Vigny M., and Koenig J. (1987) Heparan sulfate proteoglycan and laminin mediate two different types of neurite outgrowth. *J. Neurosci.* 7, 2293–2304.

Hayman E. G., Engvall E., A'Hearn E., Barnes D., Pierschbacher M., and Ruoslahti E. (1982) Cell attachment on replicas of SDS polyacrylamide gels reveals two adhesive plasma proteins. *J. Cell Biol.* 95, 20–23.

Hughes S. M., Lillien L. E., Raff M. C., Rohrer H., and Sendtner M. (1989) Ciliary neurotrophic factor induces type-2 astrocyte differentiation in culture. *Nature (Lond.)* 335, 70–73.

Hunter S. F. and Bottenstein J. E. (1990) Growth factor responses of enriched bipotential glial progenitors. *Brain Res.* 54, 235–248.

Hynes R. O. (1987) Integrins: a family of cell surface receptors. *Cell* 48, 549–554.

Juurlink B. H., Munoz D. G., and Devon R. M. (1990) Calcitonin gene-related peptide identifies spinal motoneurons in vitro. *J. Neurosci. Res.* 26, 238–241.

Kahan B. W. and Kramp D. C. (1987) Nerve growth factor stimulation of mouse embryonal carcinoma cell migration. *Cancer Res.* 47, 6324–6328.

Kato T., Ito J., and Tanaka R. (1987) Functional dissociation of dual activities of glia maturation factor: inhibition of glial proliferation and preservation of differentiation by glial growth inhibitory factor. *Brain Res.* 430, 153–156.

Kleinman H. K., Cannom F. B., Laurie G. W., Hassell J. R., Aumailley M., Terranova V. P., Martin G. R., and Dubois-Dalcq M. (1985) Biological activities of laminin. *J. Cell. Biochem.* **27**, 317–325.

Kligman D. (1982) Isolation of a protein from bovine brain which promotes neurite extension from chick embryo cerebral cortex neurons in defined medium. *Brain Res.* **250**, 93–100.

Lam A., Fuller F., Miller J., Kloss J., Manthorpe M., Varon S., and Cordell B. (1991) Sequence and structural organization of the human gene encoding ciliary neurotrophic factor. *Gene,* **102**, 271–276.

Lazarovici P., Dickens G., Kuzuya H., and Guroff G. (1987) Long-term, heterologous down-regulation of the epidermal growth factor receptor in PC12 cells by nerve growth factor. *J. Cell Biol.* **104**, 1611–1621.

Letourneau P. C., Pech I. V., Rogers S. L., Palm S. L., McCarthy J. B., and Furcht L. T. (1988) Growth cone migration across extracellular matrix components depends on integrin, but migration across glioma cells does not. *J. Neurosci. Res.* **21**, 286–297.

Levi-Montalcini R. (1952) Effects of mouse tumor transplantation on the nervous system. *Ann. NY Acad. Sci.* **55**, 330–343.

Levi-Montalcini R. (1987) The nerve growth factor 35 years later. *Science* **237**, 1154–1162.

Lillien L. E., Sendtner M., Rohrer H., Hughes S., and Raff M. C. (1988) Type-2 astrocyte development in rat brain cultures is initiated by a CNTF-like protein produced by type-1 astrocytes. *Neuron* **1**, 485–494.

Lim R., Hicklin D. J., Ryken T. C., Han X. M., Liu K. N., Miller J. F., and Baggenstoss B. A. (1986) Suppression of glioma growth in vitro and in vivo by glia maturation factor. *Cancer Res.* **46**, 5241–5247.

Lim R., Liu Y. X., and Zaheer A. (1990) Glia maturation factor beta regulates the growth of N18 neuroblastoma cells. *Dev. Biol.* **137**, 444–450.

Longo F. M., Manthorpe M., and Varon S. (1982a) Spinal cord neuronotrophic factors (SCNTFs): I. Bioassay of Schwannoma and other conditioned media. *Dev. Brain Res.* **3**, 277–294.

Longo F. M., Manthorpe M., Skaper S. D., Lundborg G., and Varon S. (1982b) Neuronotrophic activities in fluid accumulated in vivo within silicone nerve regeneration chambers. *Brain Res.* **261**, 109–117.

Longo F. M., Skaper S. D., Manthorpe M., Lundborg G., and Varon S. (1983) Temporal changes in neuronotrophic activities accumulating in vivo within nerve regeneration chambers. *Exp. Neurol.* **81**, 756–769.

Longo F. M., Hayman E., Davis G. E., Manthorpe M., Engvall E., Ruoslahti E., and Varon S. (1984a) Neurite promoting factors and extracellular matrix components accumulating in vivo within nerve regeneration chambers. *Brain Res.* **309**, 105–117.

Longo F. M., Selak I., Zovickian J., Manthorpe M., Varon S., and U. H.-S. (1984b) Neuronotrophic activities in cerebrospinal fluid of head trauma patients. *Exp. Neurol.* **84**, 207–218.

Louis J.-C., Magal E., Muir D., Varon S., and Manthorpe M. (1992) Rat brain oligodendroglial cell proliferation is inhibited by an autocrine factor. submitted.

Ludecke G. and Unsicker K. (1990) Mitogenic effect of neurotrophic factors on human IMR 32 neuroblastoma cells. *Cancer* 65, 2270–2278.

Maisonpierre P. C., Bellnisco L., Squinto S., Ip N. Y., Furth M. E., Lindsay R. M., and Yancopoulos G. D. (1990) Neurotrophin-3: A neurotrophic factor related to NGF and BDNF. *Science* 247, 1446–1451.

Manthorpe M., Skaper S. D., Adler R., Landa K. B., and Varon S. (1980) Cholinergic neuronotrophic factors: IV. Fractionation properties of an extract from selected chick embryonic eye tissues. *J. Neurochem.* 34, 69–75.

Manthorpe M., Skaper S. D., and Varon S. (1981a) Neuronotrophic factors and their antibodies: In: vitro microassays for titration and screening. *Brain Res.* 230, 295–306.

Manthorpe M., Varon S., and Adler R. (1981b) Neurite-promoting factor in conditioned medium from RN22 schwannoma cultures: bioassay, fractionation and other properties. *J. Neurochem.* 37, 759–767.

Manthorpe M., Skaper S.D., Barbin G., and Varon S. (1982a) Cholinergic neuronotrophic factors (CNTF's): VII. Concurrent activities on certain nerve growth factor-responsive neurons. *J. Neurochem.* 38, 415–421.

Manthorpe M., Longo F. M., and Varon S. (1982b) Comparative features of spinal neuronotrophic factors in fluids collected in vitro and in vivo. *J. Neurosci. Res.* 8, 241–250.

Manthorpe M., Nieto-Sampedro M., Skaper S. D., Barbin G., Longo F. M., Lewis E. R., Cotman C. W., and Varon S. (1983a) Neuronotrophic activity in brain wounds in the developing rat. Correlation with implant survival in the wound cavity. *Brain Res.* 267, 47–56.

Manthorpe M., Engvall E., Ruoslahti E., Longo F. M., Davis G. E., and Varon S. (1983b) Laminin promotes neuritic regeneration from cultured peripheral and central neurons. *J. Cell Biol.* 97, 1882–1890.

Manthorpe M. and Varon S. (1985) Regulation of neuronal survival and neuritic growth in the avian ciliary ganglion by trophic factors, in *Growth and Maturation Factors. vol. 3.* (Guroff G., ed.), John Wiley and Sons, New York. pp. 77–117.

Manthorpe M., Fagnani R., Skaper S. D., and Varon S. (1986a) An automated colorimetric microassay for neuronotrophic factors. *Brain Res.* 390, 191–198.

Manthorpe M., Skaper S. D., Williams L. R., and Varon S. (1986b) Purification of adult rat sciatic nerve ciliary neuronotrophic factor. *Brain Res.* 367, 282–286.

Manthorpe M., Rudge J., and Varon S. (1986c) Astroglial contributions to neuronal survival and neuritic growth, in *Astrocytes, vol. 2.* (Fedoroff S. and Vernadakis A., eds.), Academic Press, New York, pp. 315–376.

Manthorpe M., Pettmann B., and Varon S. (1987) Modulation of astroglial cell output of neuronotrophic and neurite promoting factors, in *Biochemical Pathology of Astrocytes* (Norenberg M. and Schousboe A., eds.), Alan R. Liss, Inc., New York, pp. 41–57.

Manthorpe M., Ray J., Pettmann B., and Varon S. (1989) Ciliary neuronotrophic factors, in *Nerve Growth Factors* (Rush R., ed.), John Wiley & Sons, New York, pp. 31–56.

Manthorpe M., Muir D., Hagg T., and Varon S. (1992) Growth promoting and inhibiting factors for neurons, in *Advances in Neural Regeneration Research* (Seil F., ed.), Alan R. Liss, New York, in press.

Martinou J. C., Le Van Thai A., Cassar G., Roubinet F., and Weber M. J. (1989) Characterization of two factors enhancing choline acetyltrans-ferase activity in cultures of purified rat motoneurons. *J. Neurosci.* **9**, 3645–3656.

Matsuda H., Coughlin M. D., Bienenstock J., and Denburg J. A. (1988) Nerve growth factor promotes human hemopoietic colony growth and differentiation. *Proc. Nat. Acad. Sci.* **85**, 6508-6512.

McCarthy K. and DeVellis J. (1980) Preparation of separate astroglial and oligodendroglial cell cultures from rat cerebral tissue. *J. Cell Biol.* **85**, 890–902.

McMorris F. A. and Dubois-Dalcq M. (1988) Insulin-like growth factor I promotes cell proliferation and oligodendroglial commitment in rat glial progenitor cells developing in vitro. *J. Neurosci. Res.* **21**, 199–209.

Montz H. P. M., Davis G. E., Skaper S. D., Manthorpe M., and Varon S. (1985) Tumor-promoting phorbol diester mimics two distinct neuronotrophic factors. *Dev. Brain Res.* **23**, 150–154.

Morrison R. S., Kornblum H. I., Leslie F. M., and Bradshaw R. A. (1987) Trophic stimulation of cultured neurons from neonatal rat brain by epidermal growth factor. *Science* **238**, 72–75.

Morrison R. S., Keating R. F., and Moskal J. R. (1988) Basic fibroblast growth factor and epidermal growth factor exert differential effects on CNS neurons. *J. Neurosci. Res.* **21**, 71–79.

Mosmann T. (1983) Rapid colorimetric assay for cellular growth and survival: Application to proliferation and cytotoxicity assays. *J. Immunol. Methods* **65**, 55–63.

Muir D., Engvall E., Varon S., and Manthorpe M. (1989a) Schwannoma cell-derived inhibitor of the neurite-promoting activity of laminin. *J. Cell Biol.* **109**, 2353–2362.

Muir D., Gennrich C., Varon S., and Manthorpe M. (1989b) Rat sciatic nerve Schwann cell microcultures: Responses to mitogens and production of trophic and neurite promoting factors. *Neurochem. Res.* **14**, 1013–1016.

Muir D., Varon S., and Manthorpe M. (1990a) An ELISA for bromodeoxyuridine incorporation using fixed microcultures. *Anal. Biochem.* **185**, 377–382.

Muir D., Varon S., and Manthorpe M. (1990b) Schwann cell proliferation is under negative autocrine control. *J. Cell Biol.* **111**, 2663–2667.

Neugebauer K. M., Tomaselli K. J., Lilien J., and Reichardt L. F. (1988) N-cadherin, NCAM, and integrins promote retinal neurite outgrowth on astrocytes in vitro. *J. Cell Biol.* **107**, 1177–1187.

Otten U., Ehrhard P., and Peck R. (1989) Nerve growth factor induces growth and differentiation of human B lymphocytes. *Proc. Nat. Acad. Sci.* **86**, 10059–10063.

Perraud F., Besnard F., Sensenbrenner M., and Labourdette G. (1987) Thrombin is a potent mitogen for rat astroblasts but not for oligodendroblasts and neuroblasts in primary culture. *Int. J. Dev. Neurosci.* **5**, 181–188.

Perraud F., Besnard F., Labourdette G., and Sensenbrenner M. (1988a) Proliferation of rat astrocytes, but not of oligodendrocytes, is stimulated in vitro by protease inhibitors. *Int. J. Dev. Neurosci.* **6**, 261–266.

Perraud F., Besnard F., Pettmann B., Sensenbrenner M., and Labourdette G. (1988b) Effects of acidic and basic fibroblast growth factors (aFGF and bFGF) on the proliferation and the glutamine synthetase expression of rat astroblasts in culture. *Glia* **1**, 124–131.

Pettmann B., Powell J., Manthorpe M., and Varon S. (1989) Biological activities of Nerve Growth Factor bound to nitrocellulose paper by Western blotting. *J. Neurosci.* **8**, 3524–3632.

Pettmann B., Janet T., Labourdette G., Sensenbrenner M., Varon S., and Manthorpe M. (1992) Biologically active basic fibroblast growth factor migrates at 27 kD in "non-denaturing" SDS-polyacrylamide gel electrophoresis. submitted.

Pringle N., Collarini E. J., Mosley M. J., Heldin C. H., Westermark B., and Richardson W. D. (1989) PDGF A chain homodimers drive proliferation of bipotential (O-2A) glial progenitor cells in the developing rat optic nerve. *Embo J.* **8**, 1049–1056.

Puro D. G., Mano T., Chan C. C., Fukuda M., and Shimada H. (1990) Thrombin stimulates the proliferation of human retinal glial cells. *Graefes Arch. Clin. Exp. Ophthalmol.* **228**, 169–173.

Rakowicz-Szulczynska E. M. and Koprowski H. (1989) Antagonistic effect of PDGF and NGF of transcription of ribosomal DNA and tumor cell proliferation. *Biochem. Biophys. Res. Commun.* **163**, 656–694.

Ridley A. J., Davis J. B., Stroobant P., and Land H. (1989) Transforming growth factors-beta 1 and beta 2 are mitogens for rat Schwann cells. *J. Cell Biol.* **109**, 3419–3424.

Rogers S. L., Letourneau P. C., Palm S. L. and Furcht L. T, (1983) Neurite extension by peripheral and central nervous system neurons in response to substratum-bound fibronectin and laminin. *Dev. Biol.* **98**, 212–220.

Rogers S. L., Letourneau P. C., Peterson B. A., Furcht L. T., and McCarthy J. B. (1987) Selective interaction of peripheral and central nervous system cells with two distinct cell-binding domains of fibronectin. *J. Cell Biol.* **105**, 1435–1442.

Rogister B., Leprince P., Bonhomme V., Rigo J. M., Delree P., Colige A., and Moonen G. (1990) Cultured neurons release an inhibitor of astroglia proliferation (astrostatine). *J. Neurosci. Res.* **25**, 58–70.

Rudge J., Davis G., Manthorpe M., and Varon, S.(1987) An examination of ciliary neuronotrophic factors from avian and rodent tissue extracts using a blot and culture technique. *Dev. Brain Res.* **32**, 103–110.

Rudge J. R., Manthorpe M., and Varon S. (1985) The output of neuronotrophic and neurite-promoting agents from brain astroglial cells: A microculture method for screening potential regulatory molecules. *Dev. Brain Res.* **19**, 161–172.

Rudkin B. B., Lazarovici P., Levi B. Z., Abe Y., Fujita K., and Guroff G. (1989) Cell cycle-specific action of nerve growth factor in PC12 cells: differentiation without proliferation. *EMBO J.* **8**, 3319–3325.

Ruoslahti E. and Piersbacher M. D. (1987) New perspectives in cell adhesion: RGD and integrins. *Science* **328**, 491–497.

Rydel R. E. and Greene L. A. (1987) Acidic and basic fibroblast growth factors promote stable neurite outgrowth and neuronal differentiation in cultures of PC12 cells. *J. Neurosci.* **7**, 3639–3653.

Saneto R. P., Altman A., Knobler R. L., Johnson H. M., and de Vellis J. (1986) Interleukin 2 mediates the inhibition of oligodendrocyte progenitor cell proliferation in vitro. *Proc. Nat. Acad. Sci.* **83**, 9221–9225.

Sawada M., Suzamura A., Yamamoto H., and Marunouchi T. (1990) Activation and proliferation of the isolated microglia by colony stimulating factor-1 and possible involvement of protein kinase C. *Brain Res.* **509**, 119–124.

Schnell L. and Schwab M. (1990) Axonal regeneration in the rat spinal cord produced by an antibody against myelin-associated neurite growth inhibitor. *Nature (Lond.)* **343**, 269–272.

Schubert D., Kimura H., LaCorbiere M., Vaughan J., Karr D., and Fischer W. H. (1990) Activin is a nerve cell survival molecule. *Nature (Lond.)* **344**, 868–870.

Schubert D., LaCorbiere M., and Esch F. (1986) A chick neural retina adhesion and survival molecule is a retinol-binding protein. *J. Cell Biol.* **102**, 2295–2301.

Schwab M. E. and Caroni P. (1988) Oligodendrocytes and CNS myelin are nonpermissive substrates for neurite growth and fibroblast spreading in vitro. *J. Neurosci.* **8**, 2381–2393.

Seilheimer B. and Schachner M. (1988) Studies on adhesion molecules mediating interactions between cells and the PNS indicate a major role for L1 in mediating sensory neuron growth on Schwann cells. *J. Cell Biol.* **107**, 341–351.

Skaper S. D., Facci L., Rudge J., Katoh-Semba R., Manthorpe M., and Varon S. (1986) Morphological modulation of cultured rat brain astroglial cells: antagonism by ganglioside GM1. *Dev. Brain Res.* **25**, 21–31.

Thoenen H., Bandtlow C., Heumann R., Lindholm D., Meyer M., and Rohrer H. (1988) Nerve growth factor: cellular localization and regulation of synthesis. *Cell. Molec. Neurobiol.* **8**, 35–40.

Thorpe L. W. and Perez-Polo J. R. (1987) The influence of nerve growth factor on the in vitro proliferative response of rat spleen lymphocytes. *J. Neurosci. Res.* **18**, 134–139.

Tomaselli K. J., Neugebauer K. M., Bixby J. L., Lilien J., Reichardt L. F. (1988) N-cadherin and integrins: two receptor systems that mediate neuronal process outgrowth on astrocyte surfaces. *Neuron* **1**, 33–43.

Vahlsing H. L., Varon S., Hagg T., Fass-Holmes B., Dekker A., Manley M., and Manthorpe M. (1989) An improved device for continuous intraventricular infusions prevents the introduction of pump-derived toxins and increases the effectiveness NGF treatments. *Exp. Neurol.* **105**, 233–243.

Varon S., Adler R., Manthorpe M., and Skaper S. D. (1983) Culture strategies for trophic and other factors directed to neurons, in *Neuroscience Approached Through Cell Culture, vol. 2*, (Pfeiffer E., ed.), CRC Press, Boca Raton, FL, pp. 53–77.

Varon S., Hagg T., and Manthorpe M. (1992) Nerve Growth Factor, in *Encyclopedia of Human Biology*, (Dulbecco R., ed.), Academic Press, Orlando, FL, in press.

Varon S., Hagg T., and Manthorpe M. (1990) Nerve growth-factor in CNS repair and regeneration, in: *Plasticity and Regeneration of the Nervous System*. Privat A., Timiras P. S., Giacobini E., Lauder J. M., and Vernadakis A., eds. Plenum, New York, in press.

Varon S., Hagg T., and Manthorpe M. (1989) Neuronal growth factors, in *Frontiers of Clinical Neuroscience, vol. 6*, (Seil F. J., ed.), Alan R. Liss, NY, pp. 101–121.

Varon S., Manthorpe M., Davis G. E., Williams L. R., and Skaper S. D. (1988) Growth factors, *Adv. Neurol.* **47**, 493–521.

Varon S., Skaper S. D., and Manthorpe M. (1981) Trophic activities for dorsal root and sympathetic ganglionic neurons in media conditioned by Schwann and other peripheral cells. *Dev. Brain Res.* **1**, 73–87.

Varon S., Skaper S. D., Barbin G., Selak I., and Manthorpe M. (1984) Low molecular weight agents support survival of cultured neurons from the central nervous system. *J. Neurosci.* **4**, 654–658.

Walicke P., Cowan W. M., Ueno A., Baird A., and Guillemin R. (1986a) Fibroblast growth factor promotes survival of dissociated hippocampal neurons and enhances neurite extension, *Proc. Natl. Acad. Sci.* **83**, 3012–3016.

Walicke P., Varon S., and Manthorpe M. (1986b). Purification of a human red blood cell protein supporting the survival of cultured CNS neurons, and its identification as catalase. *J. Neurosci.* **6**, 1114–1121.

Walicke P. A. (1990) Fibroblast growth factor (FGF): A multifunctional growth factor in the CNS, in *Advances in Neural Regeneration Research*, (Seil F., ed.), Wiley-Liss, Inc. NY, pp. 103–114.

Wewer U., Albrechtsen R., Manthorpe M., Varon S., Engvall E., and Ruoslahti E. (1983) Human laminin isolated in a nearly intact, biologically active form from placenta by limited proteolysis. *J. Biol. Chem.* **258**, 12654–12660.

Williams L. R., Vahlsing H. L., Lindamood T., Varon S., Gage F. H., and Manthorpe M. (1987). A small-gauge cannula device for continuous infusion of exogenous agents into the brain. *Exp. Neurol.* **95**, 743–754.

Wood P. (1976) Separation of functional Schwann cells and neurons from peripheral nerve tissue. *Brain Res.* **115**, 361–375.

Yavin E., Gabai A., and Gil S. (1992) Nerve growth factor mediates monosialoganglioside-induced release of fibronectin and J1/tenascin from C6 glioma cells. **56**, *in press*.

The Role of Cell Adhesion Molecules in Neurite Growth

Daren Ure and Ann Acheson

1. Introduction

Stereotyped patterns of axonal growth and axon-axon bundling leading to tract formation are common features of nervous system development in organisms as diverse as grasshoppers and humans. In grasshoppers, axons growing from individual, identified neurons follow stereotyped pathways, making turns at choice points, which suggests that the pathways bear specific guidance cues (Goodman et al., 1982; Raper et al., 1983a,b). Such path-following by growth cones has been hypothesized to result from an *adhesive preference* for molecules on the surfaces of "guidepost" cells strategically placed along the way, or for recognition molecules on other axons. These guidepost cells and previously laid down axons together are thought to create *labeled pathways* (Goodman et al., 1982; Raper et al., 1983a,b,c; Raper et al., 1984) that guide subsequent neurite growth. Labels that create adhesive preferences could be specific molecules on glial cells or axons, or could be components of the extracellular matrix. The adhesive preferences of growth cones for a given axonal or glial surface seem to be absolute rather than hierarchical, and specific guidance cues are required for both axon initiation and continued axon extension (Bastiani and Goodman, 1986; Bastiani et al., 1986; du Lac et al., 1986).

From: *Neuromethods, Vol. 23: Practical Cell Culture Techniques*
Eds: A. Boulton, G. Baker, and W. Walz ©1992 The Humana Press Inc.

In the mammalian and avian nervous systems, studies have focused primarily on regenerative neurite growth. Peripheral nervous system (PNS) neurons have the capacity to regenerate axons following injury, and to reinnervate their proper target tissues. As in insects, the substrate along which these axons extend is of critical importance. Even mammalian central nervous system (CNS) neurons, normally unable to regenerate, can do so if given a permissive substrate (David and Aguayo, 1981; Benfey and Aguayo, 1982; Smith et al., 1986). Thus, the molecular machinery a neuron needs to extend neurites is not lost after development but rather, the proper guidance cues may be lost or obscured (Smith et al., 1986; Schnell and Schwab, 1990).

1.1. How Labeled Pathways Are Created

1.1.1. Cell–Cell Adhesion Molecules

The simplest hypothesis to describe intercellular adhesion is that one set of molecules mediates axon–axon interactions and a second set mediates axon–glia interactions. The neural cell adhesion molecule (NCAM) and L1 (named after the monocolonal antibody first used in its characterization; Rathjen and Rutishauser, 1984) are two cell adhesion molecules (CAMs) that have been localized on axons and shown to contribute to axon–axon bundling both in vitro (Stallcup and Beasley, 1985; Hoffman et al., 1986; Rathjen et al., 1987) and in vivo (Thanos et al., 1984; Landmesser et al., 1988). NCAM, the most abundant and well-characterized CAM, is found on most neural cell types (Edelman, 1984; Rutishauser, 1984). NCAM exhibits homophilic binding (i.e., it acts as its own receptor; Rutishauser et al., 1982; Edelman et al., 1987; Hall et al., 1990), so that any two NCAM-bearing cells can adhere to one another. L1 (also known as NILE, NgCAM, 8D9, and G4; Grumet and Edelman, 1984; Rathjen and Rutishauser, 1984; Bock et al., 1985; Friedlander et al., 1985; Stallcup and Beasley, 1985; Lagenauer and Lemmon, 1987; Moos et al., 1988) is both homophilic and heterophilic, depending on whether it is involved in neuron–neuron or neuron–glia interactions (Grumet and Edelman, 1988; Lemmon et al., 1989; Drazba and Lemmon, 1990). A third class of CAM that mediates intercellular adhesion, N-cadherin (neural Ca^{2+}-dependent adhesion molecule), functions primarily in the formation of cell layers during morphogenesis in the nervous system and, like NCAM, is homophilic (Matsunaga et al., 1988; Takeichi, 1988). Other CAMs are also likely to be present on axons and glial cells (e.g., Rathjen et al., 1987; Hoffman et al., 1988; Mackie et al., 1988; *see* Rutishauser and Jessell, 1988 for a comprehensive review), but so far, none has been biochemically characterized as well as the molecules described herein.

1.1.2. Cell-Substrate Adhesion Molecules

Less is known about the membrane molecules that contribute to cell–extracellular matrix (ECM) adhesion. Integrin (CSAT antigen; Horwitz et al., 1985; Buck et al., 1986) mediates the interaction of chick peripheral neurons and rat PC12 pheochromocytoma cells with ECM components, including laminin, fibronectin, and type IV collagen (Bozyczko and Horwitz, 1986; Tomaselli et al., 1986; Hall et al., 1987; Tomaselli et al., 1987). However, other neuronal laminin receptors distinct from integrin have been postulated (Edgar et al., 1984, 1988; Aumailley et al., 1987). Laminin has been localized on glial cell surfaces (Cohen et al., 1987; Edgar et al., 1988); thus, axonal laminin receptors may mediate axon–glia interactions.

1.1.3. Inhibitory Molecules

Inhibitory molecules may act to restrict growth cone motility, thereby serving as "labels" for areas in which neurons should not grow (or, later in development, cannot grow) (Kapfhammer et al., 1986; Kapfhammer and Raper, 1987; Patterson, 1988; Caroni and Schwab, 1989; Schnell and Schwab, 1990). Two components of rat CNS myelin that prevent neurite growth on oligodendrocyte surfaces as well as inhibiting fibroblast spreading have been purfied and characterized (Caroni and Schwab, 1988a,b). Antibodies against these molecules, which neutralize their function, permit growth in the CNS milieu both in vitro and in vivo (Caroni and Schwab, 1988b; Schnell and Schwab, 1990). Inhibitory molecules that mediate growth cone collapse and poisoning of motility in several in vitro model systems have also been isolated from somites, tectal membranes, and brain (Cox et al., 1990; Davies et al., 1990; Raper and Kapfhammer, 1990).

1.1.4. Multiple CAMs Are Present on Neurons

A complication of the simple hypothesis that adhesive preferences are owing to CAMs on neurons is that CAMs are widely distributed and often present on the same cell. Also, L1 and NCAM, in addition to being present on neurons, are present on nonneuronal cells (Seilheimer and Schachner, 1987); both are involved in axon–glia, as well as axon–axon interactions (Lindner et al., 1983; Silver and Rutishauser, 1984; Hoffman et al., 1986). Moreover, L1, NCAM, N-cadherin, and integrin are usually present on the same neuron, rendering it unlikely that one of them alone could determine an axon's adhesive preferences. Recent studies designed to test the hypothesis that labeled pathways are created by NCAM, N-cadherin, or laminin have indeed shown that no single one of these molecules controls neurite growth along cell surfaces (Bixby and Reichardt, 1987; Bixby et al., 1987; Cohen et al., 1987; Bixby et al., 1988; Seilheimer and Schachner, 1988; Tomaselli et al., 1988; Seilheimer et al., 1989; Drazba and Lemmon, 1990; Smith et al., 1990).

2. Neuronal Cell Culture

Cell culture has been used extensively to examine adhesive phenomena involved in patterned neurite growth and the interactions of neurons with other neurons and/or nonneuronal cells. Neuronal cell survival depends on adhesion to a substrate. In explant cultures, nonneuronal cells attach to the culture surface, migrate out of the tissue, and provide a substrate on which neurons can grow. However, the advent of dissociated cell culturing necessitated the use of prepared substrates, to which a large number of purified neurons could attach. The choice of substrate represents an important variable in the study of adhesion phenomena.

2.1. The Culture Surface

Sterile, disposable, polystyrene dishes and multiwell plates in a variety of sizes are commonly used in neuronal cell culture. Tissue culture plastic differs from Petri dish plastic in that the surface is specially treated by the manufacturer to give it a positive charge so that vertebrate cells, whose membranes are net negatively charged (Weiss, 1970), will adhere to them. Glass coverslips can also be used as a culture surface. They provide better optics than Petri plastic under inverted phase contrast microscopy and facilitate preparation for electron microscopy. The coverslips must first be cleaned in boiling nitric acid, thoroughly rinsed with distilled water, and sterilized using dry heat, UV, or ethanol (Masurovsky and Bunge, 1968; Hammarback et al., 1985). The adhesive capacity of the glass can be enhanced by derivatization with the protein crosslinking agent 3-(triethoxysilyl)-propylamine. A 10% solution of the agent in DMSO for 3–4 h at 110°C following the acid wash is sufficient (Gottlieb and Glaser, 1975; Chang et al., 1987). The use of Aclar fluorocarbon plastic coverslips has also been reported to be convenient for long-term nerve tissue growth and for direct use in light and electron microscopy (Masurovsky and Bunge, 1968; Wood, 1976; Johnson and Argiro, 1983).

2.2. Preparation of the Culture Surface

Tissue culture plastic promotes better cell attachment than Petri plastic or glass, but it is still an inferior substrate for many neuronal cell types. Most culture surfaces must therefore be treated in some way prior to cell plating to enhance attachment. Pretreatment approaches have involved the application of purified proteins or protein fragments, conditioned medium, and cryostat sections. Despite the advantages of studying isolated cell types on various substrates, valuable information is still to be gained about the interactions among different cell types. To this end, many studies have focused on the coculture of neuronal and nonneuronal cells. Nonneuronal cells can replace or supplement other inert attachment-supporting substrates.

2.2.1. Coating Dishes with Purified Proteins

2.2.1.1. POLYCATIONS

Polycations were the first molecules to be used as a substrate coating that would promote attachment of a large majority of neuronal cells (Yavin and Yavin, 1974). They may be used alone or in combination with other coating substances. Anionic substrates do not promote adhesion (Letourneau, 1975). The two polycationic molecules that have been used to date are polylysine (PLYS; levo or dextro enantiomers) and polyornithine (PORN). Although PLYS has received greater use, PORN may be less toxic to cells. Both cations are available commercially and their application to culture surfaces is relatively simple.

Solution Preparation
1. Dissolve polylysine in distilled water to a final concentration of 0.1 mg/mL. Filter-sterilize. Store at 4°C.
2. Dissolve polyornithine in 0.15M borate buffer, pH 8.4, to a final concentration of 0.25 mg/mL. Filter-sterilize. Store at 4°C.

Surface Coating
1. Apply excess of substrate solution (1–2 mL in 35 mm dish) to culture surface.
2. Incubate overnight at room temperature or 4°C.
3. Aspirate excess solution and rinse 2–3 times with distilled water or PBS.

The concentrations given represent those in common practice, but investigators have reported using polycation solutions ranging from 0.005–2 mg/mL (Yavin and Yavin, 1974; Letourneau, 1975; Gundersen, 1987; Kuenemund et al., 1988). The choice of solvent (water or borate buffer) and wash buffer (water or PBS) also appears to be interchangeable. Incubation periods may be as short as 1 h when carried out at 37°C (Tomaselli et al., 1986).

2.2.1.2. TYPE I/III COLLAGEN

The desire to better mimic in vivo growth conditions and to study the function of the different CAMs has further resulted in the use of ECM molecules as growth substrates. The interstitial collagens (Types I and III) were the first to be used. They are the most abundant proteins in the vertebrate body and many cell types are able to synthesize them (Barnes, 1984). Not only do numerous cell types attach to this matrix component, but fibrillar collagen is also important for cellular differentiation and proliferation. For the purpose of coating tissue culture dishes, both of these types of collagen may be purchased in purified form, but most investigators prepare crude collagen from rat tail tendons according to the protocol by Bornstein (1958). A detailed description of the same protocol is also given by Johnson and Argiro (1983). The method makes a collagen solution that can be stored at 4°C for up to 10 mo.

Solution Preparation
1. Soak rat tails in 95% ethanol for 15 min. (Note: rat tails can be stored at –70°C before use.)
2. Dissect tendons into distilled water and tease them apart using forceps. Remove as much nontendinous material as possible.
3. Soak tendons in 0.1% (v/v) acetic acid for 2 d at 4°C.
4. Remove undigested tendon after centrifuging mixture at 1500g for 2 h. The collagen-containing aqueous phase can be stored in the refrigerator for up to 10 mo.
5. Dialyze the collagen solution against distilled, sterile water for 24 h or more. Longer dialysis periods produce more viscous solutions. The final solution can now be used directly or stored at 4°C for 1–2 wk.

Alternatives to Solution Preparation
There are minor differences in the many protocols in use. The tendon-acetic acid mixture may be gently stirred overnight in the cold and then spun at 10,000 rpm for 30 min (Campenot, personal communication). Elsdale and Bard (1972) recommend dialyzing the soluble collagen twice against 10% (v/v) Eagle's medium and adjusting the medium to pH 4.0 for the second dialysis, since the collagen solution should be more stable at low pH and ionic strength. The dialysed solution is centrifuged at 17,000 rpm for 24 h at 4°C. To simplify the procedure even more, one could omit the dialysis of the collagen-acetic acid solution (Hawrot and Patterson, 1979).

A number of methods have been described for the coating of collagen onto a culture surface. The air-drying method is the simplest procedure, but it ranks low on the collagen uniformity, ultrastructure, and lifetime that it provides (Iversen et al., 1981;

Macklis et al., 1985). The superior methods are reported to involve the use of the crosslinking agent, carbodiimide, or ammonium vapor derivitization. The latter method makes a transparent, three-dimensional collagen gel into which neuronal processes extend but cell bodies do not (Hawrot and Patterson, 1979). The gel substrate eliminates the need of a medium thickening agent but may complicate growth analysis.

Surface Coating

Air-Drying

1. Apply a thin film of collagen solution to the culture surface. It is sufficient to coat a 35 mm dish by filling then immediately removing the collagen solution from the dish.
2. Air-dry the thin collagen film overnight under sterile conditions.

Ammonium Vapor Derivitization

1. Coat the surface with the collagen solution. In a closed system, expose the coated surface to ammonia vapors by including a small container of concentrated ammonium hydroxide for 2–30 min. This step will gel the collagen into a firm, transparent layer (Bornstein, 1958).
2. Rinse the coated surface with many washes of sterile water or saline.

Hydrated Collagen Lattice

   Another method for making a three-dimensional collagen gel is that of Elsdale and Bard (1972; *see* Alternatives to Solution Preparation). By raising the ionic strength and pH of an acidic collagen solution simultaneously to physiological levels, an ordered latticework of collagen fibrils will be produced.

1. To the collagen solution (pH 4.0) on ice quickly add predetermined vols of 10X concentrated medium, serum (to reconstitute standard medium), and enough 0.142M sodium hydroxide to bring pH to 7.6.
2. Dispense solution into dishes and leave undisturbed to set for a few min.

Alternatives to Surface Coating

The carbodiimide crosslinking method by Macklis et al. (1985) is not in widespread use, although the method is not laborious and is reported to make a superior substrate. Less efficient derivitizations include exposure to UV light or salting out collagen that has been mixed with 6% NaCl (Iversen et al., 1981).

2.2.1.3. FIBRONECTIN, LAMININ, AND TYPE IV COLLAGEN

These three types of molecules, like type I/III collagen, are components of the ECM, and are thus good candidates for culture substrates. Fibronectin is a common component of serum, plasma, and the ECM. It is abundant in peripheral nerves but less so in the CNS. Laminin and type IV collagen are common constituents of the basal lamina, which surrounds a variety of cell types. These molecules possess multiple binding sites for other extracellular or cell surface molecules, and likely form important associations with them in vivo (Yamada et al., 1980).

All three ECM molecules can be purchased from suppliers or can be isolated and purified by the investigator according to described techniques. The coating procedure is similar to that of polycations, involving the adsorption of molecules onto the culture surface.

Isolation/Purification
1. Fibronectin is most often and easily purified from plasma by column chromatography, making use of fibronectin's affinity for gelatin (Engvall and Ruoslahti, 1977; Klebe et al., 1980; Ruoslahti et al., 1980; Smith et al., 1982; Smith and Furcht, 1982). Cellular fibronectin can be isolated from chick embryo fibroblasts (Yamada et al., 1980).
2. Laminin is purified most often from mouse Englebreth-Holm-Swarm (EHS) sarcoma and less often from rat yolk sac tumor (Timpl et al., 1978; Engvall et al., 1983).
3. Collagen (Type IV) can be isolated from the murine EHS tumor (Timpl et al., 1978).

Solution Preparation
1. Dilute fibronectin in 0.05M carbonate buffer to a concentration of 100–200 μg/mL and adjust to pH 9.5. Low concentration solutions can be used because fibronectin sticks to glass and polystyrene very well.
2. Dilute laminin in calcium- and magnesium-free phosphate-buffered saline (CMF-PBS) to a concentration of 2–10 μg/ 10 cm^2 tissue culture surface.

Surface Coating
   Any of the three molecules can be coated by the same procedure, but laminin alone binds very poorly to tissue culture plastic. Therefore, dishes should be precoated with a polycation (*see* above). One experiment using ^{125}I-labeled laminin showed that up to three times more laminin binds to PORN-coated dishes than to tissue culture plastic (Edgar et al., 1984). A laminin solution (10 μg/mL) left for 4–6 h at 4°C on a polylysine-coated coverslip coats laminin to a concentration of 1.7 ng/mm^2 (Lein and Higgins, 1989). The preparations and coating procedure are based on methods described by Rogers et al. (1985), Tomaselli et al. (1986), Gundersen (1987), Hall et al. (1987), Kuenemund et al. (1988), and Letourneau et al. (1990).

1. Apply an excess of the solution to the culture surface and let stand overnight at 4°C.
2. Remove excess and rinse once with CMF-PBS.

Alternatives to Surface Coating
   If one is interested in studying the effects of a specific substrate molecule, and if the cells of interest are able to attach to the untreated culture surface, it would be advantageous to block nonspecific cell–substrate attachment by postcoating the treated surface with a 0.5–1.0% bovine serum albumin (BSA) or ovalbumin (PBS) solution for at least 2 h, followed by another rinsing (Manthorpe et al., 1983; Rogers et al., 1985; Aumailley et al., 1987; Goodman et al., 1987; Hall et al., 1987).

Some investigators apply their coating solutions for as little as 30 min, but for these short periods, the adsorption should be carried out at temperatures up to 37°C (Baron-Van Evercooren et al., 1982; Edgar et al., 1984; Chang et al., 1987; Kleitman et al., 1988; Letourneau et al., 1990). It may also be desirable to completely air-dry the coating solution (Aumailley et al., 1987) or omit the rinsing step in order to leave as much coating molecule on the surface as possible (Chang et al., 1987). Occasionally, investigators will complement or even replace the precoating procedure with the addition of the coating molecule or conditioned medium directly to the experimental medium, which precipitates a fine coat of molecule on the surface (Bottenstein and Sato, 1980; Sieber-Blum et al., 1981; Hantaz-Ambroise et al., 1987; Lein and Higgins, 1989).

2.2.1.4. PROTEINS ON NITROCELLULOSE

Another method recently introduced to study the ability of purified proteins to promote neurite growth utilizes nitrocellulose, a polysaccharide widely used in chromatography and blotting procedures, which has the capacity to bind protein very tightly. This technique has allowed the study of both purified ECM components and purified CAMs as culture substrates (Lagenauer and Lemmon, 1987; Lemmon et al., 1989; Bixby and Zhang, 1990; Furley et al., 1990).

Method
1. Dissolve 5 cm^2 of nitrocellulose in 6 mL of methanol.
2. Aliquot 0.5 mL of dissolved nitrocellulose into a 60-mm Petri dish (or equivalent amounts onto other surfaces) and allow to dry in laminar flow hood.
3. Add 1–5 µL of 0.1–1.0 mg/mL solution to the nitrocellulose-coated surfaces to create discrete areas of the dish, which are coated with the protein of choice.
4. After a short time (1–15 min), rinse the substrate with BSA- or serum-containing medium to block nitrocellulose that had not bound protein. Cells will not attach to BSA-coated surfaces.

2.2.2. Coating with Conditioned Medium Components

Cells growing in vivo secrete specific factors, some of which diffuse through tissues and others that are immobilized in the ECM in the vicinity of the secreting cell. Researchers have capitalized on this property by growing primary or transformed cells in culture and using the medium conditioned by the cells to search for clues about the secreted products.

If culture dishes are exposed to conditioned medium for a brief period, some of the factors will attach to the dish. The culture surface created in this manner may promote cell attachment and growth, depending on the composition of the conditioned medium. The advantage of this surface preparation, besides being nonlaborious, is that the coating factors likely represent physiological complexes of molecules that may be more effective at promoting growth than individual purified components.

2.2.2.1. MEDIUM CONDITIONING

A number of variations have been described for the conditioning of medium, depending on the cell type and the investigator. This chapter will not detail the many protocols, but will focus on one that has proven effective in our laboratory. It involves cells of the RN22 Schwannoma cell line, which secrete a form of laminin bound likely to a proteoglycan. Media conditioned by this cell type have been studied quite extensively (Manthorpe et al., 1981,1983; Davis et al., 1985; Tomaselli et al., 1986; Muir et al., 1989).

1. When the cells have grown to half-confluency, administer fresh serum-free medium to the culture and condition it for 4 d.
2. Collect medium and adjust pH to 7.4.
3. Sterile filter the medium; this also removes any dead cells that may be present.
4. Store unused medium at –20°C.

2.2.2.2. COATING DISHES

It is common practice to coat culture surfaces first with a polycation to enhance the attachment of the secreted factors to the surface. For example, the neurite-promoting factor in RN22-

conditioned medium requires the precoating of polyornithine to the culture surface to exert its effect

1. Apply excess of conditioned medium to polycation-coated surface overnight at 4°C.
2. Remove excess and wash twice with sterile water and air-dry under sterile conditions.

2.2.3. Using Cryostat Sections as Substrates

The preapplication of cryostat sections to culture surfaces is a valuable compromise between in vitro and in vivo growth analyses (Carbonetto et al., 1987; Covault et al., 1987; Sandrock and Matthew, 1987). Not only does the bioassay afford a simplified analysis of growth characteristics, but it more closely mimics the natural pattern than assays that utilize foreign substrates. Neurites assume unique patterns on sections of different tissue types, but in general, they localize to cell surfaces or ECM regions preferentially over intracellular regions. The specific association with cell surfaces is not likely owing to mechanical forces inherent in the tissue composite, but to adhesive interactions between the neurites and specific tissue molecules. For example, antibodies to a laminin-heparan sulfate proteoglycan and NCAM both inhibit neurite elongation on cryostat sections in a tissue-specific manner (Sandrock and Matthew, 1987).

Methods
1. Mount whole or composite tissue blocks in an appropriate cryo-protectant embedding medium and section them to a thickness of 6–20 μm using a cryomicrotome. Place the sections onto glass or plastic coverslips that have been cleaned and optionally coated with a polycation.
2. Optionally, expose the section briefly (20 min) to UV light, followed by a light rinse with medium, but avoid prolonged UV exposure since extensive irradiation will denature proteins (Hammarback et al., 1985).
3. Use the prepared substrate immediately or store at –70°C.

Explants or dissociated neurons are plated onto the prepared substrates, either on or off the tissue sections. Authors have claimed viability of neurons on nonundercoated tissue substrates for up to 7 d. The use of coverslips allows the investigator to analyze tissue composition and neurite growth by both immunocytochemical and standard histological staining after cultures have been fixed.

2.2.4. Using Cultured Nonneuronal Cells as Substrates

The coculture of neurons with different nonneuronal cell types offers the advantages of studying the interaction of the cells in a more controlled manner than explant cultures. Astrocytes, Schwann cells, muscle cells, and fibroblasts are common choices of nonneuronal cells for cocultures. Whereas conditioned medium experiments involve adhesion factors secreted by other cells, coculture experiments involve adhesion factors both secreted by the nonneuronal cells or embedded in their cell membrane. The importance of studying these support cells is the belief that as a prominent part of the neural microenvironment, they regulate the capacity of peripheral neurons to regenerate and limit the ability of central neurons to grow after damage. Even fragments of cell membranes have been used as growth substrates in order to decipher the role of surface molecules that mediate regeneration (Walter et al., 1987). One cell type is usually plated and given time to attach to the substrate prior to the addition of the second cell type.

An easy method for obtaining fibroblasts involves a 20% trypsinization of newborn rat leg muscle for 10 min at 37°C, centrifugation, and plating of the cell suspension on tissue culture plastic. Depending on the study, it may be more convenient or applicable to collect fibroblasts from the sciatic nerve at the same time as Schwann cells are being isolated (Brockes et al., 1979). Typically, newborn rat or mouse sciatic nerve is trypsin/collagenase-digested, dissociated, and the cells collected by centrifugation, resuspended, and plated. If the cells are plated directly onto tissue culture plastic, only fibroblasts will attach and the remaining cells, mostly Schwann cells, can be collected. Alternatively, if the cells are plated onto a polycationic substrate, both Schwann cells and fibroblasts will attach. If only the Schwann cells are being cultured, the fibroblasts can be removed by complement-mediated immunocytolysis using specific antibodies, such as Thy 1.1, Thy 1.2, or MESA-1, followed by brief trypsinization (Seilheimer and Schachner, 1987,1988). Schwann cells can also be isolated from sciatic nerve with about 70% efficiency by an elegant method that utilizes contrasting susceptibilities of fibroblasts and Schwann cells to precisely timed applications of the antimitotic agents cytosine arabinoside and fluorodeoxyuridine (Wood, 1976).

Astrocytes are usually derived from 1-d old rat cortex following procedures by Noble et al. (1984), Fallon (1985a,b), and McCarthy and De Vellis (1980). The isolation is based on astrocyte/oligodendrocyte sorting that occurs after a specified period in culture.

3. Studying Adhesion Phenomena

Adhesive phenomena are most easily conceptualized by considering whether cells prefer to adhere to each other or to the culture surface. CAMs not only mediate the degree to which cells adhere, but they also contribute to growth characteristics of neurons, such as neurite initiation, elongation, fasciculation, and growth cone activity. Both pattern and extent of growth can be influenced by CAMs, but the extent to which these parameters change depends on the type of cell (which CAMs are present) and substrate. For example, the pattern of neurite bundling (fasciculation) generally reflects a balance between cell–cell and cell–substrate interactions. If cell–cell interactions dominate, the neurites will tend to fasciculate, but if cell–substrate adhesion dominates, the neurite will tend to spend more time on the substrate and less time growing on another axon, thus, the pattern of growth will be defasciculated (Landmesser et al., 1988; Rutishauser and Jessell, 1988).

A variety of techniques can be used to study adhesion interactions in culture. What most of them have in common is that they provide information about the adhesive forces at play by attempting to interfere with adhesion. Mechanical interference with adhesion provides relatively general information about how well a cell is attached to some other component of the culture environment. On the other hand, chemical interference with adhesion, primarily through the use of antibodies against CAMs, gives more specific information about which molecules mediate adhesive forces. The two treatments can be used alone or in combination to create a variety of assays that investigate adhesion phenomena.

3.1. Mechanical Interference with Adhesion

3.1.1. Cell Attachment

Mechanical interference with adhesion is the basis for the attachment assay. The simplest form of the assay involves plating cells from the same preparation onto a variety of substrates. After a certain interval, usually 1–24 h, the loosely attached cells are gently rinsed away with medium. The cells that remain attached are fixed and counted, and crossdish comparisons are made to assess the adhesive quality of the substrates.

The addition of mechanical shear forces allows further selection of those cells that interact specifically with the substrate. Again, cells from the same preparation are plated onto different substrates, but during the initial incubation, they are gently rotated or rocked. It is best to fix the cells at intervals in order to construct a curve representing the time-dependent adhesion characteristics on the different substrates.

There are alternative ways of estimating the number of cells that remain attached after an attachment assay. Rather than fixing the cells on the surface, they can be detached with trypsin/EDTA and counted using a Coulter counter (Goodman et al., 1987). The cells can also be prelabeled with a radioisotope, such as ^{35}S-methionine, and the attached cells counted in a scintillation counter (Cole et al., 1985). This method will not give an accurate absolute number of cells, but it is useful for crossculture comparisons. Since it is usually difficult to clearly differentiate between cell types in coculture, if one is interested in testing adhesion to a cellular substrate, it is useful to again prelabel the test cells but with a fluorescent label, such as fluorescein diacetate. After the assay, the postfixed cells can be counted using fluorescence microscopy (Keilhauer et al., 1985).

3.1.2. Growth Cone Attachment

The attachment assays previously described are not practical for assessing the attachment of growth cones to substrates. The adherence of growth cones alone has been analyzed by delivering a microjet of air (Gail and Boone, 1972; Letourneau et al., 1975) or medium (Gundersen and Barrett, 1980) at the growth cone. The pressure required to lift the growth cone is directly correlated with how well the neurite is attached to the substrate. The use of these shear forces suffers from criticisms about tearing of cellular extensions and the variability owing to cell shape and adhesion energies, but the techniques do give reproducible results.

3.2. Chemical Interference with Adhesion

3.2.1. Antibody Perturbation Studies

Chemical interference with adhesion more specifically characterizes adhesion phenomena. When antibodies specific to CAMs are used to block adhesion, it is inferred that the ligand against which the antibody is directed is playing a role in that phenomenon. Antibody blockade of CAM function can be employed alone to alter growth characteristics under otherwise normal culture conditions, or can be employed in combination with mechanical interference in an attachment assay.

Both polyclonal and monoclonal antibodies are used, but polyclonals offer a greater probability that complete function of the ligand will be blocked. In most cases, it is best to use only the variable portion of the immunoglobulin, Fab, which recognizes the epitope. Fab fragments are obtained by digesting purified IgG with papain or pepsin and separating the resultant Fab moieties from the constant Fc portion of the molecule using Protein A affinity chromatography. IgGs are disadvantageous compared to Fab fragments for two reasons. First, since each immunoglobulin has two variable regions, the molecule could give nonspecific crosslinking effects. This is especially problematic when studying abundant cell surface proteins on adjacent neurites. Second, since IgGs are large molecules, they may not be able to physically access all of the functionally important

epitopes. The use of Fab fragments should enhance epitope binding and satisfy the investigator that adhesion effects are the
result of specific blockade of CAM function.

3.2.2. Pharmacological Perturbation
and Competition Studies

Although antibody perturbation studies predominate, there
are other treatments that address which molecules are responsible for adhesion. By simply lowering the extracellular calcium
concentration to 0.1mM, it has been reported that the role of the
calcium-dependent CAM, N-cadherin, is eliminated without
degradation of the molecule (Letoureau et al., 1990). Competi-
tive binding is a workable technique because cells will adhere
only to bound substrate molecules. The addition of free molecules that compete for the CAM binding sites should reduce
cell–substrate adhesion (Tomaselli et al., 1987; Friedlander et al.,
1988; Humphries et al., 1988). Finally, enzymatic and pharmacological agents may be introduced to modify CAMs. Glycosaminoglycan lysases, such as heparitinase, chondroitinase, and
hyaluronidase, cause variable effects on neurite extension and
Schwann cell proliferation (Ratner et al., 1985; Hoffman and
Edelman, 1987; Dow et al., 1988; Hoffman et al., 1988). Beta-D-
xyloside, a blocker of proteoglycan assembly (Ratner et al., 1985;
Dow et al., 1988), and alpha-lactalbumin, a galactosyltransferase
modifier (Begovac and Shur, 1990), have also shown specific
effects in neuronal cell culture.

4. Growth Assays

The study of the relationships between cell adhesion and
growth characteristics has focused on neurite initiation, elongation, fasciculation, and branching and the dynamics of growth
cone movement. Mechanical treatments provided some of the
founding principles of the field, but chemical treatments have
been most useful in deciphering the exact roles of CAMs.
Adhesive interactions play a role in all of the previously mentioned growth characteristics, but the extent of their contributions varies.

4.1. Neurite Initiation and Elongation

Neurite initiation is a measure of the rate that neurons initiate processes under various conditions. A short window of time, anywhere from 2–24 h from cell plating, is considered the initiation period. Within that period, the fraction of neurons bearing processes longer than at least one cell diameter is determined. The phenomenon is dependent on the adhesivity of the substrate; a surface that is more adhesive enhances neurite initiation. For example, PORN- and PLYS-coated surfaces, which are more adhesive than tissue culture plastic, also promote faster neurite initiation (Letourneau, 1975). However, neurite initiation may not be directly related to adhesivity in every case. Gundersen (1987) found that neurites showed a preference for laminin, but the glycoprotein was not as adhesive as the other substrates tested.

Neurite elongation is an extension of neurite initiation. It may be quantified similarly by determining the percentage of neurons extending processes farther than defined multiples of cell body diameters, but there is room for large errors in this method (Hall et al., 1987). A more quantitative estimation can be made by making camera lucida drawings of the cells and manually measuring neurite lengths. Image analysis systems or computerized graphic tablets with morphometric software will offer the same results but with less labor. Accurate measurements can be made in this way only if neurons are plated at a sufficiently low density to allow reliable tracing of the entire neuritic arbor, as well as unequivocal identification of which neuron the neurites are growing from. One final method employs the compartmented culture system, to be discussed later in this chapter. The advantages of the system are that extension can be measured quickly, even when cells are plated at high density.

When neurons are growing in coculture, it is more difficult to visualize them. In order to measure neurite elongation it is first necessary to identify them, usually done by immunostaining with a neuron-specific marker (e.g., neurofilament). A novel approach to studying the relationship between interaxonal adhesion and elongation has been developed by Chang et al. (1987). In this assay, chick sympathetic neurites radiating from a central ganglion to an outer cellular ring form the substrate upon which a test layer of additional TRITC-stained sympathetic neurons are plated. The test neurites growing solely on the neurite base are then measured.

4.2. Neurite Fasciculation and Branching

Fasciculation and branching of neurites, in comparison to neurite elongation, are more difficult to assess. Neurite fascicles comprise numerous individual neurites. In general, wider bundles are comprised of more neurites, but variability in thickness can also result from alterations in neurite–neurite adhesion. An antibody-mediated alteration in fasciculation may be missed by light microscopy, and may only be able to be detected by electron microscopy (Rutishauser et al., 1978). A change in branching is also difficult to detect by light microscopy because new branches may remain confined to the neurite bundle. Furthermore, it is impossible to know whether branching is indeed true branching of a neurite or the divergence of two individual neurites. These drawbacks have consequently led to analysis of fasciculation and branching primarily in a qualitative manner. Fasciculation has been semiquantified by measuring bundle widths microscopically with an eyepiece and grouping the results, such as > or <3 μm (Fischer et al., 1986; Hoffman et al., 1986). Similarly, the percentage of cell aggregates with fascicles >3 μm has also represented the degree of fasciculation (Stallcup and Beasley, 1985).

4.3. Growth Cone Dynamics

Growth cone dynamics represent an important area of research because growth cones ultimately determine the branching and direction of neurite growth. Active growth cones constantly extend filopodia and lamellipodia as feelers into the microenvironment. When adhesively attractive areas are located, the contacts are stabilized, and neurite extension proceeds in the direction of the stabilized contact. Relatively nonadhesive locales are avoided. In some circumstances, the growth cone can also encounter inhibitory adhesion signals that cause retraction or collapse of active growth cones. The analyses of growth cone dynamics are at best semiquantitative. More often qualitative assessments are made by observations using videomicroscopy (Rutishauser et al., 1983; Aletta and Greene, 1988; Cox et al., 1990). Growth cone analysis suffers from the same criticisms as other in vitro assays. The growth cones under optical isolation may represent a biased subpopulation of all the growth cones.

5. Problems in Interpreting These Types of Studies

When mixed cultures of neurons and nonneuronal cells are used to study the role of CAMs in neurite growth, a major problem is the potential indirect effects of one cell type on the other. Substances secreted by the nonneuronal cell can alter the properties of the neurons, and vice versa, and adding crude tissue extracts to support neuronal survival (as are needed for ciliary, retinal, and spinal cord neurons) compounds the problem. Purified trophic factors added to keep the neuronal test populations alive may also modify CAMs, either on the neuron itself or on the nonneuronal cell surface. For example, nerve growth factor regulates L1 expression on Schwann cells in vitro (Seilheimer and Schachner, 1987). Since purified L1 is an excellent substrate for neurite growth in vitro (Lagenauer and Lemmon, 1987), NGF-regulated L1 expression on Schwann cells could be an important variable.

Cryostat sections, on the other hand, subject neurites to mixed populations of cells, and the contribution of one cell type cannot be determined except by electron microscopy. In addition, since the cells are dead, it is not possible to change their surfaces and test the relative importance of various molecules. Further, irrelevant molecules may be exposed to the neurons owing to cell damage during sectioning. Conversely, freezing and sectioning may destroy relevant cell surface molecules, particularly those containing carbohydrates.

In experiments using both mixed cultures and cryostat sections, the neurons can be plated onto the test substrates in the presence of antibodies. If the antibodies compromise neuronal attachment, neurites will not grow no matter what the substrate.

Optimal attachment is not something that can necessarily be seen with the light microscope, or even measured using attachment assays. Neurons that do not land in the right place never grow neurites, and thus, only a subpopulation of cells is being studied.

Trypsinization has to be used to dissociate the neurons before plating in any one of these experimental paradigms. Indeed, complete dissociation and low cell density are necessary so that both the number of neurons with neurites and the length of those neurites can be determined. However, some CAMs, notably NCAM, are sensitive to trypsin, and will be readily stripped from the cell surface during the dissociation procedure (Rutishauser et al., 1982). Anti-NCAM antibodies added immediately after trypsinization will not block NCAM function initially because NCAM is no longer present on the surface. During the period of resynthesis and recovery of stripped cell surface proteins, neurons will already adhere to the surfaces of the cells or cryostat sections. Once this adhesion has taken place, even if it is not very strong, anti-NCAM Fabs will no longer be able to block NCAM function (*see*, e.g., Acheson and Rutishauser, 1988). Anti-N-cadherin antibodies are unique in that they disrupt already-existing cell–cell contacts (Matsunaga et al., 1988), and N-cadherin also appears to be remobilized to the cell surface much more rapidly than NCAM. Interestingly, NCAM has not been implicated in controlling neurite growth, whereas anti-N-cadherin antibodies have strong effects. L1 appears to be the most resistant to protease treatment (Rathjen et al., 1987; and unpublished observations), and it too has strong effects in neurite growth on nonneuronal cell surfaces (Seilheimer and Schachner, 1988; Seilheimer et al., 1989; Drazba and Lemmon, 1990).

A very effective approach to studying adhesion events of cell bodies and neurites independently is by the compartmented culture system devised by Campenot (1979). In this system, neuronal cell bodies and axons are separated from one another by a fluid-impermeable barrier. Neurons plated into the center compartment extend axons into both left and right side compartments containing test substrates. Since the side compartments consist of many parallel tracks, individual axon–substrate inter-actions can be compared and there is no need to dissociate neu-rons, which was necessary in previous studies so that neurite growth could be quantified. Thus, neuronal populations can be (1) given time to recover surface proteins after dissociation or (2) studied in explant form without being exposed to enzymatic digestion. Nonneuronal cells grown in the side compartments can also be pretreated with factors that alter the expression of CAMs before neurons are added. Left and right side compartments also allow direct controls for antibody effects: antibodies can be added to only one side when both side compartments contain identical substrates. Finally, using this system, both fascicula-tion/branching patterns (a quantitative measure of the morphology of the neuron) and extent of growth (a quantitative measure of rate or linear extension) can be assessed simultaneously.

There is also some evidence that the ECM components them-selves can be active, i.e., provide information to the neurons. Laminin has been shown to potentiate NGF-mediated survival of sympathetic neurons in culture (Edgar et al., 1984), as well as potentiating the ability of BDNF to mediate survival of DRG neurons both in vitro and in vivo (Lindsay et al., 1985; Kalcheim et al., 1987). Laminin can also cause an activation of TH in chro-maffin cells (Acheson et al., 1986), implying an ability to gener-ate biochemical signals. In this respect, the choice of substrate (polycation vs ECM component) can potentially affect experi-ments designed to examine the role of CAMs in neurite growth in several ways. Active biochemical signals generated by spe-cific binding to ECM components could directly initiate growth programs that are accessory to CAM-mediated growth. Con-versely, the signals could influence neurite growth by first altering the compliment of CAMs present on the cell surface. Third, ECM-mediated changes in neuronal survival could change the population of cells being examined, and thereby change the observed effect. A lack of knowledge about which bio-chemical signals are generated makes it difficult to distinguish between the effects of passive and active adhesion on a number of growth characteristics.

6. Conclusion

Understanding the control of regeneration in the vertebrate nervous system is of central importance in devising strategies for promoting regeneration in humans. The substrate along which axons grow either permits or inhibits regeneration (David and Aguayo, 1981; Benfey and Aguayo, 1982; Smith et al., 1986; Schnell and Schwab, 1990). One element contributing to successful regeneration is the adhesivity between axon and substrate, mediated by CAMs. However, it is not known whether CAMs on the substrate or on the regenerating axon (or combinations of both) are important for growth control. Much current interest is thus focussed on understanding the molecular controls of neurite growth, and in particular, how CAMs regulate both the pattern and extent of neurite growth. However, the complexity of the problem, both in terms of evaluating growth and in terms of evaluating the relative contributions of the various CAMs, is such that in vitro model systems are necessary. Neurite initiation, elongation, and branching may all involve separate processes. Trophic factors may modulate the amount, molecular form, or function of CAMs. The first neurites to grow may use different cues than later neurites, which may simply follow along the first ones. Damage may alter CAM expression on the neurons themselves. Reentry into the cell cycle by nonneuronal cells may also change the program of CAM expression. All of these questions must be addressed systematically in vitro in order to provide information about adhesive cues in development and regeneration.

References

Acheson A. and Rutishauser U. (1988) Neural cell adhesion molecule regulates cell contact-mediated changes in choline acetyltransferase activity of embryonic chick sympathetic neurons. *J. Cell Biol.* 106, 479–486.

Acheson A., Edgar D. Timpl R., and Thoenen H. (1986) Laminin increases both levels and activity of tyrosine hydroxylase in calf adrenal chromaffin cells. *J. Cell Biol.* 102, 151–159.

Aletta J. M. and Greene L. A. (1988) Growth cone configuration and advance: a time-lapse study using video-enhanced differential interference contrast microscopy. *J. Neurosci.* 8, 1425–1435.

Aumailley M., Nurcombe V., Edgar D., Paulsson M., and Timpl R. (1987) The cellular interactions of laminin fragments. *J. Biol. Chem.* **262,** 11532–11538.

Barnes D. (1984) Attachment factors in cell culture, in *Mammalian Cell Culture, The Use of Serum-Free Hormone-Supplemented Media* (Mather J. P., ed.), Plenum Press, New York, pp. 195–237.

Baron-Van Evercooren A., Kleinman H. K., Ohno S., Marangos P., Schwartz J. P. and Dubois-Dalcq M. E. (1982) Nerve growth factor, laminin, and fibronectin promote neurite growth in human fetal sensory ganglia cultures. *J. Neurosci. Res.* **8,** 179–193.

Bastiani M. J. and Goodman C. S. (1986) Guidance of neuronal growth cones in the grasshopper embryo. I. Recognition of specific glial pathways. *J. Neurosci.* **6,** 3542–3551.

Bastiani M. J., du Lac S., and Goodman C. S. (1986) Guidance of neuronal growth cones in the grasshopper embryo. I. Recognition of a specific axonal pathway by the pCC neurons. *J. Neurosci.* **6,** 3518–3531.

Begovac P. C. and Shur B. D. (1990) Cell surface galactosyltransferase mediates the initiation of neurite outgrowth from PC12 cells on laminin. *J. Cell Biol.* **110,** 461–470.

Benfey M. and Aguayo A. J. (1982) Extenslve elongation of axons from rat brain into peripheral nerve grafts. *Nature (Lond.)* **296,** 150–152.

Bixby J. L. and Reichardt L. F. (1987) Effects of antibodies to neural cell adhesion molecule (N-CAM) on the differentiation of neuromuscular contacts between ciliary ganglion neurons and myotubes in vitro. *Dev. Biol.* **119,** 363–372.

Bixby J. L. and Zhang R. (1990) Purified N-cadherin is a potent substrate for the rapid induction of neurite outgrowth. *J. Cell Biol.* **110,** 1253–1260.

Bixby J. L., Pratt R. S., Lilien J., and Reichardt L. F. (1987) Neurite outgrowth on muscle cell surfaces involves extracellular matrix receptors as well as Ca^{2+}-dependent and -independent cell adhesion molecules. *Proc. Natl. Acad. Sci. USA* **84,** 2555–2559.

Bixby J. L., Lilien J., and Reichardt L. F. (1988) Identification of the major proteins that promote neuronal process outgrowth on Schwann cells in vitro. *J. Cell Biol.* **107,** 353–361.

Bock E., Richter-Landsberg C., Faissner A., and Schachner M. (1985) Demonstration of immunochemical identity between nerve growth factor-inducible large external (NILE) glycoprotein and the cell adhesion molecule L1. *EMBO J.* **4,** 2765–2768.

Bornstein M. B. (1958) Reconstituted rat-tail collagen used as a substrate for tissue cultures on coverslips in Maximow slides and roller tubes. *Lab. Invest.* **7,** 134–137.

Bottenstein J. E. and Sato G. H. (1980) Fibronectin and polyslysine requirement for proliferation of neuroblastoma cells in defined medium. *Exp. Cell Res.* **129,** 361–366.

Bozyczko D. and Horwitz A. F. (1986) The participation of a putative cell surface receptor for laminin and fibronectin in peripheral neurite extension. *J. Neurosci.* **6,** 1241–1251.

Brockes J. P., Fields K. L., and Raff M. C. (1979) Studies on cultured rat Schwann cells. I. Establishment of purified populations from cultures of peripheral nerve. *Brain Res.* **165,** 105–118.

Buck C., Shea E., Duggan K., and Horwitz A. F. (1986) Integrin (the CSAT antigen): functionality requires oligomeric intergrity. *J. Cell Biol.* **103,** 2421–2428.

Campenot R. B. (1979) Independent control of the local environment of somas and neurites. *Methods in Enzymol.* **28,** 302–307.

Carbonetto S., Evans D., and Cochard P. (1987) Nerve fiber growth in culture on tissue substrata from central and peripheral nervous systems. *J. Neurosci.* **7,** 610–620.

Caroni P. and Schwab M. E. (1988a) Antibody against myelin-associated inhibitor of neurite growth neutralizes nonpermissive substrate properties of CNS white matter. *Neuron* **1,** 85–96.

Caroni P. and Schwab M. E. (1988b) Two membrane protein fractions from rat central myelin with inhibitory properties for neurite growth and fibroblast spreading. *J. Cell Biol.* **106,** 1281–1288.

Caroni P. and Schwab M. E. (1989) Codistribution of neurite growth inhibitors and oligodendrocytes in rat CNS: appearance follows nerve fiber growth and precedes myelination. *Dev. Biol.* **136,** 287–295.

Chang S., Rathjen F. G., and Raper J. A. (1987) Extension of neurites on axons is impaired by antibodies against specific neural cell surface glycoproteins. *J. Cell Biol.* **104,** 355–362.

Cohen J., Burne J. F., McKinlay C., and Winter J. (1987) The role of laminin and the laminin/fibronectin receptor complex in the outgrowth of retinal ganglion cell axons. *Dev. Biol.* **122,** 407–418.

Cole G. J., Schubert D., and Glaser L. (1985) Cell-substratum adhesion in chick neural retina depends upon protein-heparan sulfate interactions. *J. Cell Biol.* **100,** 1192–1199.

Covault J., Cunningham J. M., and Sanes J. R. (1987) Neurite outgrowth on cryostat sections of innervated and denervated skeletal muscle. *J. Cell Biol.* **105,** 2479–2488.

Cox E. C., Muller B., and Bonhoeffer F. (1990) Axonal guidance in the chick visual system: posterior tectal membranes induce collapse of growth cones from the temporal retina. *Neuron* **2,** 31–37.

David S. and Aguayo A. J. (1981) Axonal elongation in peripheral nervous system "bridges" after central nervous system injury in adult rats. *Science* **214,** 931–933.

Davies J. A., Cook G. M. W., Stern C., and Keynes R. J. (1990) Isolation from chick somites of a glycoprotein fraction that causes collapse of dorsal root ganglion growth cones. *Neuron* **2,** 11–20.

Davis G. E., Manthorpe M., Engvall E., and Varon S. (1985) Isolation and characterization of rat Schwannoma neurite-promoting factor: evidence that the factor contains laminin. *J. Neurosci.* **5,** 2662–2671.

Dow K. E., Mirski S. E. L., Roder J. C., and Riopelle R. J. (1988) Neuronal proteoglycans: biosynthesis and functional interaction with neurons *in vitro*. *J. Neurosci.* **8**, 3278–3289.

Drazba J. and Lemmon V. (1990) The role of cell adhesion molecules in neurite outgrowth on Mueller cells. *Dev. Biol.* **138**, 82–93.

du Lac S., Bastiani M. J., and Goodman C. S. (1986). Guidance of neuronal growth cones in the grasshopper embryo. II. Recognition of a specific axonal pathway by the aCC neuron. *J. Neurosci.* **6**, 3532–3541.

Edelman G. M. (1984) Modulation of cell adhesion during induction, histogenesis, and perinatal development of the nervous system. *Ann. Rev. Neurosci.* **7**, 339–377.

Edelman G. M., Murray B. A., Mege R.-M., Cunningham B. A., and Gallin W. J. (1987) Cellular expression of liver and neural cell adhesion molecules after transfecting with their cDNAs results in specific cell–cell binding. *Proc. Natl. Acad. Sci. USA* **84**, 8502–8506.

Edgar D., Timpl R., and Thoenen H. (1984) The heparin-binding domain of laminin is responsible for its effects on neurite outgrowth and neuronal survival. *EMBO J.* **3**, 1463-1468.

Edgar D., Timpl R., and Thoenen H. (1988) Structural requirements for the stimulation of neurite outgrowth by two variants of laminin and their inhibition by antibodies. *J. Cell Biol.* **106**, 1299–1306.

Elsdale T. and Bard J. (1972) Collagen substrata for studies on cell behavior. *J. Cell Biol.* **54**, 626–637.

Engvall E. T. and Ruoslahti E. (1977) Binding of soluble form of fibroblast surface protein, fibronectin, to collagen. *Int. J. Cancer* **20**, 1–5.

Engvall E. T., Krusius T., Wewer U., and Ruoslahti E.(1983) Laminin from rat yolk sac tumor: isolation, partial characterization, and comparison with mouse laminin. *Arch. Biochem. Biophys.* **222**, 649–656.

Fallon J. R. (1985a) Preferential outgrowth of central nervous system neurties on astrocytes and Schwann cells as compared with nonglial cells in vitro. *J. Cell Biol.* **100**, 198–207.

Fallon J. R. (1985b) Neurite guidance by non-neuronal cells in culture: preferential outgrowth of peripheral neurites on glial as compared to nonglial cell surfaces. *J. Neurosci.* **5**, 3169–3177.

Fischer G., Kuenemund V., and Schachner M. (1986) Neurite outgrowth patterns in cerebellar microexplant cultures are affected by antibodies to the cell surface glycoprotein L1. *J. Neurosci.* **6**, 605–612.

Friedlander D. R., Grumet M., and Edelman G. M. (1985) Nerve growth factor enhances expression of neuron-glia cell adhesion molecule in PC12 cells. *J. Cell Biol.* **102**, 413–419.

Friedlander D. R., Hoffman S., and Edelman G. M. (1988) Functional mapping of cytotactin: proteolytic fragments active in cell-substrate adhesion. *J. Cell Biol.* **107**, 2329–2340.

Furley A. J., Morton S. B., Manalo D., Karagogeos D., Dodd J., and Jessell T. M. (1990) The axonal glycoprotein TAG-1 is an immunoglobulin superfamily member with neurite outgrowth-promoting activity. *Cell* **61**, 157–170.

Gail M. H. and Boone C. W. (1972) Cell-substrate adhesivity. *Exp. Cell Res.* **70**, 33–40.

Goodman C. S., Raper J. A., Ho R. K., and Chang S. (1982) Pathfinding of neuronal growth cones in grasshopper embryos, in *Developmental Order: Its Origin and Regulation* (Subtelny S. and Green P. B., eds.), Liss, New York, pp. 275–316.

Goodman S. L., Deutzmann R., and Von der Mark K. (1987) Two disinct cell-binding domains in laminin can independently promote nonneuronal cell adhesion and spreading. *J. Cell Biol.* **105**, 589–598.

Gottlieb D. J. and Glaser L. (1975) A novel assay of neuronal cell adhesion. *Biochem. Biophys. Res. Commun.* **63**, 815–821.

Grumet M. and Edelman G. M. (1984) Heterotypic binding between neuronal membrane vesicles and glial cells is mediated by a specific cell adhesion molecule. *J. Cell Biol.* **98**, 1746–1756.

Grumet M. and Edelman G. M. (1988) Neuron-glia cell adhesion molecule interacts with neurons and astroglia via different binding mechanisms. *J. Cell Biol.* **106**, 487–503.

Gundersen R. W. (1987) Response of sensory neurites and growth cones to patterned substrata of laminin and fibronectin *in vitro*. *Dev. Biol.* **121**, 423–431.

Gundersen R. W. and Barrett J. N. (1980) Characterization of the turning response of dorsal root neurites toward nerve growth factor. *J. Cell Biol.* **87**, 546–554.

Hall A. K., Nelson R., and Rutishauser U. (1990) Binding properties of detergent-solubilized NCAM. *J. Cell Biol.* **110**, 817–824.

Hall D. E., Neugebauer K. M., and Reichardt L. F. (1987) Embryonic neural retinal cell response to extracellular matrix proteins: Developmental changes and effects of the cell substratum attachment antibody (CSAT). *J. Cell Biol.* **104**, 623–634.

Hammarback J. A., Palm S. L., Furcht L. T., and Letourneau P. C. (1985) Guidance of neurite outgrowth by pathways of substratum-bound laminin. *J. Neurosci. Res.* **13**, 213–220.

Hantaz-Amroise D., Vigny M., and Koenig J. (1987) Heparan sulfate proteoglycan and laminin mediate two different types of neurite outgrowth. *J. Neurosci.* **7**, 2293–2304.

Hawrot E. and Patterson P. H. (1979) Long-term cultures of dissociated sympathetic neurons. *Methods Enzymol.* **58**, 574–584.

Hoffman S., Crossin K. L., and Edelman G. M. (1988) Molecular forms, binding functions, and developmental expression patterns of cytotactin and cytotactin-binding proteoglycan, an interactive pair of extracellular matrix molecules. *J. Cell Biol.* **106**, 519–532.

Hoffman S., Friedlander D. R., Chuong C., Grumet M., and Edelman G. M. (1986) Differential contributions of Ng-CAM and N-CAM to cell adhesion in different neural regions. *J. Cell Biol.* **103**, 145–158.

Hoffman S. and Edelman G. (1987) A proteoglycan with HNK-1 antigenic determinants is a neuron-associated ligand for cytotactin. *Proc. Natl. Acad. Sci. USA* **84**, 2523–2537.

Horwitz A. F., Duggan K., Greggs R., Dekker C., and Buck C. (1985) The cell substrate attachment (CSAT) antigen has properties of a receptor for laminin and fibronectin. *J. Cell Biol.* **101**, 2134–2144.

Humphries M. J., Akiyama S. K., Komoriya A., Olden K., and Yamada K. (1988) Neurite extension of chicken peripheral nervous system neurons on fibronectin: relative importance of specific adhesion sites in the central cell-binding domain and the alternatively spliced Type III segment. *J. Cell Biol.* **106**, 1289–1297.

Iversen P. L., Partlow L. M., Stensaas L. J., and Moatamed F. (1981) Characterization of a variety of standard collagen substrates: Ultrastructure, uniformity, and capacity to bind and promote growth of neurons. *In Vitro* **17**, 540–552.

Johnson M. I. and Argiro V. (1983) Techniques in the tissue culture of rat sympathetic neurons. *Methods Enzymol.* **103**, 334–347.

Kalcheim C., Barde Y.-A., Thoenen H., and Le Douarin, N. M. (1987) *In vivo* effect of brain-derived neurotrophic factor on the survival of developing dorsal root ganglion cells. *EMBO J.* **6**, 2871–2873.

Kapfhammer J. P. and Raper J. A. (1987) Collapse of growth cone structure on contact with specific neurites in culture. *J. Neurosci.* **7**, 201–212.

Kapfhammer J. P., Grunewald B. E., and Raper J. A. (1986) The selective inhibition of growth cone extension by specific neurites in culture. *J. Neurosci.* **6**, 2527–2534.

Keilhauer G., Faissner A., and Schachner M. (1985) Differential inhibition of neurone-neurone, neurone-astrocyte, and astrocyte-astrocyte adhesion by L1, L2, and N-CAM antibodies. *Nature* **316**, 728–730.

Klebe R. J., Bentley K. L., Sasser P. J., and Schoen R. C. (1980) Elution of fibronectin from collagen with chaotropic agents. *Exp. Cell Res.* **130**, 111–117.

Kleitman N., Wood P., Johnson M. I., and Bunge R. P. (1988) Schwann cell surfaces but not extracellular matrix organized by Schwann cells support neurite outgrowth from embryonic rat retina. *J. Neurosci.* **8**, 653–663.

Kuenemund V., Jungalwala F. B., Fischer G., Chou D. K. H., Keilhauer G., and Schachner M. (1988) The L2/HNK-1 carbohydrate of neural cell adhesion molecules is involved in cell interactions. *J. Cell Biol.* **106**, 213–223.

Lagenaur C. and Lemmon V. (1987) A L1-1ike molecule, the 8D9 antigen is a potent substrate for neurite extension. *Proc. Natl. Acad. Sci. USA* **84**, 7753–7757.

Landmesser L., Dahm L., Schultz K., and Rutishauser U. (1988) Distinct roles of adhesion molecules during innervation of embryonic chick muscle. *Dev. Biol.* **130**, 645–670.

Lein P. J. and Higgins D. (1989) Laminin and a basement membrane extract have different effects on axonal and dendritic outgrowth from embryonic rat sympathetic neurons *in vitro*. *Dev. Biol.* **136**, 330–345.

Lemmon V., Farr K. L., and Lagenauer C. (1989) L1 -mediated axon growth occurs via a homophilic binding mechanism. *Neuron* **2**, 1597–1603.

Letourneau P. C. (1975) Possible roles for cell-to-substratum adhesion in neuronal morphogenesis. *Dev. Biol.* **44**, 77–91.

Letourneau P. C., Shattuck T. A., Roche F. K., Takeichi M., and Lemmon V. (1990) Nerve growth cone migration onto Schwann cells involves the calcium-dependent adhesion molecule, N-cadherin. *Dev. Biol.* **138**, 430–442.

Lindner J., Rathjen F. G., and Schachner M. (1983) L1 mono- and polyclonal antibodies modify cell migration in early postnatal mouse cerebellum. *Nature (Lond.)* **305**, 427–430.

Lindsay R. M., Thoenen H., and Barde Y.-A. (1985) Placode and neural crest-derived sensory neurons are responsive at early developmental stages to brain-derived neurotrophic factor. *Dev. Biol.* **112**, 319–328.

Mackie E. J., Tucker R. P., Halfter W., Chiquet-Ehrismann R., and Epperlein H. H. (1988) The distribution of tenascin coincides with pathways of neural crest cell migration. *Development* **102**, 237–250.

Macklis J. D., Sidman R. L., and Shine H. D. (1985) Cross-linked collagen surface for cell culture that is stable, uniform, and optically superior to conventional surfaces. *In vitro* **21**, 189–194.

Manthorpe M., Varon S., and Adler R. (1981) Neurite-promoting factor in conditioned medium from RN22 schwannoma cultures: bioassay, fractionation, and properties. *J. Neurochem.* **37**, 759–767.

Manthorpe M., Engvall E., Ruoslahti E., Longo F. M., Davis G. E., and Varon S. (1983) Laminin promotes neuritic regeneration from cultured peripheral and central neurons. *J. Cell Biol.* **97**, 1882–1890.

Masurovsky E. B. and Bunge R. P. (1968) Fluoroplastic coverslips for long-term nerve tissue culture. *Stain Technol.* **43**, 161–165.

Matsunaga M., Hatta K., and Takeichi M. (1988) Role of N-cadherin cell adhesion molecules in histogenesis of neural retina. *Neuron* **1**, 289–295.

McCarthy K. D. and De Vellis J. (1980) Preparation of separate astroglial and oligodendroglial cell cultures from rat cerebral tissue. *J. Cell Biol.* **85**, 890–902.

Moos M., Tacke R., Scherer H., Teplow D., Frueh K., and Schachner M. (1988) Neural adhesion molecule L1 as a member of the immunoglobulin superfamily with binding domains similar to fibronectin. *Nature (Lond.)* **334**, 701–703.

Muir D., Engvall E., Varon S., and Manthorpe M. (1989) Schwannoma cell-derived inhibitor of the neurite-promoting activity of laminin. *J. Cell Biol.* **109**, 2353–2362.

Noble M., Fok-Seang J., and Cohen J. (1984) Glia are a unique substrate for the *in vitro* growth of central nervous system neurons. *J. Neurosci.* **4**, 1892–1903.

Patterson P. (1988) On the importance of being inhibited, or saying no to growth cones. *Neuron* **1**, 263–267.

Raper J. A. and Kapfhammer J. P. (1990) The enrichment of a neuronal growth cone collapsing activity from embryonic chick brain. *Neuron* **2**, 21–29.

Raper J. A., Bastiani M. J., and Goodman C. S. (1983a) Pathfinding by neuronal growth cones in grasshopper embryos. I. Divergent choices made by growth cones of sibling neurons. *J. Neurosci.* **3**, 20–30.

Raper J. A., Bastiani M. J., and Goodman C. S. (1983b) Pathfinding by neuronal growth cones in grasshopper embryos. II. Selective fasciculation onto specific axonal pathways. *J. Neurosci.* **3**, 31–41.

Raper J. A., Bastiani M. J., and Goodman C. S. (1983c) Guidance of neuronal growth cones: selective fasciculation in the grasshopper embryo. *Cold Spring Harbor Symp. Quant. Biol.* **48**, 587–598.

Raper J. A., Bastiani M. J., and Goodman C. S. (1984) Pathfinding by neuronal growth cones in grasshopper embryos. IV. The effects of ablating the A and P axons upon the behavior of the G growth cone. *J. Neurosci.* **4**, 2329–2345.

Rathjen F. G. and Rutishauser U. (1984) Comparison of two cell surface molecules involved in neural cell adhesion. *EMBO J.* **3**, 461–465.

Rathjen F. G., Wolff J. M., Frank R., Bonhoeffer F., and Rutishauser U. (1987) Membrane glycoproteins involved in neurite fasciculation. *J. Cell Biol.* **104**, 343–353.

Ratner N., Bunge R. P., and Glaser L. (1985) A neuronal cell surface heparan sulfate proteoglycan is required for dorsal root ganglion neuron stimulation of Schwann cell proliferation. *J. Cell Biol.* **101**, 744–754.

Rogers S. L., McCarthy J. B., Palm S. L., Furcht L. T., and Letourneau P. C. (1985) Neuron-specific interactions with two neurite promoting fragments of fibronectin. *J. Neurosci.* **5**, 369–378.

Ruoslahti E., Hayman E., Pierschbacher M., and Engvall E. (1980) Fibronectin: purification, immunochemical properties and biological activities. *Methods Enzymol.* **82A**, 803–831.

Rutishauser U. (1984) Developmental biology of a neural cell adhesion molecule. *Nature (Lond.)* **310**, 549–554.

Rutishauser U. and Jessell T. M. (1988) Cell adhesion molecules in vertebrate neural development. *Physiol. Rev.* **68**, 819–857.

Rutishauser U., Gall W. E., and Edelman G. (1978) Adhesion among neural cells of the chick embryo IV. Role of the cell surface molecule CAM in the formation of neurite bundles in cultures of spinal ganglia. *J. Cell Biol.* **79**, 382–393.

Rutishauser U., Grumet M., and Edelman G. M. (1983) Neural cell adhesion molecule mediates initial interactions between spinal cord neurons and muscle cells in culture. *J. Cell Biol.* **97**, 145–152.

Rutishauser U., Hoffman S., and Edelman G. M. (1982) Binding properties of a cell adhesion molecule from neural tissue. *Proc. Natl. Acad. Sci. USA* **79**, 685–689.

Sandrock A. W. and Matthew W. D. (1987) Identification of a peripheral nerve neurite growth-promoting activity by development and use of an *in vitro* bioassay. *Proc. Natl. Acad. Sci. USA* **84,** 6934–6938.

Schnell L. and Schwab M. E. (1990) Axonal regeneration in the rat spinal cord produced by an antibody against myelin-assocaited neurite growth inhibitors. *Nature (Lond.)* **343,** 269–272.

Seilheimer B. and Schachner M. (1987) Regulation of neural cell adhesion molecule expression on cultured mouse Schwann cells by nerve growth factor. *EMBO J.* **6,** 1611–1616.

Seilheimer B. and Schachner M. (1988) Studies of adhesion molecules mediating interaction between cells of peripheral nervous system indicate a major role for L1 in mediating sensory neuron growth on Schwann cells in culture. *J. Cell Biol.* **107,** 341–351.

Seilheimer B., Persohn E., and Schachner M. (1989) Antibodies to the L1 adhesion molecule inhibit Schwann cell ensheathment of neurons *in vitro. J. Cell Biol.* **109,** 3095–3103.

Sieber-Blum M., Sieber F., and Yamada K. M. (1981) Cellular fibronectin promotes adrenergic differentiation of quail neural crest cells *in vitro. Exp. Cell Res.* **133,** 295–385.

Silver J. and Rutishauser U. (1984) Guidance of optic axons *in vivo* by a preformed adhesive pathway on neuroepithelial endfeet. *Dev. Biol.* **106,** 485–499.

Smith D. E., Mosher D. F., Johnson R. B., and Furcht L. T. (1982) Immunological identification of two sulfhydryl-containing fragments of human plasma fibronectin. *J. Biol. Chem.* **257,** 5831–5838.

Smith D. E. and Furcht L. T. (1982) Localization of two unique heparin binding domains of human plasma fibronectin with monoclonal antibodies. *J. Biol. Chem.* **257,** 6518–6523.

Smith G. M., Miller R. H., and Silver J. (1986) Changing role of forebrain astrocytes during development, regenerative failure, and induced regeneration upon transplantation. *J. Comp. Neurol.* **251,** 23–43.

Smith G. M., Rutishauser U., Silver J., and Miller R. H. (1990) Maturation of astrocytes *in vitro* alters the extent and molecular basis of neurite outgrowth. *Dev. Biol.* **138,** 377–390.

Stallcup W. B. and Beasley L. (1985) Involvement of the nerve growth factor-inducible large external glycoprotein (NILE) in neurite fasciculation in primary cultures of rat brain. *Proc. Natl. Acad. Sci. USA* **82,** 1276–1280.

Takeichi M. (1988) The cadherins: cell-cell adhesion molecules controlling animal morphogenesis. *Development* **102,** 639–655.

Thanos S., Bonhoeffer F., and Rutishauser U. (1984) Fiber-fiber interactions and tectal cues influence the development of the chicken retinotectal projection. *Proc. Natl. Acad. Sci. USA* **81,** 1906–1910.

Timpl R., Martin G. R., Bruckner P., Wick G., and Wiedemann H. (1978) Nature of the collagenous protein in a tumor basement membrane. *Eur. J. Biochem.* **84,** 43–52.

Tomaselli K. J., Damsky C. H., and Reichardt L. F. (1987) Interactions of a neuronal cell line (PC12) with laminin, collagen IV, and fibronectin: identification of integrin-related glycoproteins involved in attachment and process outgrowth. *J. Cell Biol.* **105,** 2347–2358.

Tomaselli K. J., Reichardt L. F., and Bixby J. L. (1986) Distinct molecular interactions mediate neuronal process outgrowth on nonneuronal cell surfaces and extracellular matrices. *J. Cell Biol.* **103,** 2659–2672.

Tomaselli K. J., Neugebauer K. M., Bixby J. L., Lilien J., and Reichardt L. F. (1988) N-Cadherin and integrins: two receptor systems that mediate neuronal process outgrowth on astrocyte surfaces. *Neuron* **1,** 33–43.

Walter J., Kern-Veits B., Huf J., Stolze B., and Bonhoeffer F. (1987) Recognition of position-specific properites of tectal cell membranes by retinal axons *in vitro. Development* **101,** 685–696.

Weiss L. (1970) The cell periphery. *Int. Rev. Cytol.* **26,** 63–105.

Wood P. M. (1976) Separation of functional Schwann cells and neurons from normal peripheral nerve tissue. *Brain Res.* **115,** 361–375.

Yamada K. M., Kennedy D. W., Kimata K., and Pratt R. M. (1980) Characterization of fibronectin interactions with glycosaminoglycans and identification of proteolytic fragments. *J. Biol. Chem.* **255,** 6055–6063.

Yavin E. and Yavin Z. (1974) Attachment and culture of dissociated cells from rat embryo cerebral hemispheres on polylysine-coated surface. *J. Cell Biol.* **62,** 540–546.

Three-Dimensional Organ Culture Systems

B. Rogister, J. M. Rigo, P. P. Lefebvre, P. Leprince, P. Delree, D. Martin, J. Schoenen, and G. Moonen

1. Introduction

Normal brain functions are to the highest degree dependent on the cytoarchitecture and the intercellular relationships that govern both the nervous system metabolism (for instance, the integrity of the blood-brain barrier) and specific functions (synaptic transmission, glioneuronal relationships). Thus, the study of the biology of the nerve cells should be approached using experimental designs that preserve the histological structure in which these cells take part in vivo.

Tissue culture disrupts these structural and functional relationships to various degrees according to the method used, as illustrated in Fig. 1. Furthermore, new and possibly abnormal relationships might be reestablished during the cultivation procedure. On the other hand, disruption of the tissue organization increases the accessibility to experimental manipulations, allowing, for instance, the study of the effect of medium conditioned by one cell type on another cell type or somatic cell hybridization.

In this chapter, we review different culture systems implying various degrees of tissue disruption to illustrate the advantages and limitations of these methods. For the purposes of clarity, specific examples will be given with description of the

From: *Neuromethods, Vol. 23: Practical Cell Culture Techniques*
Eds: A. Boulton, G. Baker, and W. Walz ©1992 The Humana Press Inc.

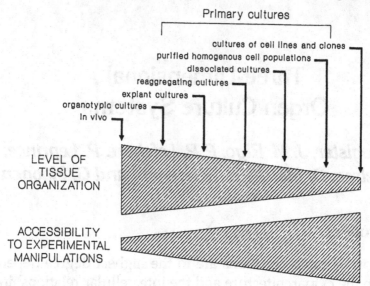

Fig. 1. Methods for culturing nervous tissue have been classified as a function of the level of tissue organization and the accessibility to experimental manipulation. For instance, tissue organization is higher when primary dissociated cultures contain several cell-types as compared to culture techniques allowing the separation of homogenous cell types. Cell lines or clones allow genetic studies or somatic hybridization that are not possible using primary (i.e., not subcultured) cultures.

dissection and culture procedures. Some results obtained using these culture methods will also be mentioned as illustrations of the possibilities offered by the various techniques.

2. Organotypic Cultures of the Organ of Corti and of the Spiral Ganglion

2.1. Introduction

Cell survival in organotypic cultures depends on adequate passive diffusion of metabolites from the organ to the culture medium and from the medium to the organ. Such a culture method is thus only possible when the considered organ has a size lower than well-defined limits (roughly, 1 mm thickness). In our laboratory, we have used organotypic cultures to study the trophic and toxic interactions that take place in the

developing inner ear, since these interactions might be implicated in the pathophysiology of several types of deafness.

Cochlear hair cells, which are responsible for the transduction of the sound stimulus into a bioelectric signal, are divided
in two subpopulations: the outer hair cells (OHC) organized in
three rows and the inner hair cells distributed in one row, parallel to the outer hair cells. Two different afferent neuronal systems, radial (originating from Type I ganglion neurons) and spiral
(originating from Type II neurons) are connected respectively
to the inner and outer hair cells (Berglund and Ruygo, 1987;
Kiang et al., 1982).

The innervation pattern of the two systems is different—
many Type I neurites contact a single inner hair cell, whereas
a single Type II neurite makes synapses with many outer hair
cells. To preserve the complexity of the architecture of the
cochlea, we have developed an organotypic culture system of
the mammalian inner ear. The 5-d-old rat was chosen in these
experiments because the cochlea is not completely ossified at
that age, whereas the development of the sensory and neuronal
structures is nearly achieved.

2.2. Methodology

2.2.1. Dissection Procedure

Five-day-old Wistar rats are killed by cervical dislocation.
The base of the skull is turned facing the operator. Using curved
forceps and small scissors, the mandibles are sectioned and
removed. The bullae are exposed and are easily recognized since
they are limited by a white ring that corresponds to the ossify-
ing bulla tympani. The temporal bone is excised by four cuts:
sagitally along the midline (1), laterally to the bullae (4) and
transversally, rostral (2) and caudal (3) to the bullae (Fig. 2A).
The tissue is then transferred into a glass Petri dish (diameter 60
mm) in warmed (37°C) phosphate buffered saline (PBS). Under
a dissecting stereomicroscope (magnification 10X), using sharp
watchmaker forceps, the bony ring is removed opening the bullae tympani where the cochlea is easily recognized. The cochlea
is freed from the surrounding tissues. Particular attention is paid

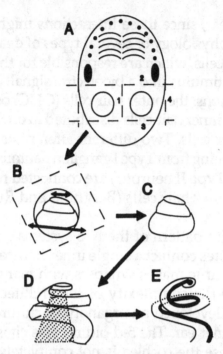

Fig. 2. Schematic dissection of rat cochlea with the organ of Corti and the spiral ganglion (not drawn to scale for the purpose of clarity). (A) The head of the 5-d-old rat is turned upside-down and the upper mandible is seen forward. Tympani bullae are dissected by four incisions (anterior (2), posterior (3), lateral (4) and medial (1)). (B) The lower part of the temporal bone which contains the cochlea is excised. (C) The cochlea is freed from newly formed bone that is very soft. (D) The organ of Corti is freed from surrounding fibrous tissue. (E) The spiral ganglia (large black spiral) is separated from the organ of Corti.

to the apex where the surrounding tissues stick to the bony capsule (Fig. 2B). The bony cochlea now completely exposed is separated from the labyrinth block and transferred in another glass Petri dish containing warm PBS (Fig. 2C). The bony shell is removed (Fig. 2D). The stria vascularis and spiral ligament (outer wall of the cochlea duct) are stripped away as a single piece from the base to the apex. The organ of Corti is stripped from the columella, which constitutes a central axis (Fig. 2E). Isolated organs of Corti are usually cut in three sectors before being plated as explants.

The same procedure is used to prepare combined explants of the organ of Corti and of the spiral ganglion (dark line on Fig. 2E). Very carefully, the spiral ligament is removed, together with the stria vascularis as above. With very sharp forceps, the columella is then cut transversally to its long axis between each turn.

2.2.2. Tissue Culture

Plastic Petri dishes (35 mm) (NUNC) are coated by overnight incubation at room temperature with 1 mL of a 0.1 mg/mL solution of polyornithine in 15 mM borate buffer (pH 8.4). Before use, the dishes are rinsed twice with distilled water and once with culture medium.

The explants are seeded in Eagle's Minimum Essential Medium (MEM) supplemented with bovine insulin (5 μg/mL) and glucose (final concentration: 6 g/L). A volume of 600 μL culture medium is required to maintain the explants on the substratum and to prevent cultures from drying. After 1 h, when the explants adhere to the substratum, 66 μL of Fetal Calf Serum (FCS) (final concentration 10% v/v) are added to the culture medium for 24 h. After that time, medium is renewed using MEM added with the N1 cocktail that contains transferrin (5 μg/mL), putrescine (100 μM), bovine insulin (5 μg/mL), progesterone (0.02 μM), and selenite (0.03 μM) (Bottenstein and Sato, 1979; *see also* the chapter of Bottenstein). Under the phase contrast microscope, the outer hair cells are recognized by their stereocilia organized in V or W (Fig. 3). They can be observed in these cultures from the time of explantation up to 7 d (time during which the cultures remain organotypic). The mean number of OHC (305 ± 45/mm) remains relatively stable for at least 7 d. The inner hair cells are more difficult to recognize at the phase-contrast microscopy level. Ultrastructural studies performed in these cultured organs of Corti show that both outer and inner hair cells remain intact (Lefebvre et al., 1990a).

The explants containing both the organ of Corti and the spiral ganglion are seeded in culture as described above but in Dulbecco's modified essential medium (DMEM) supplemented

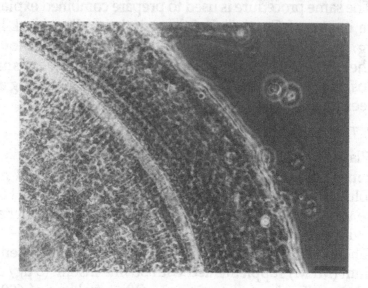

Fig. 3. Phase-contrast microphotograph of 5-d-old rat organ of Corti grown in vitro for 3 d. The three rows of outer hair cells (OHC) can be easily recognized. Scale bar: 50 μM. (Reprint from Lefebvre et al., 1990a.)

with the N1 cocktail. In order to observe the neurons and their neuritic processes, an immunocytochemical staining of cytoskeletal elements is performed (Lefebvre et al., 1991), using antineurofilament antibodies (Fig. 4).

2.3. Applications of Organotypic Cultures of the Organ of Corti and of the Spiral Ganglion

2.3.1. Explants of the Organ of Corti as an In Vitro Model to Investigate the Pathophysiology of Inner Ear Diseases

Ménière's disease is characterized by recurrent attacks of vertigo and tinnitus and progressively leads to deafness. The primary, yet misunderstood anomaly is an increased pressure of the high K^+-containing endolymphatic fluid that leads to the breakdown of the endoperilymphatic barrier and to an abnormal increase of the K^+ concentration in the perilymph. That

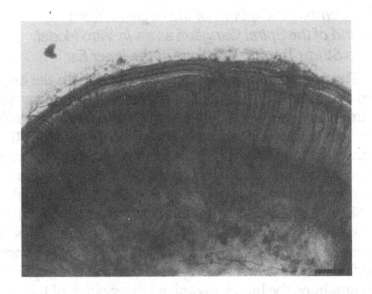

Fig. 4. Microphotograph of a 5-d-old rat explant containing both the organ of Corti and the spiral ganglion. Immunohistochemical staining using antineurofilament antibodies shows the afferent auditory neurites and the peripheral innervation in the organ of Corti. Scale bar: 100 μM.

increase of extracellular potassium concentration is thought to play a key role in determining the progressive loss of cochlear hair cells and auditory neurons (Zenner, 1986). In order to try to understand the pathological mechanisms leading to the death of the sensory and neuronal structures in the cochlea in Ménière's disease, we used as a bioassay the serum-free three-dimensional culture preparation of cochlear hair cells of 5-d-old rat and reported the release by the cochlea, in response to an increase of extracellular potassium concentration, of a low-mol-wt cytotoxic activity for hair cells and afferent auditory neurons (Lefebvre et al., 1990a). Since a high potassium concentration was demonstrated to have no direct cytotoxic activity on the sensory and neuronal structures of the inner ear in vitro (Zenner, 1986), we have suggested that this cochlea-derived toxic activity might play a role in the pathophysiology of the hearing loss that occurs during the progression of Ménière's disease.

2.3.2. Combined Explants of the Organ of Corti
and of the Spiral Ganglion as an In Vitro Model
to Study Trophic Relations in the Inner Ear

The preparation of explants containing both the organ of Corti and the spiral ganglion has been developed to investigate the trophic relationships between the hair cells and the afferent auditory neurons. Indeed, we were able to demonstrate that the organ of Corti releases diffusible factors that allow the survival of, and induce the neuritogenesis by, the auditory neurons. One of the molecular vectors of this neurite promoting effect is the Nerve Growth Factor (NGF) (Lefebvre et al., 1990b). When the explants are incubated in the presence of antibodies raised against nerve growth factor, the innervation pattern is completely modified, further demonstrating the key role of this trophic factor in maintaining the highly organized innervation of the cochlea.

Moreover, if a sector of the organ of Corti is destroyed, a neuronal degeneration is observed near the lesion, indicating that the presence of the organ of Corti in the explant is a necessary condition for the survival of neurons of the spiral ganglion. Extrapolating such trophic interrelationships to inner ear diseases might lead to a possible explanation of the retrograde degeneration of the first auditory neurons that follows an injury of the Organ of Corti.

3. Microexplant Cultures
(Cerebellum and Hippocampus)

3.1. Introduction

The microexplant culture system was originally designed to obtain long-term survival of cerebellar macroneurones, including Purkinje cells in conditions suitable for intracellular recording and injection. It was subsequently adapted to serum-free conditions and used for investigations on neuronal migration, trophic interactions, and neuritogenesis in the developing cerebellum (Moonen et al., 1982a; Neale et al., 1982; Gibbs et al., 1982; McDonald et al., 1982). In this method, although the histological organization of the donor organ is lost, close cellular interactions are preserved because of partial dissociation.

We shall describe the methodology we used in a recently developed application of that microexplant method aimed at the study of astrogliosis.

3.2. Microexplant Culture of Developing Cerebellum and Hippocampus

3.2.1. Dissection Procedure

Neonatal rats (less than 12 h of extra-uterine life) are killed by cervical dislocation. The skull is open and the entire brain (from the medulla to the rostral part of the frontal lobes) is lifted using incurvated forceps to cut the cranial nerves. The brain is transferred in a glass Petri dish containing PBS warmed at 37°C.

Under a stereomicroscope (magnification 10X), and using sharp watchmaker forceps, the cerebellum is then dissected from the brain by two incisions made below and beyond·it and by then separating it from the underlying pons (Fig. 5). The cerebella are then freed of surrounding meninges and choroid plexus and transferred in another glass Petri dish containing 200 μL of PBS. Using iridectomy scissors, they are cut in little pieces of tissue of about 200 μM of diameter.

The two hemispheres are separated by a medial longitudinal section (Fig. 6). Viewing the hemispheres from the inner face, thalamus and neighboring deep structures are excised. The hippocampi are then seen as a large comma-like structure. Meninges and choroid plexi are removed and the hippocampi are dissected and then processed as described for cerebellar microexplants.

3.2.2. Culture Method

Microexplants are washed with DMEM supplemented with 10% (v/v) Horse Serum (HS). Microexplants from two hippocampi or from one cerebellum are then placed in a 20 mL glass Erlenmeyer culture flask containing 2 mL of HS-DMEM per flask. Each flask is then flushed with 95% air/5% CO_2, tightly closed, and shaken at 100 rpm for 24 h at 37°C on an orbital gyratory shaker.

After 24 h, the microexplants are transferred into 24-wells NUNC multiwell dishes previously coated with polyornithine as described for the cultures of the organ of Corti. The content of

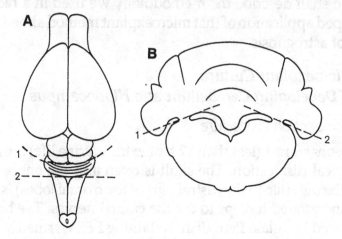

Fig. 5. Schematic dissection of the cerebellum. **(A)** Brain is excised from the skull and deposited in a glass Petri dish on its ventral face. Stripped lines show where the incisions have to be performed to obtain the cerebellum with the pons. **(B)** Cerebellum is freed from the pons by cutting cerebellum pediculi (stripped lines).

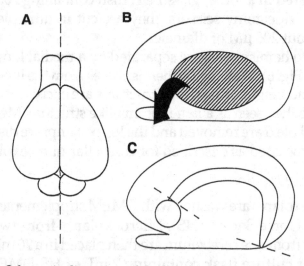

Fig. 6. Schematic dissection of the hippocampus. **(A)** The portion of the brain just above the dissected cerebellum is cut sagitally into the interhemispheres scissura. **(B)** Viewing the hemisphere from its internal face, all the hatched part has to be removed (thalamus and neighboring deep structures). **(C)** Hippocampus is then revealed like a large comma structure. Stripped line shows where the cuts have to be performed to free the hippocampus from the hemispheres.

one Erlenmeyer flask is equally divided into two wells. Cultures are kept in a 95% air/5% CO_2 incubator at 37°C. After 24 h, microexplants have attached to the substratum and the medium is renewed; after a new 48-h period, medium is replaced with DMEM supplemented with the N1 components (described in Section 2.2.2.) for an additional 24-h period.

At this time of culturing, microexplants are spreading over the substratum, radial neurite outgrowth is vigorous and, especially in cerebellar microexplants, glia fibrillary acidic protein (GFA) positive cells (Bignami and Dahl, 1974; Bignami et al., 1972) can be seen migrating out. A few neuronal cells (neurofilament- or neuron-specific enolase-positive) can also be seen migrating out of the two types of microexplants, mostly along radial GFA(+) processes (Selak et al., 1985).

3.3. An In Vitro System for the Study of Postlesional Gliosis Using Microexplants

The reaction of astroglia after a CNS injury (traumatic, toxic or ischemic) is a well-documented phenomenon that occurs both in developing and in adult mammal nervous system. It consists of two cellular events; proliferation and hypertrophy with accumulation of various cytoskeletal protein such as Glial Fibrillary Acidic protein (GFA) or vimentin. Although the astrogliosis is histologically well documented, little is known about the molecular aspects of its induction and regulation (Eng, 1988). Tissue culture allows the study of putative regulatory molecules released during astrogliosis because biochemical analysis of the culture media conditioned by dying neurons or reactive astrocytes are easily performed.

After 3 d in vitro, the neuronal populations of cerebellar and hippocampal microexplants respond differently to the glutamate-related agonist kainic acid. Indeed, in hippocampal microexplant preparations, most of the neurons are killed, whereas most remain unaffected in cerebellar microexplants as revealed by histological studies as well as by quantification of the release of the lactate dehydrogenase (LDH) in the medium as an index of cell death. It has to be pointed out that these observations are valid only in the conditions mentioned in the

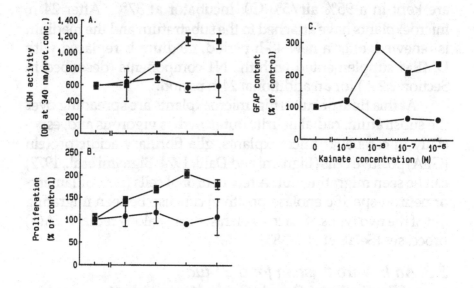

Fig. 7. (A) Lactate-Dehydrogenase (LDH) activity release by hippocampal (■) and cerebellar (●) microexplants exposed to various concentrations of kainic acid for 24 h. After the Kainic acid exposition, CM were collected and assayed for protein concentration as described by Bradford (Bradford, 1976). LDH activity release was assayed spectrophotometrically using a modification of the method of Morgenstern (Morgenstern et al., 1965) and described in (Morgan et al., 1988). The substrate solution was lactic acid (10 ml of a 85% solution) in 1 L of 0.67M 2-amino-2-methyl-1-propanol adjusted to pH 9 by addition of 5 M HCl. NAD (4 mg/mL) was added to the ice-cold buffered substrate immediately before use. 250 µL of CM were mixed with an equal vol of the substrate solution and incubated for 30 min at 37°C. The reaction was stopped by addition of glycine-sodium hydroxyde buffer, pH 10.5 (0.5 M; 250 µL/assay tube) and the optical density was read in a LKB spectrophotometer at 340 nm. (B) Incorporation of [³H]-thymidine into hippocampal (■) and cerebellar (●) microexplants exposed to various concentrations of kainic acid. Kainic acid was added in the culture medium with 1 µCi/well of [³H]-thymidine. After 24 h, cultures were washed three times with PBS and digested in 0.5 mL of sodium hydroxide (0.1 N) for 10 min in 37°C. Twenty µL of the digestion products were assayed for protein concentration as described by Bradford (Bradford, 1976) and the remaining vol was buffered with 10 µL of HCl (16.3 N) and transferred into counting vials. Ten mL of scintillation liquid (Aquasol) were added and incorporated radioactivity was counted in a LS 3801 Beckmann scintillation counter. Results obtained in DPM were normalized for protein concentration and expressed as percentage of the control value. (C) Glial Fibrillary Acidic (GFA)

previous section and should not be extrapolated to different conditions, for instance, for longer culture duration, without appropriate analysis.

Thus, incubation with increasing concentrations of kainic acid is followed, in hippocampal microexplants, by a dose-dependent neuronal death quantified by an increase of the release of lactatedehydrogenase (LDH) in the culture medium (Fig. 7A). Only in hippocampal, but not in cerebellar preparations, a cellular proliferation was measured using an [^{3}H]-thymidine incorporation assay, proliferation that was also dependent on the kainic acid concentration (Fig. 7b). DNA-synthesizing cells were identified as astroglial cells using a double immunolabeling with anti-GFA and anti-BromodeoxyUridine (BrdU) antibodies after a BrdU incorporation. Moreover, 2 d after a 24-h exposure to kainic acid, we observed an increase of GFA protein content which again, was only found in hippocampal microexplants (Fig. 7C).

protein content measured in hippocampal (■) and in cerebellar (●) microexplants exposed to various concentrations of kainic acid. After 24 h of exposition, medium was changed for 48 h with N1-supplemented DMEM to allow the accumulation of GFA. After 48 h, cultures were lysed for 10 min at room temperature in 250 μL of the following buffer: 50 mM Tris-HCl pH 6.8, 2 mM EDTA, 2 mM phenyl methylsulfonylfluoride, and 0.5% Triton-100 (Parisi et al., 1986). The cells were scrapped with a rubber policeman and homogenized. The homogenate was centrifuged at 12,000 g for 10 min at 4°C. The cytoskeletal pellet was dissolved in sample buffer (10 mM Tris-HCl, pH 6.8, and Sodium-Dodecyl-Sulfate (SDS)). One aliquot was removed for protein determination according to the method of Lowry (Fryer et al., 1986). β-mercaptoethanol and glycerol (5% final concentration) were added and samples were immersed in a boiling water bath for 2 min. Four μL of samples were run on a 12.5% polyacrylamide-SDS gel (Laemmli, 1970). After electrophoresis, the migrated proteins were transferred onto Immobilon membrane according to Towbin (Towbin et al., 1979). After saturation with bovine serum albumine fraction V (5% in PBS), the membrane was incubated for 2 h in 1/300 dilution of rabbit polyclonal anti-GFA antibodies. Bound antibodies were revealed with biotinylated swine antirabbit Ig antibody and Avidin-Biotin-Horseradish peroxydase complex (ABComplex-HRP). Peroxydase was revealed with chloronaphtol and immunostained bands were scanned using a Laser densitometer LKB 2202 Ultroscan. Results are expressed as percentage of the control value.

The use of such an in vitro model could permit characterization of the molecular(s) signal(s) that induce the astrocytic reaction following a neuronal injury. Such a characterization would be more tedious using in vivo models of CNS injury that constantly include an alteration of the blood-brain barrier, allowing inflammatory cells to penetrate the CNS and thus, greatly increasing the number of molecular influences on astrocytic cells (Rogister et al., 1990) (*see,* however, O'Callaghan et al., 1990).

4. Dissociated Cells in Coculture or Cultured in the Presence of Conditioned Media

4.1. Introduction

Potential paracrine influences between identified cell types can be unraveled using purified primary (or even clonal) cell cultures by cultivating two purified cellular populations in conditions that allow direct cell-to-cell contacts, or in bicompartimental culture devices in which communication between the two cell types occurs only through diffusible substances, or, finally, by treating one cell type with a medium conditioned by the other cell type.

This kind of approach has been extensively used to analyze multiple aspects of cellular communication in the nervous system including glioneuronal interactions.

We shall illustrate it for adult rat dorsal root ganglia (DRG) neurons. As in the previous sections, we shall first describe the culture methodology. Thereafter, some of the results obtained with such cultures will be briefly exposed.

4.2. Purification and Culture Method of Adult Rat DRG Neurons

4.2.1. Dissection Method

A 3- to 6-mo-old male Wistar rat is killed by ether exposure followed by cervical dislocation. The spinal column is dissected and placed in a glass Petri dish. The column is sagitally transected with curved scissors. The two hemisected columns are cut in four pieces, which are transferred into Petri dishes containing PBS. Under a stereoscopic microscope, the spinal cord is removed, and 40–50 ganglia are dissected. The ganglia are carefully freed of connective tissues, and the roots are cut close to the ganglion.

4.2.2. Dissociation of the Tissue and Purification of the Neurons

All these steps are schematically summarized in Fig. 8 (Delrée et al., 1989). Cleaned ganglia are incubated in 1 mL of a 0.5% collagenase solution for 45 min at 37°C while submitted to agitation in a gyratory shaker (100 rpm). Collagenase is thereafter replaced by 1 mL of 0.25% trypsin for 30 min. Ganglia are washed with DMEM supplemented with 10% (V/V) Fetal Calf Serum (FCS) and centrifuged for 5 min at 300 g. Supernatant is discarded, and the pellet is resuspended in 2 mL of DMEM-FCS.

The ganglia are mechanically dissociated by successive cycles of up and down aspiration through a Pasteur pipet. After five cycles, undissociated material is allowed to sediment and the supernatant is removed and saved. Two additional mL of DMEM-FCS are added to the pellet, and the dissociation is further achieved using a Pasteur pipet whose caliber at the tip has been flame-narrowed (tip diameter, 200 μM).

The pooled cell suspension is filtered through a 150 μm nylon sieve. This step allows the elimination of a substantial amount of myelin debris and of the undissociated fragments of the ganglia. To discard the remaining myelin debris, the filtrate is then layered onto 10 mL of a freshly prepared isosmotic Percoll solution (final concentration: 26%, density, 1.040) in a 50 mL conical tube and centrifuged (800g, 20 min at room temperature) as described by Goldenberg and De Boni (1983).

The pellet is recovered with a Pasteur pipet and layered onto a 10 μM nylon sieve placed above a 60 mm Petri dish previously filled with 5 mL of DMEM supplemented with N1 components (*see* Section 2.2.2.). Most of the neurons, whose diameter is greater than 15 μM are retained by the sieve, whereas most of nonneuronal cells, the diameter of which is below 10 μM, go through it.

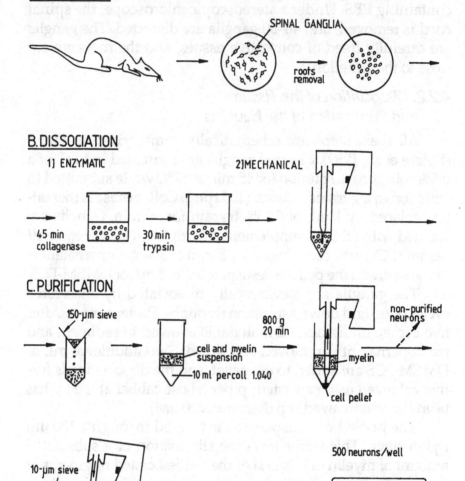

Fig. 8. Summary of the purification procedure for Dorsal Root Ganglia neurons (for explanations, *see* text). (Reprinted from Delrée et al., 1989.)

4.2.3. Culture of Neurons Purified from Adult Rat DRG

Recovery of neurons is achieved by reverting the sieve onto a 60 mm diameter Petri dish containing 5 mL of DMEM-N1. This neuronal suspension is centrifuged (200g for 10 min). The supernatant is discarded, and the pellet is resuspended in 1 mL of DMEM-N1. After counting in a hemocytometer, 500 neurons are seeded in a 6 mm microwell (96 microwell plates, NUNC) containing 50 µL of DMEM-N1. Microwells precoated with 50 µL of polyornithine as described above, are used after additional treatment with 50 µL of laminine (10 µg/mL in PBS, 1 h incubation). Ten hours after seeding, one-half of the culture medium is renewed and thereafter, such a change is performed every 48 h. DMEM-N1 allows less than 20% of neurons to survive at 48 h in vitro (Fig. 9) (Delrée et al., 1989). To improve neuronal survival, the cultures are supplemented with a 50% dilution of conditioned medium from C6 cells, astrocytes, or Schwann cells (those conditioned media allow a neuronal survival of about 70% at 48 h in vitro) (Delrée et al., 1989). Such cultures can be maintained for several days.

4.3. Plasticity of Neurotransmitter Expression in Adult Rat DRG Neurons

We have observed striking modifications of the neurotransmitters phenotype of cultured adult rat DRG neurons (Fig. 10) (Schoenen et al., 1989). Some neurotransmitters are expressed in lower percentage of neurons in vitro than in vivo (β-endorphin), others remain stable (CGRP), and still others that are not, or only very sparsely, expressed in vivo are expressed in high percentage of the cultured neurons (serotonin or TRH). We have investigated a possible regulation of the expression of serotonin and TRH in adult DRG neurons by various conditions of culture.

Figure 11 shows that coculture of adult rat DRG neurons with astrocytes, but not with Schwann cells or fibroblasts, promotes the expression of serotonin. This effect is mimicked by addition of conditioned medium of rat glioma C6 (data not shown). The expression of TRH seems to be insensitive to coculture or to the presence of various conditioned media. TRH expression seems indeed to be repressed in vivo by the connection between DRG and the spinal cord since in vivo proximal but not distal rhizotomies are followed by a dramatic increase of TRH expression in adult DRG neurons *in situ* (in preparation).

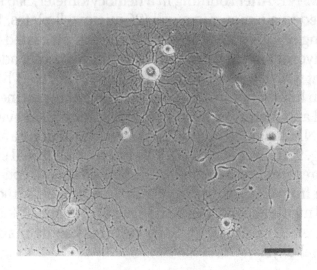

Fig. 9. Phase-contrast microphotograph of purified adult rat DRG neurons after three days in vitro in 50% C6 glioma cells CM, 50% DMEM (v/v). Notice the multipolarity of these neurons. Scale bar: 100 μM. (Reprinted from Delrée et al., 1989.)

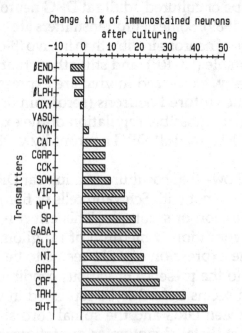

Fig. 10. Diagram illustrating for each antiserum the difference between the percentage of positively stained neurons in cultured and freshly dissociated DRG. (The origin of antisera is detailed in Schoenen et al., 1989.)

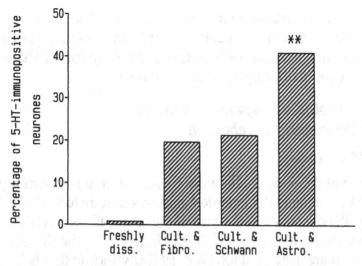

Fig. 11. Nonpurified adult DRG neurons were cultured for 48 hours in conditioned media of different cell types: 3T3 fibroblasts, Schwann cells and astrocytes. After 2 d, the cultures were fixed and the percentage of serotonin-immunoreactive neurons was determined and compared to the percentage of serotonin-positive neurons in a preparation of freshly dissociated but not cultured DRG neurons. It appears that culture by itself induces the appearance of serotonin-immunoreactivity in a significant percentage of DRG neurons. Moreover, this percentage is modulated by the cellular environment and is particulary high in astrocyte conditioned medium.

5. Aggregate Cultures

5.1. Introduction

In the previously described systems, cells or explants were plated in "static" conditions in which cells interact not only with each other but also with the substratum (tissue culture dish plastic, polycationic substratum, collagen, laminine). Methods have also been designed in which the cells–culture substratum interactions are no longer possible through permanent agitation of the culture device.

Again, two different procedures can be used, that is, without or with an initial step of tissue dissociation. The first method could be referred to as organotypic suspension culture and the second as reaggregation culture. We and others have used the first method to study the developmental neuronal migration. The second method initially described by Moscona (Moscona,

1965, 1973) has been used for several studies including neuronal differentiation and myelinization. Confrontation of aggregates of normal and tumoral cells permits investigation of the biochemical events underlying tumor invasion.

5.2. Organotypic Suspension Culture of Developing Cerebellum

5.2.1. Methodology

Postnatal d-7 cerebella were dissected, freed of meninges, and cut in transversally oriented explants that include the outer surface. Whole identified parts of the cerebellum such as the paraflocculi can also be used. Immediately after the dissection, the explants are incubated for 1 h in [^{3}H]-thymidine that is incorporated by the proliferating neuroblasts of the external granular layer. An aliquot of the explants is then fixed to be processed for autoradiography, permitting counting of the number of labelled cells at the external granular layer level. The remaining explants are transferred in 20 mL Erlenmeyer flasks and incubated for 2 or 3 days on a gyratory shaker (80 rpm). The medium used is MEM supplemented with the N1 cocktail. These conditions do not allow further proliferation in the external granular layer as demonstrated by [^{3}H]-thymidine incorporation and scintillation counting (Moonen et al., 1987). The postmitotic granule neurons migrate inward along the radial glia processes (Bergman glia) (Rakic, 1981). When the explants are fixed after 2 or 3 d and processed for autoradiography, labeled postmitotic, postmigratory neurons can be seen at the level of the internal granular layer, whereas the external granular layer has virtually disappeared.

5.2.2. Results

It has been demonstrated conclusively that the developmental neuronal migration results from a complex interaction between the migrating, postmitotic neurons and the radially oriented processes of an early subset of astroglia, the radial glia (Rakic, 1985). Using the in vitro model illustrated in the present section, some of the biochemical steps that are involved in the complex sequence of events that occur during that critical period of nervous system development have been characterized.

These processes include proteolysis (Moonen et al., 1982b; Lindner et al., 1986), as well as cell-to-cell interaction involving the L1 member of the immunoglobulin superfamily (Lindner et al., 1983; Edelman, 1984), as well as a recently described endogenous lectin (cerebellar soluble lectin) (Lehmann et al., 1990).

5.3. Multicellular Tumor Spheroids

5.3.1. Introduction

Three-dimensional culture systems in which the confrontation of multicellular spheroids from tumor origin and of brain cell aggregates occurs are useful to study the biochemical mechanisms underlying tumor invasion (Lund-Johansen, 1990; Backlund and Bjerkvig, 1989). The use of this method often requires three steps:

1. Normal brain cells aggregates formation;
2. Multicellular tumor spheroids production; and
3. Confrontation.

At this last step, immunocytochemical techniques coupled with image analysis can be used to study the invasion of the normal tissue by tumor cells. The effects of different substances on the extent and the kinetic of invasion can therefore be studied.

5.3.2. Methodology

5.3.2.1. BRAIN CELLS AGGREGATES (PULLIAM ET AL., 1988)

Brain from E15 to E20 fetuses are used. Dissociation is performed in DMEM by passages through two nylon sieves of, respectively, 210 and 130 μm mesh. Cells are washed three times with fresh DMEM and then resuspended in DMEM with 10% (v/v) FCS.

Viable cells are then counted with an hemocytometer and 4 mL of a cell suspension containing approximately $4-5 \times 10^7$ viable cells are placed into 25 mL Erlenmeyer flasks that have been siliconized before sterilization to reduce cell adhesion to the sides of the flasks. The flasks are rotated at 76 rpm at 37°C in a humidified atmosphere of 5% CO_2. After 2–3 d, aggregates have been formed. They are transferred to 50 mL siliconized Erlenmeyer flasks, and 5 mL of DMEM supplemented with 15% FCS are added to the suspension. Five mL of medium are removed every 2–3 d and replaced with fresh medium. The cells are cultured in suspension for 10 d to allow them to form stable aggregates.

5.3.2.2. MULTICELLULAR TUMOR SPHEROIDS (MTS) (YUHAS ET AL., 1977)

Tumor cell lines of interest (for instance, gliomas) are first cultured in their appropriate medium (which could be MEM, DMEM, or F12), supplemented with 10% FCS in monolayer cultures. When they reach confluence, they are harvested by mild trypsinization (0.25% v/v trypsin, 3–5 min at 37°C) or, even scraped off the surface. At that time, approx 10^6 cells in 10 mL of medium are added to 100 mm plastic Petri dishes that have been coated (thickness 2–3 mm) with 0.5% Agar dissolved in complete medium.

5.3.2.3. CONFRONTATION

After 10–30 d (time necessary to allow MTS production), brain cell aggregates are added to the MTS preparations. The medium can be either DMEM supplemented with 10% FCS or a chemically defined medium (such as Costar SF-X hybridoma medium).

5.3.3. Results

This model offers two advantages for the study of the mechanisms of tumour invasion when compared to two-dimensional models. 1) They are well suited for long-term experiments and are easy to manipulate. 2) They present closer homology to *in situ* brain. For example, MTS of human glioma present a more extensive extracellular matrix under three-dimensional geometry than in monolayer cultures (Glimelius et al., 1988). This method can also be used to study the efficacy and the toxicity of antitumor treatments.

6. Conclusion

In the sense used in this chapter, tridimensional culture systems refers to highly organized in vitro preparation that are allowed to survive, differentiate, and regenerate as opposed to ex vivo short term preparation such as the brain slices (*see,* however, the chapter by Wray). Indeed, historically, tridimensional cultures were the first ones to be used when Harrisson (1907) explanted fragments of the spinal cord of frogs to analyze neuritic outgrowth and help to solve the controversy that appeared at that time between Golgi and Cajal about the nature of nerve fibers.

As pointed out in the introduction, tridimensional culture systems and indeed, all the culture systems, can be classified according to the level of organization, or in other words, to the complexity and sophistication of intercellular relationships. These relationships are either maintained and allowed to complexify, owing to partial tissue dissociation, or are reestablished like in the aggregate cultures. One should not, however, underestimate those cellular relationships that are reestablished in dissociated cell cultures that usually result in a so-called monolayer of cells. Many examples could be given, such as formation of functional gapjunctions between cells (Moonen and Nelson, 1978; Ransom and Kettenmann, 1990; Mugnaini, 1986), synaptogenesis (Nelson and Lieberman, 1981), auto-conditioning of the medium (Grau-Wagemans et al., 1984), and so on. Regarding this last point, it is a common experience in nerve cell culture that the density of seeded cells is a critical factor for survival, and indeed bioassays of neuronal-survival promoting factors using dissociated cells require very low cell densities in order to minimize such cellular interactions (Manthorpe et al., 1986).

7. In Vivo Veritas?

The experimentor naturally wishes to know whether the cellular interactions that are unraveled or defined using such in vitro cultured material actually happen in vivo, either in the developing or in the adult animal. Indeed, it is our opinion that to oppose in vitro and in vivo methods is misleading. We believe that if a given process is for the first time described using in vitro methods, its in vivo relevance has to be investigated. On the other hand, if such a process is first discovered in vivo, the use of in vitro methods can and most often does help to understand it at a more mechanistic level. The experimental strategy has then to be planned as a continuous in vitro to in vivo back and forth oscillation. For instance, the effect of NGF on cholinergic central neurons was first described in aggregate cultures (Honegger and Lenoir, 1982) and led to a large number of in vivo works on the trophic interactions between cortical structures (e.g., hippocampus) and subcortical cholinergic nuclei (e.g., septal nucleus) that are mediated by NGF (Hefti, 1986; Williams et al., 1986; Rosenberg et al., 1988).

Along this line, it is worth to point out that integration of tissue culture and in vivo experiments somehow culminates in the use of cultured material for grafting in the central nervous system. The last decade has seen a tremendous increase of the number of experimental works that use grafting methods. In most instances, "acute" embryonic material is grafted in developing or in adult animals. An increasing number of works also deal with the grafting of cultured cells, including aggregates of engineered cells or even artificial montages such as encapsulated aggregates. These experimental strategies nicely illustrate the complementarity between tissue and cell culture on one hand and in vivo experiments on the other hand. The therapeutic potential of these methods could be enormous.

Acknowledgments

B. R., P. L., and J. S. are, respectively, Research Assistant, Research Associate, and Senior Research Associate of the Fonds National de la Recherche Scientifique (Belgium).

This work was supported by grants from the Fonds National de la Recherche Scientifique (Belgium), Fonds Médical Reine Elisabeth (Belgium) and from the Research Fund of the Faculty of Medecine, University of Liège, Belgium.

The authors would thank P. Ernst-Gengoux and A. Brose for their expert technical assistance.

References

Backlund E. O. and Bjerkvig R. (1989) Stereotactic biopsies as a model for studying the interaction between gliomas and normal brain tissue in vitro. *J. Neurosurg. Sci.* **33,** 31–33.
Berglund A. M. and Ruygo D. K. (1987) Hair cell innervation by spiral ganglion neurons in the mouse. *J. Comp. Neurol.* **255,** 560–570.

Bignami A., Eng L. F., Dahl D., and Uyeda C. T. (1972) Localization of the glial fibrillary acidic protein in astrocytes by immunofluorescence. *Brain Res.* **43,** 429–435.

Bignami A. and Dahl D. (1974) Astrocyte-specific protein and neuroglial differentiation. An immunofluorescence study with antibodies to the glial fibrillary acidic protein. *J. Comp. Neurol.* **153,** 27–38.

Bottenstein J. E. and Sato G. (1979) Growth of rat neuroblastoma cell line in serum-free supplemented medium. *Proc. Natl. Acad. Sci. USA* **76,** 514–517.

Bradford M. M. (1976) A rapid and sensitive method for the quantification of microgram quantities of protein utilizing the principal of protein-dye binding. *Anal. Biochem.* **72,** 248–255.

Delrée P., Leprince P., Schoenen J., and Moonen G. (1989) Purification and culture of adult rat dorsal root ganglia neurons. *J. Neurosci. Res.* **23(2),** 198–206.

Edelman G. M. (1984) Modulation of cell adhesion during induction, histogenesis, and perinatal development of the nervous system. *Ann. Rev. Neurosci.* **7,** 339–377.

Eng L. (1988) Regulation of glial intermediate filaments in astrogliosis, in Biochemical Pathology of Astrocytes, (Norenberg M. D., Hertz L., and Schousboe A., eds.), Liss, New York, pp. 79–90.

Fryer H. J. C., Davies G. E., Manthorpe M., and Varon S. (1986) Lowry protein assay using an automatic microtiter plate spectrophotometer. *Anal. Biochem.* **153,** 262–266.

Gibbs W., Neale E. A., and Moonen G. (1982) Kainic acid sensitivity of mammalian Purkinje cells in monolayer cultures. *Devel. Brain. Res.* **4,** 103–108.

Glimelius B., Norling B., Nederman T., and Carlsson J.(1988) Extracellular matrices in multicellular spheroids of human glioma origin: Increased incorporation of proteoglycans and fibronectin as compared to monolayer cultures. *APMIS* **96,** 433–444.

Goldenberg S. and De Boni U. (1983) Pure populations of viable neurons from rabbit dorsal root ganglia, using gradients of Percoll. *J. Neurobiol.* **14,** 195–206.

Grau-Wagemans M-P, Selak I., Lefebvre P. P., and Moonen G. (1984) Cerebellar macroneurons in serum-free cultures: evidence for intrinsic neuronotrophic and neuronotoxic activities. *Devel. Brain. Res.* **15,** 11–19.

Hefti F. (1986) Nerve growth factor promotes survival of septal cholinergic neurons after fimbrial transection. *J. Neurosci.* **6,** 2155–2162.

Honegger P. and Lenoir D. (1982) Nerve Growth Factor (NGF) stimulation of cholinergic telencephalic neurons in aggregating cultures. *Dev. Brain Res.* **3,** 229–238.

Kiang N. Y. S., Rho J. M., Northrop C. E., Lieberman M. C., and Ruygo D. K. (1982) Hair cell innervation by spiral cells in adult cats. *Science* **217,** 175–177.

Laemmli U. K. (1970) Cleavage of structural proteins during the assembly of the head of bacteriophage T4. *Nature* 227, 680–683.

Lefebvre P. P., Weber T., Leprince P., Rigo J.-M., Delrée P., Rogister B., and Moonen G. (1991) Kainate and NMDA toxicity for cultured developing and adult rat spiral ganglion neurons: further evidence for a glutamatergic excitatory neurotransmission at the inner hair cell synapse. *Brain Res.* 555, 75–83.

Lefebvre P. P., Weber T., Rigo J.-M., Delrée P., Leprince P., and Moonen G. (1990a) Potassium-induced release of an endogenous toxic activity for outer hair cell and auditory neurons in the cochlea: a new pathophysiological mechanism in Meniere's disease? *Hearing Res.* 47, 83–94.

Lefebvre P. P., Leprince P., Weber T., Rigo J.-M., and Moonen G. (1990b) Neuronotrophic effect of developing otic vesicle on cochleovestibular neurons: evidence for nerve growth factor involvement. *Brain Res.* 507(2), 254–260.

Lehmann S., Kuchler S., Theveniau M., Vincendon G., and Zanetta J.-P. (1990) An endogenous lectin and one of its neuronal glycoprotein ligands are involved in contact guidance of neuron migration. *Proc. Natl. Acad. Sci. USA* 87, 6455–6459.

Lindner J., Rathjen F. G., and Sachner M. (1983): L₁ mono- and polyclonal antibodies modify cell migration in early postnatal mouse cerebellum. *Nature* 305, 427–430.

Lindner J., Guenther J., Nick H., Zinser G., Antonicek H., Schachner M., and Monard D. (1986) Modulation of granule cell migration by a glia-derived protein. *Proc. Natl. Acad. Sci. USA* 83, 4568–4571.

Lund-Johansen M. (1990) Interactions between human glioma cells and fetal rat brain aggregates studied in a chemically defined medium. *Invasion Metastasis* 10, 113–128.

Manthorpe M., Fagnani R., Skaper S. D., and Varon S. (1986) An automated colorimetric assay for neurotrophic factors. *Dev. Brain Res.* 25, 191–198.

McDonald R. L., Moonen G., Neale E. A., and Nelson P. G. (1982) Cerebellar macroneurones in microexplant cell culture. Postsynaptic amino acid pharmacology. *Dev. Brain Res.* 5, 77–88.

Moonen G., Neale E. A., McDonald R. L., Gibbs W., and Nelson P. G. (1982a) Cerebellar macroneurons in microexplant cell culture. Methodology, basic electrophysiology and morphology after horseradish peroxidase injection. *Dev. Brain Res.* 5, 59–73.

Moonen G., Grau-Wagemans M.-P., and Selak I. (1982b) Plasminogen activator-plasmin system and neuronal migration. *Nature* 298, 753–755.

Moonen G. and Nelson P. G. (1978) Some physiological properties of astrocytes in primary cultures, in *Dynamic Properties of Glial Cells* (Schoffeniels E., Franck G., Tower D. B., and Hertz L. eds.), Pergamon, Elmsford, NY, pp. 389–393.

Moonen G., Selak I., and Grau-Wagemans M.-P. (1987) In vitro analysis of glial-neuronal communication during cerebellum ontogenesis, in *Glio-Neuronal Communication in Development and Regeneration.* (Althaus H. H. and Seifert W., eds.), Springer Verlag, Berlin, pp. 324–338.

Morgan D. L. M., Clover J., and Pearson J. D. (1988) Effects of synthetic polycations on leucine incorporation, lactate dehydrogenase release, and morphology of human umbilical vein endothelial cells. *J. Cell Sci.* **91,** 231–238.

Morgenstern S., Flor R., Kessler G., and Klein B. (1965) Automated determination of NAD-coupled enzymes. Determination of lactate dehydrogenase. *Anal. Biochem.* **13,** 149–161.

Moscona A. A. (1965) Recombination of dissociated cells and the development of cell aggregates in *Cells and tissues in culture,* (Willmer E., ed.), Academic Press, New York, pp. 489–529.

Moscona A. A. (1973) Cell aggregation, in *Cell biology in medicine,* (Bittar E., ed.), Wiley, New York, pp. 571–591.

Mugnaini E. (1986) Cell junctions of astrocytes, ependyma and related cells in the mammalian central nervous system with emphasis on the hypothesis of a generalized functional syncytium of supporting cells, in *Astrocytes Development, Morphology and Regional Specialization of Astrocytes.* (Fedoroff S. and Vernadakis A., eds.), Academic, London, pp. 329–371.

Neale E. A., Moonen G., McDonald R. L., and Nelson P. G. (1982) Cerebellar macroneurons in microexplant cell culture ultrastructural morphology. *Neuroscience* **7,** 1879–1890.

Nelson P. G. and Lieberman M. (1981) *Excitable Cells in Tissue Culture.* Plenum, New York, **1,** 310.

O'Callaghan J. P., Miller D. B., and Reinhard J. F. (1990) Characterization of the origins of astrocyte response to injury using the dopaminergic neurotoxicant, 1-methyl-4-phenyl-1,2,3,6,-tetrahydropyridine. *Brain Res.* **521,** 73–80.

Parisi G., Tropea R., Giuffrida S., Lombardo M., and Giuffré F. (1986) Cystic meningiomas. Report of seven cases. *J. Neurosurg.* **64,** 35–38.

Pulliam L., Berens M. E., and Rosenblum M. L. (1988) A normal human brain cell aggregate model for neurobiological studies. *J. Neurosci. Res.* **21,** 521–530.

Rakic P. (1981) Neuronal-glial interaction during brain development. *Trends Neurosci.* July, 184–187.

Rakic P. (1985) Mechanisms of neuronal migration in developing cerebellar cortex, in *Molecular Basis of Neural Development.* (Edelman G. M., Gall W. E., and Cowan W. M. eds.), Wiley, New York, pp. 139–160.

Ransom B. R. and Kettenmann H. (1990) Electrical coupling, without dye coupling, between mammalian astrocytes and oligodendrocytes in cell culture. *Glia* **3,** 258–266.

Rogister B., Leprince P., Delrée P., Van Damme J., Billiau A., and Moonen G. (1990) Enhanced release of plasminogen activator inhibitor(s) but not of plasminogen activators by cultured rat glial cells treated with interleukin-1. *Glia* **3**, 252–257.

Rosenberg M. B., Friedmann T., Robertson R. C., Tuszynski M., Wolff J. A., Breakefield X. O., and Gage F. H. (1988) Grafting genetically modified cells to the damaged brain: restorative effects of NGF expression. *Science* **242**, 1575–1578.

Schoenen J., Delrée P., Leprince P., and Moonen G. (1989) Neurotransmitter phenotype plasticity in cultured dissociated adult rat dorsal root ganglia: an immunocytochemical study. *J. Neurosci. Res.* **22(4)**, 473–487.

Selak I., Foidart J.-M., and Moonen G. (1985) Laminine promotes cerebellar granule cells migration in vitro and is synthesized by cultured astrocytes. *Devel. Neurosci.* **7**, 278–285.

Towbin H., Staehelin T., and Gordon J. (1979) Electrophoretic transfer of protein from polyacrilamide gels to nitrocellulose sheets: procedure and some applications. *Proc. Natl. Acad. Sci. USA* **76**, 4350–4354.

Williams L. R., Varon S., Peterson G. M., Wictorin K., Fischer W., Bjorklund A., and Gage F. H.(1986) Continuous infusion of nerve growth factor prevents basal forebrain neuronal death after fimbria fornix transection. *Proc. Natl. Acad. Sci. USA* **83**, 9231–9235.

Yuhas J. M., Li A. P., Martinez A. O., and Ladman A. J. (1977) A simplified method for production and growth of multicellular tumour spheroids. *Cancer Res.* **37**, 3639–3643.

Zenner H. P. (1986) K^+-induced motility and depolarization of cochlear hair cells. Direct evidence for a new pathophysiological mechanism in Meniere's disease. *Arch. Otorhinolaryngol.* **243**, 108–111.

Organotypic Slice Explant Roller-Tube Cultures

Susan Wray

1. Introduction

Within the past decade, organ cultures have reemerged as an important complement to in vivo studies for examining regulatory mechanisms in various neuronal systems. The principle behind organ cultures is to maintain a tissue explant in a state as close as possible to that found in vivo, i.e., to preserve some aspect of the spatial, structural, and/or synaptic organization of the original tissue. As a result, such cultures are often referred to as *organotypic* and are valuable tools to study organized, yet relatively isolated neuronal systems.

All organ cultures are generated from tissue fragments or slices. However, there are many ways to maintain organ cultures, and the manner in which this is done, e.g., in a slide chamber, flask, Petri dish, or test tube, has been used to define the organ culture method (Masurovsky et al., 1971; Bornstein, 1973; Paul, 1975; Toran-Allerand, 1978; Feldman et al., 1982; Gähwiler, 1981a,1984a,1988; Fowler and Crain, 1985; Baldino and Geller, 1986; Lasnitzki, 1986; Romijn et al., 1988). Three important factors directly influencing final thickness, maximal survival time, and organotypic nature of the culture are: (1) how the explant tissue is incubated; (2) the material used to adhere the slice to the supporting surface; and (3) the supporting surface itself. Obviously, culture thickness, survival, and organotypic

From: *Neuromethods, Vol. 23: Practical Cell Culture Techniques*
Eds: A. Boulton, G. Baker, and W. Walz ©1992 The Humana Press Inc.

nature are interrelated and directly influence the extent to which individual cells can be visualized. Each of these parameters will be discussed separately in later sections of this chapter.

In all organ cultures, the orientation (sagittal, coronal, and so on) of the section taken for culturing can be varied. In this way, one can systematically control the anatomically defined population of neurons under study, as well as their environment. In addition, the source of tissue used in organ cultures can be either postnatal (starting material that is already substantially differentiated) or embryonic. This allows the investigator to compare various developmental aspects of the system being studied. For example, one can compare cultures of neuronal areas before synaptic input occurs to cultures derived from tissue in which synapses are already established. In such an investigation, morphologic, cellular, and molecular changes associated with neuronal circuit development can be addressed.

The roller-tube technique is one we have found highly effective in preparing organ cultures. Chicken plasma has been used historically as the adherent in these roller-tube cultures and glass slides as the supporting surface (Gähwiler, 1988). Roller-tube cultures prepared in this manner maintain many organotypic features and thin substantially, allowing *individual cells* to be visualized throughout the cultured tissue. Therefore, if individual, single-cell analysis is required to examine cellular features in organotypically organized tissue, organ cultures generated using the roller-tube technique are recommended.

Because of the thinness of the final tissue, a wide variety of techniques are compatible with slice explant roller-tube cultures. In specific neuronal subtypes, the cascade of events, starting with receptor-mediated stimulation, through transcription, precusor processing, and ending with product secretion (neuropeptide, neurotransmitter, and so forth), can be readily studied in a controlled, yet synaptically organized environment. Regulation of these events in identical neuronal cell types can be compared in slices:

1. Obtained from different brain regions;
2. Generated from different aged animals; and
3. With the same cells but obtained from a different anatomic orientation.

Hence, by combining different technical assays with different slice parameters, one can achieve enormous flexibility. These features make slice explant roller-tube cultures a viable and attractive model system for addressing questions on development and regulation in the central nervous system (CNS).

2. Slice Explant Roller-Tube Cultures

The slice explant roller culture technique (Sobkowicz et al., 1974; Geller and Woodward, 1979; Gähwiler, 1981a,b,1984a,1988; Geller, 1981b; Wray et al., 1988,1989a) allows long-term culturing of postnatal or embryonic neurons, which permits easy visualization of individual cells, and maintenance of organotypic intercellular relationships (interneurons and glia) in the cultured tissue (Figs. 1 and 2). It is important to understand that embryonic/neonatal brain tissue is extremely fragile. Therefore, in preparing these cultures, one must acquire the skills necessary to minimize the damage done to the tissue being plated before success or failure of the technique may be judged.

2.1. Ability to Identify Individual Cells

In optimal slice explant roller-tube cultures, the tissue thins to a quasi-monolayer, 1–3 cells thick. The thinning process takes place within the first few days and is owing to cell death and cell movement. The thinness of these slice explant cultures is one of the most important features of this culture technique. This thinning allows individual *living* cells in the culture to be visualized using phase contrast microscopy or Nomarski optics. In addition, single cells can be identified from specific brain regions based on morphologic features and microinjections of dyes and/or electrophysiological recordings can be done (Gähwiler, 1980b; Geller, 1981b; Baldino and Wolfson, 1985; Baldino and Geller, 1986). Equally important, the thinness of the tissue permits *single-cell analysis* of *fixed* cultures using light and electron microscopic immunocytochemical and immunohistochemical methods (Zimmer and Gähwiler, 1984; Baldino et al., 1985; Sofroniew et al., 1988; Frotscher and Gähwiler, 1988; Wray et al., 1988,1989a). More recently, *in situ* hybridization histochemistry has been performed on these cultures (Fig. 1; Wray et al., 1989a,1991).

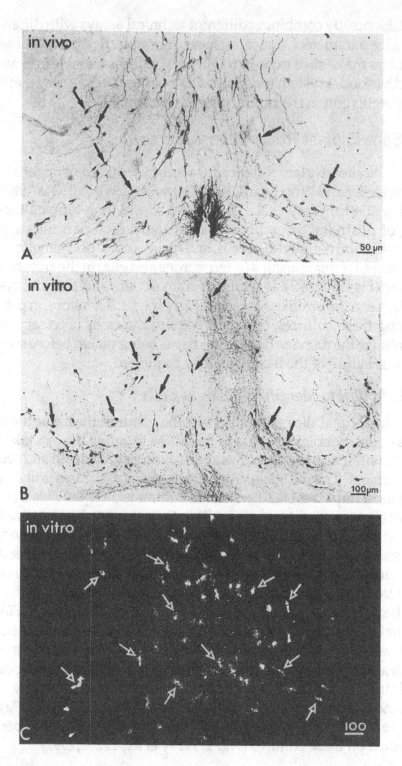

2.2. Organotypic Organization

In a single slice, these cultures contain relatively normal interneuronal and glial interactions and anatomic distribution of various cell types (Gähwiler 1981a,1984b; Baldino et al., 1985; Sofroniew et al., 1988; Wray et al., 1988,1989a). If one has several slices through a given neuronal region, the organotypic nature of these cultures expands to include relative neuronal topography. This means that cell populations found in several neuronal regions may maintain the same relationship in vitro, i.e., occurring in particular slice cultures in a manner proportional to that seen in vivo (Wray et al., 1988).

It is important to keep in mind, however, that the final organization of the culture slice is dependent on the initial tissue taken and the age of the animal from which it is obtained. In general, immature tissue shows greater distortion of the original cytoarchitecture than older tissue. The term immature here can apply to either embryonic or postnatal tissue and relates to when the structure studied develops (cell differentiation, synapse formation, and so on). Such distortion is most likely due to enhanced cell migration, proliferation of undifferentiated cell types (both glial and neuronal), and lack of synaptic input. Thus,

Fig. 1 *(opposite page)*. Maintenance of topographic distribution in vitro of a dispersed but identifiable neuronal population. Comparison of luteinizing hormone releasing hormone (LHRH) cells found in vivo (A) and in vitro after 18 d in slice culture (B and C). (A). A 50 μm vibratome section from a 4 d-old rat immunocytochemically stained for LHRH (this section is at level of the organum vasculosum lamina terminalis). (B). Culture section derived from a 4 d-old animal, from the same anatomic level as that shown in *A*, immunocytochemically stained for LHRH after 18 d in vitro. (C) Culture section as in *B*, but this section was maintained for the last 6 (of 18) d in defined medium and, following fixation, labeled for cells containing LHRH mRNA using *in situ* hybridization histochemistry. Arrows point to a few of the immunopositive LHRH cells (A and B) as well as to those cells expressing LHRH mRNA (C). Note that the LHRH neuronal distribution is similar in vivo and in vitro, with the LHRH cells forming an inverted V around the midline, but that the in vitro tissue has spread to almost twice the surface area originally present in vivo (*see* scale lines).

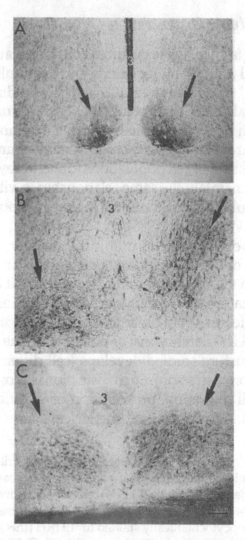

Fig. 2. Maintenance of nuclear organization of identifiable neuronal populations located within discrete nuclei. The suprachiasmatic nucleus is a discrete nucleus, containing numerous neuropeptides in vivo (A). After 18 d in slice explant cultures (B and C), although the surface area of the nucleus increases and can be distorted (B) or normal (C) in shape, the neuropeptide cells are found in discrete regions, similar to that seen in vivo. (A). A 50 μm vibratome section, at the level of the suprachiasmatic nucleus, from a 4-d-old rat immunocytochemically stained for vasoactive intestinal polypeptide (VIP, arrows) (B and C). Culture sections derived from a 4-d-old animal, from the same anatomic level as that shown in A, immunocytochemically stained for VIP (B) and neurophysin (C) after 18 d in vitro.

it is perhaps inaccurate to say that immature tissue distorts more than mature tissue, but rather that this technique does not encourage tissue to become organized if it has not yet developed these features when plated. In either case, the age of the tissue from which the cultures are generated is an extremely important variable. Initial studies on early postnatal tissue (4–6 d) are recommended. This postnatal age group maintains organotypic characteristics and shows substantial survival and thinning from a variety of brain regions, including hippocampus, anterior hypothalamus, cerebellum, and brainstem, as well as survival of phenotypically different cell types (Fig. 3). Once this tissue is growing successfully, the ideal age of the tissue to be explanted can be determined.

Having the ability to change the synaptic environment via slice orientation and age of initial animals can yield important information about the cells studied. In addition, the ability to reintroduce specific inputs experimentally by coculturing in the presence of tissue containing afferent systems opens up a variety of investigations concerning axonal choice and mechanisms of ingrowth (Gähwiler and Hefti, 1984; Gähwiler et al., 1987; Zimmer and Gähwiler, 1987; Wray et al., 1988). Slice explant roller-tube cocultures are extremely useful since they can be used to study whether *direct* or *indirect* interactions occur experimentally (*see* Section 5.2.). Specifically, tissues may be plated onto the same coverslip and inserted into a test tube (direct contact possible) or onto two separate coverslips and inserted back-to-back in the same test tube (indirect interactions possible via secretion into the media).

Therefore, the slice explant roller culture system is an in vitro model system particularly appropriate to examine the intrinsic properties of identifiable single neurons, neuronal subpopulations, and total populations (monitoring cellular activity, gene and gene products) and the ways in which these neuronal subtypes respond to extrinsic factors (both humoral and synaptic).

2.3. Tissue Types Successfully Cultured

A variety of neuronal and nonneuronal tissue types have been studied using slice explant cultures (Figs. 3 and 4). Embryonic tissues from rat and mouse survive and thin similarly. However, tissues of *neonatal* rat brain seem to survive and

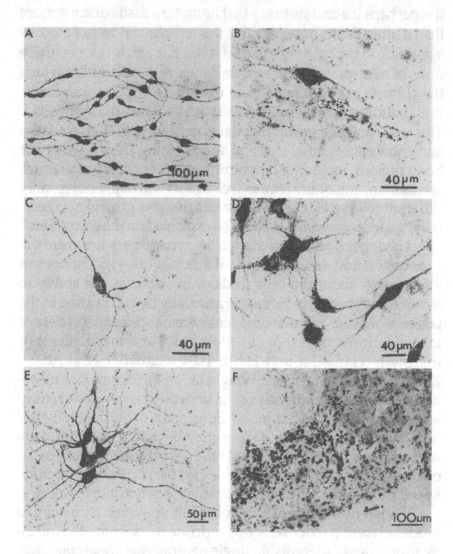

Fig. 3. Examples of various cell types maintained in organotypic explant slice cultures for 17–21 d. All cultures were derived from postnatal day 4 rats. **(A)** Tyrosine hydroxylase (the first enzyme in the catecholaminergic pathway, TH) positive cells located along the midline, just lateral to the third ventricle in tissue derived from the caudal hypothalamus. TH-containing cells formed long bands, often containing 50–100 immunopositive cells. TH-positive cells in these cultures generally possessed a simple, bipolar morphology, with the majority of cells oriented along the same axis. **(B)** Neurotensin cells cultured from hypothalamic regions averaged 10 µm in diameter and occasionally had immunopositive, varicose, axonal

maintain an organotypic structure better than neonatal mouse (personal observation). This is in part a reflection of size differences that make the neonatal mouse brain more difficult to work with. Slice explant cultures have been routinely obtained from the following brain regions of postnatal rat: hippocampus (Gähwiler, 1980b,1981b,1984a,b; Zimmer and Gähwiler, 1984; Frotscher and Gähwiler, 1988), cerebellum (Gähwiler et al., 1973; Gähwiler, 1981b,1984a), septum (Gähwiler and Brown, 1985; Gähwiler et al., 1987), preoptic area/hypothalamus (Geller and Woodward, 1979; Geller, 1981a,b; Baertschi et al., 1982; Gähwiler, 1984a; Baldino et al., 1985; Baldino and Wolfson, 1985; Baldino and Geller, 1986; Wray et al., 1988,1989a), and brainstem (Wray et al., 1988; Braschler et al., 1989). Both pituitary (Fig. 3, Baertschi et al., 1982; Gähwiler et al., 1984; Wray et al., 1988) and spinal cord tissues (Delfs et al., 1989) survive well in these cultures. Although spinal cord tissues do not thin as much as those previously listed, single cells are still discernible by immunocytochemistry. Our laboratory has been able to maintain various nonneuronal tissues in slice explant cultures. These include neonatal ovaries, testes, and embryonic nasal regions (unpublished data; Fig. 4).

processes studding their somas and proximal dendrites. (C). A multipolar corticotropin releasing hormone (CRH) cell with many branching processes was seen in the periventricular region. (D) Culture from the level of the paraventricular nucleus immunocytochemically stained for neurophysin (NP). Immuno-positive NP cells were usually found in clusters (5–10 cells, averaging 20–25 μm in diameter). Often large numbers of NP immunoreactive cells remained grouped in close proximity to the lateral border of the third ventricle comparable to their position in vivo. These cells averaged 20-30 μm in diameter and were often multipolar. (E) TH-containing cells immunohistochemically detected in cultures from brainstem tissue at the level of the locus coeruleus. Within this slice, clusters of large stained cells were observed. These clusters consisted of 3–30 cells whose average diameter was 25–35 mm. These cells were often multipolar with long dendritic processes. (F) Gonadotropes (stained for luteinizing hormone) as well other anterior pituitary cell types (not shown) were maintained in slice explants.

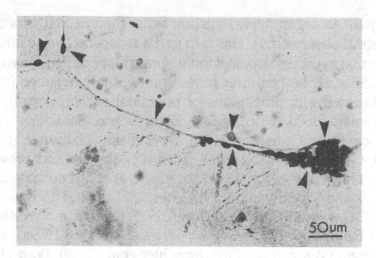

Fig. 4. Slice explant roller tube culture derived from embryonic mouse tissue. This slice was derived from a 11.5 d old mouse embryo and includes nasal and brain tissue. The slice was cultured for 7 d and then immunocyto-chemically stained for LHRH. Numerous immunopositive cells (arrowheads) are detected in "tracks," resembling the organization of this system, at this early embryonic stage, in vivo (*see* Wray et al., 1989b).

3. Slice Explant Protocol

3.1. Generating Brain Slice Explant Cultures from Postnatal Animals

3.1.1. General Considerations

The success of slice explant cultures depends greatly on treating the initial tissue as *gently* as possible and maintaining a *clean* environment. Failure to exercise sterile technique quickly manifests itself through culture contamination. Unfortunately, rough handling of the tissue before culturing is more difficult to detect. Typically, the cultures will die after a few days, leaving the investigator with few clues as to why. Many factors can contribute to culture death: poor media quality, animals suffering from malnutrition, and/or the lot of chicken plasma used. Although these variables affect culture survival, treatment of the tissue prior to culturing is often a variable that is overlooked, and is critical for tissue survival. The key, particularly for the novice, is to perfect the dissecting technique before evaluating the reasons for success or failure of the culture procedure. Thus, as mentioned earlier, starting out with tissue from early postnatal animals that has been previously shown to do well is recommended. This could include cerebellum, hippocampus, or anterior hypothalamic regions from postnatal 4–6 d old rats.

3.1.2. Setting Up

The area where the initial dissections, cutting, and plating of the slice cultures are performed should be as sterile as possible. In our laboratory, these procedures are done in a tissue enclosure hood in an area of limited traffic. In addition, it is helpful to wear a surgical mask until the cultures are plated. (**Note**: a laminar flow hood is used only when the cultures are inserted into the test tubes [before rotation] and during feeding [*see* Section 3.4.4.]).

The following instruments are routinely used during the dissection and plating procedures: dissecting microscope, fiber optic lighting, tissue chopper, repeating Eppendorfs, alcohol burner, small oven, large surgical scissors (decapitation), small scissors (to remove the brain from the calvarium), No. 5 forceps (removal of the pia, blood vessels, and the like and manipulation of the tissue slices), polished flat surfaced small spatulas (transfer of the blocked tissue and tissue slices), razor blade holders and breakable blades (to dissect out the area of interest), and aclar film (plastic discs used on tissue chopper that can be sterilized and discarded after using).

All tools, with the exception of the fine forceps and small scissors, are heat-sterilized in a small oven before culturing. These tools are resterilized throughout the plating procedure by rinsing in a series of alcohol washes followed by flaming them in an alcohol lamp. The finer tools are cleaned through a series of alcohol washes only (water, 50, 95, and 100%). Aclar discs are used on the tissue chopper as disposable, sterile cutting surfaces. They are cut from aclar film (Allied Fibers and Plastics, Pottsville, PA), washed several times in alcohol, air-dried, and placed in packages that are then gas-sterilized and set aside for at least 2 wk to remove any harmful components of gases.

3.1.3. Animals

As discussed earlier, the age of the animal from which the tissue is obtained can critically affect the survival, thinness, and/or organotypic characteristics of the final culture. Hence, the age of the initial animal should be chosen carefully while taking into account the neuronal area, cell type, and so forth to be studied. Successful slice explant cultures have been obtained from either postnatal or embryonic animals (*see* Section 3.1.7.). However, we recommend that one initially tries the technique on early post-natal animals as described herein.

In our laboratory, lactating rats with 2–5 d-old pups (Sprague-Dawley, Taconic Farms, Germantown, NY) are received on a weekly schedule and housed for at least 2 d before using. It is important to remember that the health of the pups from which the cultures are made can greatly influence the survival of the cultured material. Animals received from an outside facility should be allowed to recover for a few days to regain a nutritionally healthy condition. Poor cultures are invariably obtained when rat pups appeared to be underfed or in poor health.

In our hands, 4–5-d old pups have consistently yielded good cultures. The oldest animals used in our laboratory have been 9–10 d of age. Whereas extended fiber outgrowth and survival of neuropeptidergic cells is typically seen in material from 9–10-d old rats, this tissue consistently becomes more necrotic after 13 d in culture than tissue generated from younger animals. Our observations agree with previous accounts (Gähwiler, 1988): The older the animal, the less the tissue survives or the greater cell death.

3.1.4. Dissection and Slice Preparation (Fig. 5)

To gain access to the brain, rats are quickly decapitated using large scissors, under aseptic conditions, in a tissue enclosure hood (Labconco, via Fisher Scientific, Pittsburgh, PA). Small scissors are placed in the spinal opening and a cut is made toward the ear, keeping the scissors pulled outward so as not to damage the tissue with too deep a cut. The scissors are moved rostrally within this cut and a second cut is made toward the eyes. A third cut is then made directly across the skull. At this point, the scissors are removed and replaced in the spinal opening and a single cut is made in the skull in the opposite direction as the first cut,

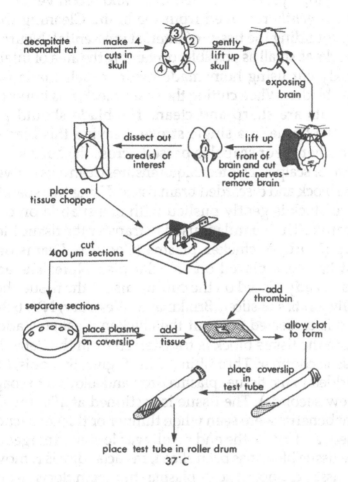

Fig. 5. Schematic of protocol for generation of slice explant roller-tube cultures. *See* Section 3.1.4.

toward the ear. The scissors are removed and the flap of skull is lifted up, exposing the dorsal aspect of the brain. The brain is gently lifted upward, the optic nerves are carefully cut, and the brainstem is freed from the surrounding skull. At this point, the brain is removed from the skull and placed directly in a Petri dish containing Geys balanced salt solution (GBS, Gibco; for slides, Gibco BRL, Gaithersburg, MD) enriched with glucose (5mg/mL). Once the brain is removed, it is easy to access the pituitary, if this is a tissue of interest. Remember, the tissue to be cultured must be kept moist (GBS + glucose) at all times.

Using forceps, the pia, meninges, and blood vessels are quickly *but gently* removed from the brain. Cleaning those areas surrounding the tissue of interest is essential. Neuronal tissue blocks as small as possible, containing the area of interest are quickly cut using homemade razor scapels made from breakable blades. When cutting the tissue block, it is important that the cuts are sharp and clean. The blade should pass through the tissue in a single stroke. Leaving this blade in place, another razor scapel is passed across the back side of the first razor scalpel. This technique ensures clean cuts between the tissue block and discarded brain tissue. Using two spatulas, the tissue block is gently pushed with one spatula onto the other spatula. This spatula is used to transfer the tissue block to a drop (20 μL) of chicken plasma (Cocalico, Reamstown, PA) that has been placed on an aclar disc. **Note:** The aclar disc was already placed on the cutting disc of the tissue chopper (McIlwain tissue slicer, Brinkmann, Westbury, NY) before the brain is dissected out, but the chicken plasma is added just before the tissue block is transferred from the Petri dish to the tissue chopper. Thrombin (20 μL, Sigma, St. Louis, MO) is then added to the tissue/plasma drop and allowed to coagulate (a few seconds). The tissue is sectioned at 400 μm. (No apparent benefits were seen when thinner or thicker sections were used, and often, the end result was less advantageous). After the tissue block has been sliced, the aclar disc is removed from the tissue chopper, excess plasma/thrombin clot is cleaned away from the tissue, and the sliced tissue block is gently pushed off the aclar disc (take care to keep the block intact, so as not to lose the order of the slices) into a drop of GBS + glucose in a clean Petri dish. Using a spatula and forcep, the individual slices containing the specific area(s) to be investigated are separated, trimmed if necessary, placed in fresh GBS + glucose in a new Petri dish, and refrigerated. The entire process from decapitation to refrigeration for each brain should take about 5–10 min. (If you are working with a small piece of tissue, it can be manually dissected rather than sliced on the tissue chopper.)

3.1.5. Plating and Incubation of the Tissue Slices

After all the tissue sections have been collected, the dissection area is cleaned quickly with an alcohol-soaked cloth. Precleaned coverslips (Gold seal, 12 × 24 mm, cleaning procedure listed below), a few at a time, are arranged in a sterile Petri dish and 20 µL of chicken plasma (Cocalico, Reamstown, PA) are spread over the coverslip surface. (The plasma/thrombin used for plating is made fresh on the day of culturing. The plasma/thrombin clot used to adhere the tissue block while cutting may be frozen from the fresh material used on previous culture days). Tissue slices are transferred to the plasma-coated coverslips using a spatula. Thrombin (0.4 NIH U/20 µL GBS + glucose, Sigma) is added to the coverslip to coagulate the plasma. Coagulation should start immediately, adhering the explant to the coverslip. At this point, the Petri dishes are covered and placed in a sterile hood for ~20 min to assure sticking of the explant to the slide. Still using a sterile hood, the coverslips are placed in 15 mL plastic tubes (Falcon, Becton Dickinson, Lincoln Park, NJ) containing 1 mL of media (media recipe, *see below*), tightly capped, and placed upright in a test tube rack. This is the first point at which to determine whether the plasma/thrombin clot is working. The explants in tubes are left upright in this moist environment for ~60 min. The tissue slices should not change position on the coverslip. The test tubes are then inserted in a roller drum (Bellco, Vineland, NJ) and rotated at approx 10 revolutions/h at 37°C.

3.1.6. Cocultures

Two types of cocultures can be used. The first type involves incubating different tissue types on the same coverslip surface; the second incubates different tissue types on separate coverslips, back-to-back, in the same tube. The latter procedure does not allow direct contact between the different tissue types to occur, but allows for humoral interactions (*see* Section 5.2.). For both coculturing procedures, we have used different tissue types from the same litter, but not necessarily from the same animal. The rest of the procedure used for the cocultures is identical to that previously described for single cultures.

3.1.7. Embryonic Animals

Timed-pregnant females are received no earlier than 6–8 d after fertilization. The animals should acclimate for at least 2 d before using. The females are anesthetized by inhalation of carbon dioxide. The uterus is removed and placed in cold GBS + glucose (as above). The embryos are quickly dissected from the uterus and placed in a Petri dish containing fresh GBS + glucose. This dish is then transferred to the culturing area. Embryos are removed from their amniotic sac and the area of interest is removed (*see* Fig. 4). If this region is small enough, e.g., less than 400 µm, no further chopping is necessary. If a tissue chopper is required, the procedure is the same as described above for postnatal animals.

3.2. Support Surfaces Used for Plating Cultures

Extensive preliminary investigations should be done before opting for a particular supporting surface. We recommend that initial studies be done on glass coverslips before attempting to alter the substrate material.

3.2.1. Glass Coverslips

Glass coverslips have been used most often as substrate surfaces for slice explant cultures. The only requirement for these coverslips is that they fit snuggly into the roller tubes, and are clean and sterile before using. Coverslips are cleaned by placing them in a beaker with 95% alcohol for 1 h. Several 1-h soakings are recommended, after which the coverslips are put into fresh alcohol, covered, and left overnight. The following day, the alcohol is replaced by deionized, distilled water, and the coverslips are boiled for 20–30 min. After cooling, the water is replaced by several alcohol rinses. The coverslips are stored in fresh alcohol until the day of use. On the morning of culturing, the coverslips are placed into a cleaned, glass Petri dish and heated in a small oven (Fisher, Isotemp 500 series, Pittsburgh, PA) at ~200°C for 1 h and then cooled.

3.2.2. Nonglass Coverslips

Nonglass and porous coverslips are potentially attractive because they could circumvent two major problems of glass coverslips: penetration of oxygen and nutrients from top surfaces only; and inability of growing cultures to attach directly to the glass surface. New products are available that allow penetration of oxygen and fluid from both surfaces, present the tissue with a surface for cell adhesion, migration, and axonal growth, and are compatible with many histochemical procedures (Romijn et al., 1988; Millicell from Millipore, Bedford, MA, and Anocell from Anotek, New York, NY). Although many of these products still need to be tested, these porous membranes may provide a good alternative to the glass coverslips used routinely. However, tissue types can behave differently on these membranes. For example, we have found that neuronal areas from the middle and caudal hypothalamus, which are difficult to maintain on glass surfaces, survive, produce dramatic fiber outgrowth, and thin sufficiently for single cell analysis on coverslips made from Anocell (personal observation, Fig. 6). On the other hand, rostral forebrain areas as well as preoptic/anterior hypothalalmic regions are maintained better on glass coverslips. The exact mechanism behind surface preference is unclear. A number of factors seem to contribute to this phenomenon. The presence of ventricular surfaces with ependymal cells and beating cilia influences survival on glass. Slices with such cells seem to "stick" less to the culture surface. These slices may do better on surfaces that allow anchoring, such as Anocell, as opposed to glass. Another factor that may influence the choice of the support surface may be the ratio of cell bodies to axonal tracts. Tissue regions with low fiber tract areas may adhere well on glass, whereas areas with high large tract regions may require surfaces that allow cells to anchor quickly, since large areas of degenerating processes could cause cell shifts to occur, resulting in reorganization or tissue loss. However, no certain formula exists for predicting how tissue areas will do on any particular support surface, and hence must be worked out empirically.

3.3. Substances Used to Adhere Cultures to Support Surfaces

Of all the substances used to adhere the tissue slice to the support surface (the adherent), *cultures grown in the presence of chicken plasma/thrombin clots thin most readily while maintaining an*

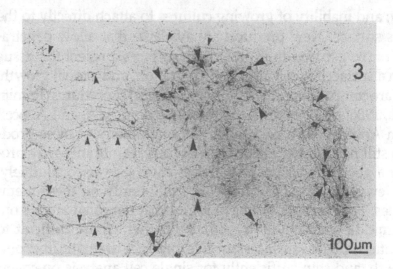

Fig. 6. Example of a slice explant culture grown on nonglass coverslip (Anocell). This culture was derived from postnatal day 5 rat pups and cultured for 13 d. The explant is from the hypothalamus at the level of the paraventricular nucleus. The culture was immunocytochemically stained for oxytocin, a neuropeptidergic neuron known to be located in this area in vivo. The third ventricle is to the right (3). Numerous oxytocin cells (large arrowheads) and fibers (small arrowheads) are visible.

organotypic nature. This does not mean alternative adherents are not available or worth trying, but one must understand that the adherent used must:

1. Be strong enough to maintain the culture relatively stably through continual rolling;
2. Permit thinning; and
3. Allow survival of neural tissue and potentially promote fiber outgrowth.

The choice of adherent is influenced by the project's goal, and one must decide before choosing an adherent (*see below*) whether it is critical to visualize single cells and how much reorganization of the tissue is allowable.

After plating, treatment of the tissue slices depends on the adherent used. If the adherent bonds the tissue to the coverslip quickly, e.g., plasma/thrombin clot, the slides are placed in test tubes and rotated (approx within 1 h for plasma/thrombin clots).

If, however, the adherent takes time to bond the tissue to the coverslip, as with polyornithine and/or collagen, the coverslips are set in sterile humid chambers, 40–50 µL of media added on top of the tissue, and the chambers placed flat in incubators with a moist environment. The cultures are maintained in a flat orientation for ~24 h, or until the tissue has adhered, and then inserted in test tubes and rotated as indicated in Section 3.1.5.

3.3.1. Plasma/Thrombin Clot

Plasma clots are one of the most common substances on which to grow tissue fragments, specifically in roller-tube cultures. Unfortunately, like serum, plasma is not as defined as one would like, and thus, it is difficult to make consistent. Usually, the plasma is derived from chicken blood. Whereas many commercial sources of chicken plasma are available, most have been unsuitable; one problem is the anticoagulant used. Heparin appears to be the most compatible anticoagulant for tissue survival, but the concentration used during plasma preparation is critical (*see* Paul, 1975 for review of plasma preparation). Since not everyone has facilities for chickens available, we have established a commercial source, Cocalico (Reamstown, PA), to produce a plasma that is well-suited for our organotypic roller cultures. Cocalico chicken plasma is received frozen, lyophilized, and stored at –20°C. On the day of culturing, the plasma is reconstituted with distilled water. Both plasma and thrombin are centrifuged at 6000 rpm for ~20 min before use.

The plasma obtained from Cocalico maintains healthy cultures for at least 2 wk. Longer growth periods may result in some thickening and reorganization of the tissue, but this effect is tissue-dependent. We have found that this plasma maintains a variety of tissue types, including forebrain (hypothalamus and preoptic areas), cerebellum, and pituitary from postnatal rats, as well as brain and nasal areas from embryonic mice.

The source of thrombin is less of a problem, due to standardization; 0.4–0.5 U of thrombin are applied in vol equal to that of plasma. In our laboratory, thrombin (Sigma, 100 U) is reconstituted with 5 mL of Geys + glucose (same solution as in Section 3.1.4.), centrifuged (*see above*), and 20–25 µL applied to an equal vol of plasma.

3.3.2. Polyornithine

Slice explant cultures can be maintained on polyornithine (or polylysine)-coated coverslips. Use of this adherent removes factors present in the plasma/thrombin clot however optimal thinning of the culture rarely occurs. To prepare polyornithine-coated coverslips, a solution of 0.3–0.5 µg/mL polyornithine (Sigma) in distilled water is used. This solution is placed on cleaned coverslips and allowed to set for 2–3 h in a sterile hood. The coated coverslips are then rinsed 3X in distilled water and air-dried. These coverslips can be prepared several days in advance and stored under sterile conditions. Coverslips must be rinsed thoroughly because unbound polyornithine is toxic to cells.

3.3.3. Collagen

Slice explant cultures can be maintained on collagen-coated coverslips with thinning dependent on tissue type and concentration of collagen used. Collagen is available from many commercial sources or can be made by the investigator (Paul, 1975; Maurer, 1986; Toran-Allerand, 1990). We have had some success using collagen obtained commercially (Sigma, rat tail type I). The final concentration of collagen applied to the coverslips is important, and we have found that 2–6 mg/mL is a good starting point. The collagen is diluted in 0.01–0.05M acetic acid, placed on the coverslips, and allowed to dry. (The coverslips may be stored in a balanced salt solution media.)

3.4. Media

Without question, the choice of media is critical to the success of the cultures. If possible, it is advantageous to initially select media that are known to: (1) grow the cells one is interested in, and (2) work well in slice explant cultures. In some instances, a compromise medium may be necessary. We have found that the presence of serum, specifically horse serum, is important for thinning of these cultures. We have found that calf serum is not as effective as horse serum for tissue thinning and survival. To date, we have not had cultures thin to the extent required for single cell analysis if nonserum-containing media was present from the time of plating. Others have had some success in culturing in defined media from the onset (Romijn et al., 1988). If extensive thinning is not required, then serum-free media (*see below*) can be used throughout culturing. Alternatively, after thinning, cultures can be transferred into defined media, allowing one to work in a controlled environment.

3.4.1. Serum-Containing Media

The serum-containing media outlined below are used by many investigators (Gähwiler, 1981a; Llano et al., 1988; Wray et al., 1988, 1989a) and supports a variety of tissues including forebrain, cerebellum, brainstem, pituitary, gonads, and even embryonic tissue.

The serum-containing media used consist of 25% inactivated horse serum (heat at 56°C for 30 min in a water bath), 50% Eagle's basal medium, and 25% Earles' balanced salt solution. The media is supplemented with 5.0–7.5 mg/mL glucose, 2mM glutamine, penicillin (25 µg/mL), streptomycin (25 µg/mL), and neomycin (50 µg/mL) (PSN antibiotic mixture, *see* Section 3.4.3.).

The main concern in selecting media constituents is product consistency. Most of the above products are relatively easy to obtain from companies who sell standard culture products. Unfortunately, horse serum presents its own problem. We pretest horse serum lots on dissociated cultures from brain and dorsal root ganglion prior to using it on slice explant cultures. Typically, horse serum from three different sources is tested. The lot in which neurons are maintained the best (number of cells, neurite outgrowth, thinness of final slice culture, axonal outgrowth) is ordered in quantity (enough for at least 1 yr of culturing). Serum-containing media is changed 1–2 times/wk depending on the type and amount of tissue being cultured.

3.4.2. Defined Media (Serum-Free)

Cellular properties of neurons in culture should preferentially be studied on cultures maintained in defined media, thereby eliminating unknown factors present in serum. However, we have found that for optimal thinning to occur, cultures are best grown first in the presence of serum and then changed into a serum-free media. In contrast, Romijn et al. (1988) report maintenance and thinning (to approx eight cells thick) of cultures

grown in defined media from the day of plating. Our cultures
survive well in serum-free media for at least 6 d (Wray et al.,
1991). The present feeding schedule used in our laboratory is as
follows: for the first 11–12 d, the cultures are grown in serum-
containing media (described above) , followed by defined
(serum-free) media for a maximum of 6 d (Fig. 1C). All experimental manipulations are done only after two changes in
defined media.

Our defined media recipe is based on previous formulations (Maurer, 1986). It is composed of 50% Eagle's basal Media
(Gibco), 50% Ham's F-12 nutrient mixture (Gibco), 10 mg/mL
bovine serum albumin (Sigma); $1 \times 10^{-4}M$ putrescine (Sigma), 5
µg/mL insulin (Sigma), 100 µg/mL transferrin (Sigma), 2 mM
glutamine (Gibco), 5 µl/mL PSN antibiotic mixture (Gibco, *see*
Section 3.3.1.), $3 \times 10^{-8}M$ selenium (Sigma), 5.0–7.5 mg/mL glucose (Sigma), and 38 µM ascorbic acid. Defined medium is
changed more often than serum-containing medium—at least
every 2–3 d.

3.4.3. Culture Supplements:
Phenol Red, Antibiotics, and Antimitotics

Initially, our cultures were grown in the presence of phenol
red, but once the feeding schedule was established it was no
longer added to the media. Although contamination of short-
term cultures should not be a problem, the PSN antibiotic mixture (for formula, *see* Section 3.3.1.) ensures long-term culture
survival. Since we have not seen any deleterious effects of this
antibiotic mixture on the cultures, the PSN antibiotic mixture is
routinely used for all our cultures. In our laboratory, cultures
have been successfully grown for as long as 11–53 d. Most
experiments in our lab, however, are performed on explants cultured for 17–21 d (Wray et al., 1988,1989a,1991).

For slice explant cultures derived from postnatal animals,
antimitotics can be used to reduce the number of nonneuronal
cells. Most investigators have not found overgrowth to occur in
these cultures, and thus, do not routinely use antimitotics. In
contrast, when slice explant cultures are derived from embryonic tissue, antimitotics are often necessary because of the large
number of proliferating cells. If antimitotics are required, we have

used fluorodeoxyuridine ($8 \times 10^{-5}M$, Sigma). The antimitotic is added a few days after the cultures are plated; the length of exposure is dependent on the tissue being cultured.

3.4.4. Feeding

Feeding the cultures should be done in a sterile environment, i.e., in a laminar flow hood. The media should be filtered before use and allowed to warm to room temperature. Test tubes are gently removed from the roller drum and placed in a test tube rack. To change the media, simply remove the test tube cap, turn the test tube upside down, and pour old medium into a beaker. Replace with appropriate fresh medium, recap test tube tightly, and replace in roller drum. One never removes all the old medium. If changing to defined medium, we recommend at least two changes, separated by 1–2 d, in order to assure removal of the original serum.

4. Methodologies Compatible with Slice Explant Cultures

Slice explant cultures are compatible with a variety of existing techniques: these include examination of both the living culture and fixative-preserved culture. To date, most investigations have coupled slice explant cultures with electrophysiologic techniques and/or histochemical/immunocytochemical techniques.

In the next sections, some of the issues addressed and techniques used with slice cultures are briefly described. For exact procedural details, the reader is urged to refer to the original papers. It should be clear that when a variety of techniques are combined, a large amount of information can be obtained from any one culture.

4.1. Examination of Living Tissue

4.1.1. Identification of Cell Type Based on Neuronal Morphology, Location, and/or Vital Dyes

Because of the thinness of these slice explant cultures, it is often possible to identify a cell type by its morphologic features, e.g., cell shape (Purkinje cells in the cerebellum; Gähwiler et al., 1973), size (magnocellular neurons in the supraoptic nucleus;

Gähwiler et al., 1978), and/or location in neuronal region (pyramidal neurons in the hippocampus; Gähwiler, 1980b). In addition, detailed morphologic aspects of individual living cells have been obtained using injections of various dyes after the culture has thinned. Often, these dyes can be used to identify axonal pathways (cocultures) as well as cellular morphology. Lucifer yellow has been very effective in delineating cellular features, including dendritic spines and neuritic arborizations without being toxic (Gähwiler, 1980a). We have also used DiI (Honig and Hume, 1986) and fluorescent latex microspheres (Katz et al., 1984) to retrogradely label cells already in culture.

An alternative approach is to label cells prior to culturing. We have been able to label hypothalamic cells using Fast Blue. In these experiments, Fast Blue was injected into the tail vein of 2–3-d old rat pups; tissues from these animals were cultured 2–3 d later, as previously described. Upon plating, the slices were quickly viewed under a fluorescent microscope and numerous labeled cells were detected. Using extremely sterile techniques, these cells can be monitored during culturing (unpublished data). Another compound, used by others in labeling cells prior to culturing, is Fluoro-Gold (Fluorochrome Inc., Engelwood, CO). The ability to label cells with nontoxic markers not only gives information on morphology and circuitry but enhances the possibility of being able to record directly from the cells of interest.

4.1.2. Cellular Activity

Once a cell type has been identified as indicated above, cellular activity (both ionic and receptor-mediated) can be monitored using various electrophysiologic techniques, including extracellular (Gähwiler et al., 1978; Geller and Woodward, 1979; Baldino and Wolfson, 1985), intracellular (Gähwiler 1980a,b; Gähwiler and Brown, 1985), and patch-clamp recordings (Llano et al., 1988). Neuronal activity has been characterized in a number of different tissue types maintained in slice explant cultures (Gähwiler et al., 1973; Gähwiler, 1980b; Geller, 1981a,b, *see also* refs. above).

Our laboratory has begun such investigations on hypothalamic tissue. For these studies, we used a modified patch electrode (Hamill et al., 1981) to record intracellular activities. Briefly, the coverslip was transferred to a recording chamber and placed onto the stage of an inverted microscope. The slice chamber was perfused with oxygenated control bathing solution containing 140 mM NaCl, 4.7 mM KCl, 2.5 mM CaCl$_2$, 1.13 mM MgCl$_2$, 10 mM HEPES, and 10 mM glucose. The recording pipet was filled with 150 mM KCl, 2 mM MgCl$_2$ and 5 mM HEPES. Recordings were carried out at room temperature and the sodium channel blocker tetrodotoxin blocked virtually all spontaneous activity (Wray et al., 1991).

4.1.3. Neuronal Secretion

Most explant studies on neuronal secretion involve the use of acute or short-term cultures (Morris et al, 1986; Stern et al., 1986). However, the few secretion studies performed using long-term roller-tube cultures indicate the usefulness of this technique in studying neurotransmitter/neuropeptide release and local-ization of synthesis sites (Baertschi et al., 1982; Gähwiler et al., 1984). These studies involved measurement of secreted product in the media using standard radioimmunoassay methods or indirect assays (*see* Section 5.2.)

4.2. Examination of Fixative-Preserved Tissue

At present, examination of cells in living cultures is limited by the lack of appropriate cell markers that are compatible with cellular viability. When such markers are unavailable, examina-tion of fixed material, as in vivo, has been used to study tissue organization, neuronal structure, function, and regulation.

4.2.1. Neuronal Synapses, Pathways, and Phenotype and Regulation of Gene Expression

One substance that is especially compatible with light and electron microscopic (EM) analysis of synaptology and fiber ingrowth is horseradish peroxidase (HRP). Although HRP is injected into neurons when the tissue is still living, processing and visualization of the product requires fixed material (Zimmer and Gähwiler, 1984; Frotscher and Gähwiler, 1988). The HRP is processed using standard methods, e.g., hydrogen peroxide and 3,3'diaminobenzidine as the chromagen. At the light microscopic level, HRP can delineate neuronal morphology and pathways.

Examination at the EM level can determine cellular structure, synaptic relationships, and the degree of connective reorganization. These techniques are also applicable to cocultures (Zimmer and Gähwiler, 1984).

If individual cells cannot be identified and/or isolated in living cultures, immunohistochemical, immunocytochemical, and/or *in situ* hybridization histochemical assays may be performed on fixed cultures (*see* Section 8 for immunocytochemistry and *in situ* hybridization histochemistry procedural details). These methods provide valuable data concerning survival of cell types (Baldino et al., 1985; Sofroniew et al., 1988; Wray et al., 1988), organotypic organization of the slices (Wray et al., 1988), afferent connections (Gähwiler and Brown, 1985), efferent projects (Gähwiler and Hefti, 1984; Gähwiler et al., 1987; Wray et al., 1988), functional activity (Wray et al., 1988), and gene transcription (Wray et al., 1989a).

The compatibility of slice explant roller-tube cultures and *in situ* hybridization histochemistry largely due to thinning (Wray et al., 1989a,1991) enables one to examine:

1. Developmental onset of phenotypic expression;
2. Receptor-mediated changes in transcription and translation of specific neuronal genes in differentiated cells;
3. Second messenger systems that directly influence gene expression;
4. Regulation of gene expression in embryonic and/or nonsynaptically associated cells; and
5. The sequence of genes activated that lead to the final event, e.g., secretion of a product. Use of molecular probes is a very new and exciting aspect to investigations using slice explant roller-tube cultures.

Thus, together, both histochemical and immunocytochemical techniques yield data with high anatomic resolution that can be used to determine how individual neurons and/or organized neuronal circuits, albeit "simplified," respond to specific intrinsic and extrinsic stimuli.

4.2.2. Product Synthesis and Processing

Depending on the markers available, immunocytochemical methods can be used to determine the gene products synthesized and processed in the cells of interest. Alternatively, radioimmunoassays (RIAs) on homogenized tissue could be performed, but with inherent loss of single cell information. Single cell resolution could be achieved using isotopic pulse chase methods and autoradiography, if antibodies or other histochemical markers are not available.

4.2.3. Receptor Localization

To map the distribution of specific receptors yet still maintain cellular resolution, slice explant cultures can be analyzed, after application of radiolabeled markers, using autoradiography (Hösli and Hösli, 1990; Toran-Allerand, 1978) and this approach can be combined with immunocytochemical analysis (Lerea and McCarthy, 1990).

5. Examples of Slice Explant Systems

Slice explant roller-tube cultures are particularly good models with which to examine neurons within the CNS, specifically hippocampus, cerebellum, brainstem, and preoptic/hypothalamic areas. Tissues cultured by this method thin to quasimonolayers during the first 2 wk in vitro, but maintain several aspects typical of the tissue in vivo.

5.1. Slice Explant Cultures: A Model System for Examination of Neuropeptide-Expressing Cells

Neuropeptidergic cells are extremely diverse, and even cells expressing the same neuropeptide gene and gene products can be composed of functionally distinct subpopulations based on the heterogeneity of:

1. Projections;
2. Afferents;
3. Coexistence with other neurotransmitter and/or neuropeptide products;

4. Receptors; and
5. Access to humoral environment.

This enormous complexity has made examination of molecular and functional properties of neuropeptide cells in vivo extremely difficult. For these reasons, our laboratory has chosen to examine the regulation of these cell types in slice explant roller-tube cultures.

5.1.1. Maintenance of Topography of a Dispersed Neuropeptide Population (Fig. 1)

Neuropeptide-expressing cells can be widely distributed throughout the brain. The luteinizing hormone releasing hormone (LHRH) system is an example of such population. In rats, it is composed of approx 1300 cells that are distributed in a continuum throughout the forebrain (Wray and Hoffman, 1986). The anatomic distribution of this system makes studying the entire LHRH population in vivo extremely difficult.

We have reported (Wray et al., 1988,1989a) that 20–30% of LHRH neurons survive, express their appropriate gene and gene products, and maintain an organotypic distribution for long periods of time in vitro using a slice explant roller culture technique. The majority of the LHRH neurons are contained in a few slices, making regulatory studies on this system in vitro much more feasible than in vivo. In slice explant cultures, LHRH neurons responded to a stimulus known to affect the system in vivo. Using *in situ* hybridization histochemistry and single-cell analysis procedures to monitor mRNA level changes in individual cells, we found that a specific population of LHRH cells responded to estradiol as had been reported in vivo (Zoeller et al., 1988). Thus, in slice explant cultures, LHRH cells are maintained in a dispersed, yet anatomically correct, distribution pattern and respond appropriately to a humoral agent. For the LHRH system and other neuronal systems that are anatomically dispersed, slice explant cultures offer a unique opportunity to examine large numbers of cells from distinct anatomic regions under controllable environmental conditions.

5.1.2. Maintainence of Topography of Neuronal Populations Located in Discrete Nuclei (Fig. 2)

Neuropeptide-expressing cells can be maintained in relatively discrete nuclei but can also be intermingled with numerous other cell types. The suprachiasmatic nucleus, an example of such a neuropeptide-containing structure, is composed of heterogeneous neuropeptide populations with numerous intranuclear synaptic connections (Gainer et al., 1989). We have reported that the suprachiasmatic nucleus in slice explant cultures can be maintained organotypically for 13–21 d (Wray et al., 1988; Gainer et al., 1989). Interestingly, a slice culture containing this nucleus may become disorganized dorsally but remain considerably organized ventrally, the area housing the nucleus proper, with large numbers of intranuclear synapses. Three different peptides localized in this nucleus in vivo have been shown to be maintained in vitro using slice explant cultures (Gainer et al., 1989).

5.2. Cocultures

As mentioned earlier, cocultures using the slice explant roller-tube method offer a unique opportunity to compare direct interactions (Fig. 7) between tissue types vs humoral interactions (Gähwiler et al., 1984; Wray et al., 1988). Such questions can be addressed by plating the tissue onto two coverslips rather than one. Furthermore, slice explant cocultures have been useful in studying fiber ingrowth and axonal target selection (Gähwiler and Hefti, 1984; Gähwiler and Brown, 1985; Gähwiler et al., 1987; Zimmer and Gähwiler, 1987; Wray et al., 1988) as well as synaptic formation and induction of morphologic changes, e.g., dendritic spine formation (Wray et al., 1988).

To date, there are few membrane markers to identify specific neuronal cell types while the cell is still alive. This makes electrophysiologic recordings of cells not distinctly anatomically organized or morphologically discernible extremely difficult. How then does one determine whether the cells of interest are

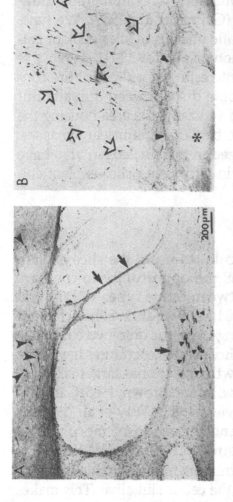

Fig. 7. Axonal growth examined using cocultures. (A) 18-d-old preoptic area/hypothalamus-brainstem coculture immunohistochemically double-stained for LHRH and TH. Tissue was originally taken from a postnatal day 4 animal. Both LHRH cells (large arrowheads) within the forebrain tissue and TH-containing cells (small arrowheads) within the brainstem tissue are visible. Interestingly, when these tissues were cocultured, TH-containing fibers were rarely seen to cross into the forebrain, areas to which they normally project. However, LHRH fibers were found to cross over into the brainstem tissue (arrows), an area rarely receiving LHRH projections in vivo. (B) 18 d old preoptic area/hypothalamus-brainstem-anterior pituitary triple cocultures immunohistochemically stained for LHRH (open arrows). Closed arrowheads indicate anterior pituitary/hypothalamus border and single asterisk shows the middle of the anterior pituitary tissue. LHRH cells are seen in preoptic/hypothalamic tissue. Black arrows indicate border between the hypothalamus and the brainstem (mainly out of field of photograph). Note: As in vivo, LHRH fiber plexi in vitro are primarily located along the ventral border of the tissue (black arrowheads). In these tricultures, LHRH fibers entered the anterior pituitary tissue but did not enter the brainstem. Although the anterior pituitary is the main target tissue for LHRH secretion, LHRH terminals end within the median eminence in vivo, never actually entering the anterior pituitary directly.

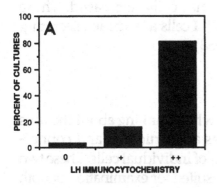 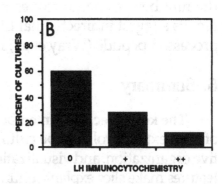

Fig. 8. Evidence for LHRH secretion in vitro shown using cocultures. Effect of preoptic/hypothalamus on LH cell staining in cocultured anterior pituitary tissue. **(A)**. Hypothalamus-anterior pituitary coculture ($n = 32$). **(B)**. Coculture in the presence of an LHRH antagonist in the culture media ($n = 25$). LH staining pattern of a pituitary that was classified as ++ had greater than 30 cells heavily stained whereas an anterior pituitary classified as + had less than 30 cells lightly stained, and 0 indicates no LH positive cells (x-axis). Y-axis: percent of anterior pituitary cocultures found in 0, +, ++ category. Many heavily stained pituitaries were found when cocultured with preoptic/hypothalamic tissue containing LHRH cells. The hypothalalmus did not have to be in contact with the anterior pituitary (back-to-back coverslips in the same test tube) for LH cells to be detected. The effect of the hypothalalmus on LH cell detectablity was abolished by the addition of a highly specific LHRH antagonist to the culture media **(B)**.

functional? Certainly the presence of gene and gene products is helpful, but not conclusive. Instead, one would like to show the end result, i.e., secretion of an appropriate neurotransmitter and/ or neuropeptide. We are currently developing a modified RIA that will be sensitive enough to detect the release of LHRH in our culture media. However, using cocultures, an alternative approach to demonstrate LHRH secretion was performed (Fig. 8). Since LHRH cells normally influence gonadotropes in the anterior pituitary, we cocultured forebrain-containing LHRH cells with pituitary pieces. These cocultures included both tissues on the same coverslip as well as each tissue on a separate coverslip, back-to-back, in the same test tube. We found that in either case, gonadotropes were detectable in the pituitary tissue. However, when an LHRH antagonist was added to the media, the number of gonadotropes dramatically decreased. These results suggest indirectly that LHRH cells are, in fact, secreting processed peptide (Wray et al., 1988).

6. Summary

The key aspects to remember when thinking about the uses of slice explant roller-tube cultures are: structural and connective organization, and visualization of individual cells. These two features make slice explant cultures suited for examination of both intrinsic properties of neuronal function and extrinsic factors that influence these properties. In addition, analysis of single neurons, neuronal subtypes, and total neuronal populations in slice explant cultures may identify the events that lead to developmental and homeostatic regulation of neuronal gene expression. Thus, this model system, because of its organotypic nature, should provide a way to address the relationship between neuronal phenotype, anatomic circuitry, and neuronal functions.

7. Products and Suppliers

Note: Most of these products are available from many different sources.

Aclar Film (gage 7.80 mils)	Allied Corporation, Engineer Plastics Department, P.O. Box 1205, Pottsville, PA, 17901, (800)233-0251
Albumin, bovine # A-0281	Sigma, P.O. Box 14508, St. Louis, MO 63160, (800)325-3010
Anocell	Anotec, 226 East 54th St., New York, NY 10022, (212)751-3770
Basal Media Eagle # 320-1010 or 1017	Gibco Laboratories, 3175 Staley Rd., Grand Island, NY 14072, (800)828-6686
Blade Breakers and blades	Surgimed-MLB, 22 S. State St., Newtown, PA 18940, (215)968-3186

Chicken Plasma # 30-0390	Cocalico Biologicals, Inc., 449 Stevens Rd., P.O. Box 265, Reamstown, PA 17567, (215)267-7548
Collagen Type 1	# C 7661 Sigma Chemical (for address *see* Albumin)
Earle's BSS	#310-4010 or 4015 Gibco Lab. (for address *see* Basal Media)
F-12 Nutrient Mix.	#320-1765AG Gibco Lab. (for address *see* Basal Media)
Geys BSS	#310-4260 (for slides) Gibco Lab. (for address *see* Basal Media)
Glass Coverslips #3312 Gold Seal (12 × 24)	Fisher Scientific, P.O. Box 1768, Pittsburgh, PA 15230, (412)562-8300
Horse Serum	Source varies
Insulin, bovine	#I - 5500 Sigma Chemical (for address *see* Albumin)
L-glutamine	#320-5030AG (200 m*M*) Gibco Lab. (for address *see* Basal Media)
McIlwain Tissue Chopper	Brinkmann, Cantiague Rd. Westbury, NY 11590, (800)645-3050
PSN Antibiotic Mix. #600-5640	Gibco Lab. (for address *see* Basal Media)
Putrescine	#P - 7505 Sigma Chemical (for address *see* Albumin)
Roller Drum #7736	Bellco Glass Biotechnology, 340 Edrudo Rd., P.O. Box B, Vineland, NJ 08360, (800)257-7043
Sodium Selenite	#S - 1382 Sigma Chemical (for address *see* Albumin)
Thrombin	#T - 4265 Sigma Chemical (for address *see* Albumin)
Tissue Culture Water	Source varies
Transferrin, bovine	#T - 5761 Sigma Chemical (for address *see* Albumin)

8. Protocols

8.1. Immunocytochemistry

For immunocytochemical assays, we made only a few changes from the protocol used in our laboratory to process tissue sections. However, antibodies must be checked in vivo before use on cultured slices. This allows one to establish the initial concentration to stain the cultures. In general, we found that slice cultures tend to have a slightly higher background than tissue sections. Thus, the signal-to-noise ratio must be critically evaluated for each antibody used.

Cultures are fixed (4% paraformaldehyde-0.2% picric acid for 1–2 h), rinsed in PBS (2 × 10 min) to remove excess fixative, and blocked in 10% normal goat serum with detergent (0.2% Triton X-100, RPI, Elkgrove Village, IL) for 1 h. Cultures are rinsed in 10% normal goat serum (2 × 30 min), and then incubated in primary antiserum overnight at 4°C. The next day, the cultures are processed using standard immunocytochemical methods. We have obtained sensitive staining using the avidin-biotin-horseradish peroxidase complex (Vector Laboratories, CA). Double-label immunocytochemistry (Fig. 7A; Wray et al., 1988), as well as fluorescent procedures are also compatible with slice explant roller-tube cultures.

8.2. In Situ *Hybridization Histochemistry*

We have found that a good signal is achieved using synthetic deoxynucleotide probes approx 50 nucleotides in length (Wray et al., 1989a,1991), as well as ribonucleotide probes (unpub. data). We have not tried the new nonradioactive methods on slice cultured material. However, most techniques are expected to be compatible with slice explant tissue by modification of recipes used in vivo.

Explant cultures are processed for *in situ* hybridization histochemistry by a modified procedure previously used on tissue sections (Zoeller et al., 1988; Wray et al., 1989). Briefly, cultures are fixed with 4% formaldehyde (15–20 min), rinsed in PBS (2 × 5 min), placed in 0.25% acetic acid/0.1M triethanolamine hydrochloride-0.9% NaCl (10 min), rinsed in 2XSSC (2×1.5 min), dehydrated (70, 80, 95, and 100% ethanol [2 min/wash]), delipidated in chloroform (7 min), rinsed (1 min each) in 100% than 95% ethanol, and air-dried.

Synthetic 48–50 base deoxynucleotide probes (5pmol) are 3'end-labeled with [^{35}S]dATP to a specific acitivity of 7,000–15,000 Ci/mmol. Labeled probe is applied to individual cultures in 25 µL (500,000 cpm) of hybridization buffer. The cultures are hybridized overnight at 37°C in humid chambers. The following day, cultures are rinsed in 2XSSC, washed in 2XSSC/50% formamide at 40°C (4 × 15 min), rinsed in 2XSSC at room temp (2 × 1 h), and then allowed to air dry. After drying, the cover-slips are dipped in NTB3 (Eastman Kodak) and exposed. The emulsion covered cultures are developed, counterstained, dried, and then mounted culture side down onto microscope slides (Wray et al., 1989a,1991).

Quantification of *in situ* hybridization histocytochemistry can be done when a radioisotope-labeled probe is used. Integrated densities of areas of silver grains over individual labeled cells and the cell areas enclosing these grains are measured using an image analysis system (Wray et al., 1989a,1991). This density/cell value reflects the level of mRNA/cell. Hence, the direction of change in mRNA levels and the statistical significance of these changes can be determined (Wray et al 1989).

Acknowledgments

I would like to thank S. Key for technical assistance, Devera Schoenberg for editorial comments, and W. Hayes and H. Gainer for their suggestions and review of this manuscript.

References

Baertschi A. J., Beny J-L., and Gähwiler B. H. (1982) Hypothalamic paraventricular nucleus is a privileged site for brain-pituitary interaction in long-term tissue culture. *Nature* 295, 145–147.

Baldino F., Jr. and Geller H. M. (1986) Hypothalamic explant tissue cultures in pharmacological research, in *Modern Methods in Pharmacology, vol. 3, Electrophysiological Techniques in Pharmacology* (Geller H. M., ed.), Liss, New York, pp. 103–119.

Baldino F., Jr., Higgins G. A., Moke M. T., and Wolfson B. (1985) Primary explants as a model of the hypothalamus *in situ. Peptides* 6, 249–256.

Baldino F., Jr. and Wolfson B. (1985) Postsynaptic actions of neurotensin on preoptic-anterior hypothalamic neurons *in vitro*. *Brain Res.* **325**, 161–170.

Banker G. A. and Cowan W. M. (1977) Rat hippocampal neurons in dispersed cell culture. *Brain Res.* **126**, 397–425.

Bornstein M. B. (1973) Organotypic mammalian central and peripheral nerve tissue, in *Tissue Culture: Methods and Applications*. (Kruse P. F. and Patterson M. K. eds.), Academic Press, New York, pp. 86–92.

Bottenstein J. E. and Sato G. (eds.) (1985) *Cell Culture in the Neurosciences*. Plenum Press, New York.

Braschler U. F., Iannone A., Spenger C., Streit J., and Lüscher H-R. (1989) A modified roller tube technique for organotypic cocultures of embryonic rat spinal cord, sensory ganglia and skeletal muscle. *J. Neurosci. Methods* **29**, 121–129.

Clarke M. J. O., Lowry P., and Gillies G. (1987) Assessment of corticotropin-releasing factor, vasopressin and somatostatin secretion by fetal hypothalamic neurons in culture. *Neuroendocrinology* **46**, 147–154.

Delfs J., Friend J., Ishimoto S., and Saroff D. (1989) Ventral and dorsal horn acetylcholinesterase neurons are maintained in organotypic cultures of postnatal rat spinal cord explants. *Brain Res.* **488**, 31–42.

Fedoroff S. and Hertz L. (eds.) (1977) *Cell, Tissue and Organ Cultures in Neurobiology*. Academic Press, New York.

Feldman S. C., Dreyfus C. F., and Lichtenstein E. S. (1982) Somatostatin neurons in the rodent hippocampus: an in vitro and in vivo immunocytochemical study. *Neurosci. Lett.* **33**, 29–34.

Fowler J. and Crain S. M. (1985) Long-term organotypic cultures of neonatal mouse hippocampal explants in plastic multiwell plates. *Proc. Eur. Tissue Soc.* **33**, 23.

Freshney R. I. (ed.) (1986) *Animal Cell Culture*. IRL Press, Oxford.

Frotscher M. and Gähwiler B. H. (1988) Synaptic organization of intracellularly stained CA3 pyramidal neurons in slice cultures of rat hippocampus. *Neuroscience* **24**, 541–551.

Gähwiler B. H. (1980a) Labelling of neurons within CNS explants by intracellular injection of lucifer yellow. *J. Neurobiol.* **12**, 187–191.

Gähwiler B. H. (1980b) Excitatory action of opioid peptides and opiates on cultured hippocampal pyramidal cells. *Brain. Res.* **194**, 193–203.

Gähwiler B. H. (1981a) Organotypic monolayer cultures of nervous tissue. *J. Neurosci. Methods* **4**, 329–342.

Gähwiler B. H. (1981b) Morphological differentiation of nerve cells in thin organotypic cultures derived from rat hippocampus and cerebellum. *Proc. Roy. Soc. Lond. B.* **211**, 287–290.

Gähwiler B. H. (1984a) Slice cultures of cerebellar, hippocampal and hypothalamic tissue. *Experientia* **40**, 235–243.

Gähwiler B. H. (1984b) Development of the hippocampus *in vitro*: cell types, synapses and receptors. *Neuroscience* **11**, 751–760.

Gähwiler B. H. (1988) Organotypic cultures of neural tissue. *Trends Neurosci.* **11**, 484–489.

Gähwiler B. H. and Brown D. A. (1985) Functional innervation of cultured hippocampal neurones by cholinergic afferents from cocultured septal explants. *Nature* 313, 577–579.

Gähwiler B. H., Enz A., and Hefti F. (1987) Nerve growth factor promotes development of the rat septo-hippocampal cholinergic projection *in vitro. Neurosci. Lett.* 75, 6–10.

Gähwiler B. H. and Hefti F. (1984) Guidance of acetylcholinesterase-containing fibers by target tissue in cocultured brain slices. *Neuroscience* 13, 681–689.

Gähwiler B. H., Mamoon A. M., and Tobias C. A. (1973) Spontaneous bioelectric activity of cultured cerebellar purkinje cells during exposure to agents which prevent synaptic transmission. *Brain Res.* 53, 71–79.

Gähwiler B. H., Marbach P., and Baertschi A. J. (1984) The hypothalamo-hypophyseal system in culture in *Neuronal Comminications* (Meyer B. J. and Kramer S., eds.), A. A. Balkema, Rotterdam, pp. 145-154.

Gähwiler B. H., Sandoz P., and Dreifuss J. J. (1978) Neurones with synchronous bursting discharges in organ cultures of the hypothalamic supraoptic nucleus area. *Brain Res.* 151, 245–253.

Gainer H., Castel M., and Wray S., (1989) Immunocytochemical analysis of the suprachiasmatic nucleus in organotypic cultures. *Soc. Neurosci. Abstr.* 15, 431.9.

Geller H. (1981a) Histamine actions on activity of cultured hypothalamic neurons: evidence for mediation by H1 and H2-histamine receptors. *Dev. Brain Res.* 1, 89-101.

Geller H. (1981b) Electrophysiological pharmacology of hypothalamic neurons in explant tissue culture, in *Advances in Physiological Sciences, vol. 14.* (Stark E., Makara G. B., Haluze B., and Rappay G. Y. eds.), Akademiai Kiado, Budapest, pp. 107-111.

Geller H. M. and Woodward D. J. (1979) Synaptic organization of tuberal hypothalamus in tissue culture: Effects of electrical stimulation and blockers of synaptic transmission. *Exp. Neurol.* 64, 535–552.

Hamill O. P., Marty A., Neher E., Sakmann B., and Sigworth F. J. (1981) Improved patch-clamp techniques for high-resolution current recording from cells and cell-free membrane patches. *Pflügers Arch.* 391, 85–100.

Hösli E. and Hösli L. (1990) Autoradiographic localization of binding sites for neurotransmitters in explant culutres of rat central nervous system, in *A Dissection and Tissue Culture Manual of the Nervous System.* (Shahar A., de Vellis J., Vernadakis A., and Haber B., eds.), Liss, New York, pp. 343–345.

Honig M. G. and Hume R. I. (1986) Fluorescent carbocyanine dyes allow living neurons of identified origin to be studied in long-term cultures. *J. Cell Biol.* 103, 171–187.

Katz L. C., Burkhalter A., and Dreyer W. J. (1984) Fluorescent latex microspheres as a retrograde neuronal marker for *in vivo* and *in vitro* studies of visual cortex. *Nature* 310, 498–500.

Lasnitzki I. (1986) Organ Culture, in *Animal Cell Culture* (Freshney R.I., ed), IRL Press, Oxford, pp. 149–182.

Lerea L. S. and McCarthy K. D. (1990) Analysis of receptor expression on cultured cells via combined receptor autoradiography and immunocytochemistry, in *A Dissection and Tissue Culture Manual of the Nervous System*. (Shahar A., de Vellis J., Vernadakis A., and Haber B. eds.), Liss, New York, pp. 346–350.

Llano I., Marty A., Johnson J. W., Ascher P., and Gähwiler B. H. (1988) Patch-clamp recording of amino acid-activated responses in "organotypic" slice cultures. *Proc. Natl. Acad. Sci. USA* **85**, 3221–3225.

Masurovsky E. B., Benitez H. H., and Murray M. R. (1971) Synaptic development in long-term organized cultures of murine hypothalamus. *J. Comp. Neurol.* **143**, 263–278.

Maurer H. R. (1986) Towards chemically-defined, serum-free media for mammalian cell culture, in *Animal Cell Culture: A Practical Approach* (Freshney R.I., ed.), IRL Press, Oxford, pp. 13–32.

Morris M., Eskay R. L., and Sundberg D. K. (1986) A tissue culture model for the study of peptide synthesis and secretion from microdissected hypothalamic explants. *Methods Enzymol.* **124**, 359–371.

Paul J. (1975) *Cell and Tissue Culture*. Churchill Livingstone, New York, NY.

Romijn H. J., de Jong B. M., and Ruijter J. M. (1988) A procedure for culturing rat neocortex explants in a serum-free nutrient medium. *J. Neurosci. Methods* **23**, 75–83.

Sobkowicz H. M., Bleier R., and Monzain R. (1974) Cell survival and architectonic differentiation of the hypothalamic mamillary region of the newborn mouse in culture. *J. Comp. Neurol.* **155**, 355–376.

Sofroniew M. V., Dreifuss J. J., and Gähwiler B. H. (1988) Slice explants of rat hypothalamus examined by immunohistochemical staining for neurohypophyseal peptides and GFAP. *Brain Res. Bull.* **20**, 669–674.

Stern J. E., Mitchell T., Herzberg V. L., and North W. G. (1986) Secretion of vasopressin, oxytocin, and two neurophysins from rat hypothalamo-neurohypophyseal explants in organ culture. *Neuroendocrinology* **43**, 252–258.

Toran-Allerand C. D. (1978) Culture of hypothalamic neurons: Organotypic culture. *Biologie Cellulaire des Processes Neurosecretoires Hypothalamiques*, Coll. Internation. du CNRS. **280**, 759–776.

Toran-Allerand C. D. (1990) Long-term Organotypic Culture of the CNS in Maximow Assemblies, in *Methods in Neuroscience, vol. 2, Cell Cultures* (Conn P. M., ed.), Academic Press, New York. (in press)

Wray S. and Hoffman G. (1986) A developmental study of the quantitative distribution of LHRH neurons within the central nervous system of postnatal male and female rats. *J. Comp. Neurol.* **252**, 522–531.

Wray S., Gähwiler B. H., and Gainer H. (1988) Slice cultures of LHRH neurons in the presence and absence of brainstem and pituitary. *Peptides* **9**, 151–1175.

Wray S., Zoeller R. T., and Gainer H. (1989a) Differential effects of estrogen on luteinizing hormone-releasing hormone gene expression in slice explant cultures prepared from specific rat forebrain regions. *Mol. Endocrinol.* **3**, 1197–1206.

Wray S., Grant P., and Gainer H. (1989b) Evidence that cells expressing luteinizing hormone releasing hormone mRNA in the mouse are derived from progenitor cells in the olfactory placode. *Proc. Nat. Acad.
Sci. USA* **86,** 8132–8136.

Wray S., Kusano K., and Gainer H. (1991) Maintenance of LHRH and Oxytocin Neurons in slice explants cultured in serum-free media; effects
of tetrodotoxin on gene expression. *Neuroendo.* **54,** 327–339.

Zimmer J. and Gähwiler B. H. (1984) Cellular and connective organization
of slice cultures of the rat hippocampus and fascia dentata. *J. Comp.
Neurol.* **228,** 432–446.

Zimmer J. and Gähwiler B. H. (1987) Growth of hippocampal mossy fibers:
a lesion and coculture study of organotypic slice cultures. *J. Comp.
Neurol.* **264,** 1–13.

Zoeller R. T., Seeburg P. H., and Young W. S., III. (1988) *In situ* hybridization
histochemistry for messenger ribonucleic acid (mRNA) encoding
gonadotropin-releasing hormone (GnRH): effect of estrogen on cellular levels of GnRH mRNA in female rat brain. *Endocrinology* **122,**
2570–2577.

Neurons

Jeffrey R. Buchhalter and Marc A. Dichter

1. Introduction

There are several preparations that can be utilized to study the development of the vertebrate nervous system and the physiology and pharmacology of individual vertebrate neurons. These include both in vivo and in vitro preparations. The latter include brain slices (both conventional "thick" and new "thin" slices) as well as organotypic central nervous system (CNS) cultures and primary dissociated cell cultures. Each of these preparations can be appropriately utilized for particular kinds of studies and no single one of these is best for all neurobiological studies. This chapter will focus on the system of primary dissociated cell cultures of different parts of the mammalian CNS. These preparations have become more and more popular for studying issues related to the growth and development of the nervous system, the expression of neuron specific properties, and cellular physiology and pharmacology of mammalian CNS function.

Dissociated cell cultures are derived from specific regions in the fetal or postnatal CNS and can be prepared from almost any part of the brain. This preparation allows cells to develop and differentiate in a relatively controlled, although nonphysiologic environment. Factors that regulate survival, growth, and differentiation of neurons can be examined. In addition, the relative influence of genetic vs environmental factors on the growth, differentiation, and ultimate phenotype of the cells can be examined. As the cells mature in culture, regionally specific

From: *Neuromethods, Vol. 23: Practical Cell Culture Techniques*
Eds: A. Boulton, G. Baker, and W. Walz ©1992 The Humana Press Inc.

features develop that can be examined in detail both at the individual cell and at the small network levels. Using a variety of analytic techniques, these preparations have proven to be particularly valuable for the study of trophic influences on growth and differentiation, regionally specific expression of neuronal phenotypes, the detailed biophysical analysis of membrane channels, and the physiology and pharmacology of synaptic connections that form between neurons. Dissociated cell cultures, however, do suffer from several significant handicaps. The neurons are grown in an artificial environment, and it is virtually impossible to reproduce all of the factors that may be necessary for "normal" growth and differentiation in vivo. In addition, the dissociation procedure disrupts normal cell-to-cell contacts and may result in aberrant intercellular communication. The dissociation and culturing procedures themselves, may be somewhat harsh and could contribute to decreased neuronal viability and an inhibited differentiation. For physiological studies, these preparations do not maintain the normal anatomic relationships found in the intact CNS or in slices removed from the intact CNS; neurons cannot easily be identified by criteria that are considered routine *in situ*. For example, it is currently impossible to determine that a neuron was a hippocampal CAl pyramidal cell or dentate gyrus granule cell unless that subregion was selectively dissociated. Alternatively, it is possible to determine that a neuron with a pyramidal morphology may or may not contain a neurotransmitter or neuropeptide by immunohistochemical staining after an acute recording.

2. General Considerations for Producing Primary Cultures of Mammalian CNS

CNS neurons can be grown in a variety of configurations (*see below*), and the specific kind of culture used should be dependent on the nature of the experimental question being asked. For example, electrophysiological studies are facilitated by a small number of well-dispersed, well-developed neurons that can be individually examined. For analyses of synaptic interactions, it is appropriate to have neurons at a density high enough to insure the formation of synapses, but sufficiently low so that the synaptic interactions are not too complex. The analyses of neurotransmitter uptake or release, or other types of biochemical studies, are facilitated by larger numbers of neurons to produce a high signal-to-noise ratio. It must be appreciated, however, that the neurons from the same CNS region may have different properties under different culture conditions.

Throughout this chapter, different aspects of tissue culture techniques will be discussed as individual topics. However, it is clear that significant interactions among the different techniques can occur. For example, the density of plating will influence the substrate used for coating the growing surface, and the number of glia in the culture system will influence the allowable concentration of serum and excitatory amino acids in the medium (Rosenberg and Aizenman, 1989).

The successful growth of central neurons is highly dependent upon the presence of glia and/or glia-derived factors. Thus, in general, it is much more difficult to produce predominantly neuronal cultures with prolonged survival than to grow neurons together with glial cells.

2.1. Cultures of Different CNS Regions

Many regions of the mammalian CNS have been successfully cultured (Shahar et al., 1989). The ease with which an area can be isolated from fetal brain is a function of its size, clarity of anatomic boundaries, and location. For example, the cerebral cortex is a large structure that is clearly delineated and easily accessible on the surface of the removed brain. The hippocampus is much smaller and requires careful exposure in the dorsal-caudal aspect of the exposed hemisphere, but it has clearly delimited borders under the dissecting microscope (Peacock et al., 1979b). Brainstem nuclei are very small regions that require transverse cuts to be made above and below the region of interest followed by removal of the area with a punch biopsy technique (Buse, 1985). The latter strategy can also be applied to obtain neurons from a subfield of a structure that was removed en bloc. Thus, the CA1 and CA3 regions of hippocampus, dentate gyrus, or the ventral horn of the spinal cord can be isolated (Kay and Wong, 1986). However, this technique is most easily applied to postnatal tissue, which is usually not satisfactory for long-term cultures.

Another means of isolating specific neuron subgroups from a region is to take advantage of selective growth conditions. For example, cultures can be enriched for cerebellar Purkinje cells by coating coverslips with anti-Thy-1 antibody and using postnatal day 1 animals (Messer et al., 1984); whereas coverslips coated with polylysine alone and use of postnatal day 7 pups are optimal for granule cells (Meier et al., 1984).

Each area of the CNS appears to have a "critical" period for removal to optimize survival. It is generally believed that this period coincides with the migration of neuroblasts from the germinal matrix zone into the structure of interest and comes before terminal mitosis of the neurons in question. It had previously been thought that neurons do not divide in culture. However, it has been demonstrated that immature neuroblasts removed from various central structures can continue to divide in culture and then differentiate to form mature neurons (Kriegstein and Dichter, 1983).

Neurons from many CNS regions, including spinal cord, hypothalamus, striatum, hippocampus, and neocortex can be removed from embryonic brains and survive for 4–8 wk or longer. In general, however, it has not been possible to maintain postnatal neurons in primary dissociated cell culture for periods beyond several days. It is possible that improvements in tissue culture techniques will allow the long-term survival of such neurons in the future. Table 1 shows the times indicated in the literature as optimal for various parts of the CNS.

2.2. Methods of Dissociation

Neurons can be dissociated by enzymatic and/or mechanical techniques. Once the tissue is dissected, it can be cut into small pieces, incubated in a calcium-free buffer solution, and vigorously triturated in order to produce a suspension of isolated cells. This procedure can produce many cells from a given region but usually produces a suspension with a relatively low viability. However, those viable neurons can be cultured successfully. More commonly, enzyme treatment for 15–120 min followed by trituration with Pasteur pipets is employed for cell dissociation. By use of these techniques, suspensions of cells with greater than 90% viability (as assessed by trypan blue exclusion) can be obtained. The choice of enzymes and number of triturations is determined empirically. Trypsin in concentrations of 0.025–0.25% has been the most commonly used enzyme. Other enzymes used include: papain, collagenase, hyaluronidase, and DNAase. The number of triturations required is related to the diameter of the pipet tip, which is adjusted by fire polishing. The possibility that an enzyme used for dissociation may effect one or more ion channels should be considered, particularly in experiments in which neurons are used within several hours of dissociation (Kay and Wong, 1986). This concern may be less important with long-term cultures, although it is possible that once proteins are removed from cell membranes by trypsin, they will not be regenerated.

2.3. Substrate

The choice of the material onto which the neurons will be plated is largely determined by the intended application. Small, round, glass cover slips are often used when neurons are desired for electrophysiological, morphological, developmental, and immunohistochemical studies. Large, plastic Petri dishes or flasks are used for pharmacological and neurochemical projects.

The composition of the two-dimensional surface onto which the neurons are plated is of different relevance depending on the type of application. In the most common application, neurons are grown for the study of electrophysiological, morphological, or biochemical properties. Thus, whatever substrate produces the optimal density of neurons with the least difficulty and expense is chosen. Other studies are concerned with questions of the importance of growth substrates such as extracellular matrix proteins on the promotion of survival and process elongation, and these studies will obviously need to focus more directly on the substrates used for growth.

Table 1
Regional CNS Cultures*

Region	Species	Age	Substrate	Serum	Special features	References
Spinal cord	Mouse	E12–14	Collagen	FCS-10 HS-10 HS-5, NuSer	Motor neurons	Ransom et al., 1977 De Deyn and Macdonald, 1987 Smith, 1989
	Rat	E13–15	Glia	RS-4, CEE 1	Serotonin neurons	Yamamoto et al., 1981
Brainstem	Mouse	E13–15	Coll., fibroblasts PL, Coll. gel	RS-7.5		
Brainstem (and cerebellum)	Rat	E19–20	PL	SF		Ahmed et al., 1983
Locus ceruleus	Mouse	P1–5	Fibronectin Glia	FBS-10 HS-10	Noradrenergic neurons	Masuko et al., 1986
Mesencephalon	Mouse	E13	Collagen	SF	DA and GABA neurons	Dal Toso et al., 1988
Hypothalamus	Rat	E16–18	PL	FCS-10	Peptide hormone secretion	Whatley et al., 1981
Striatum	Rat	P1–5, P4–5	Collagen	FCS-25		Wilkinson et al., 1974
	Rat	P1–2	Glia	NuSer-10		Dubinsky, 1989
	Rat	E-19	Glia	HS-2		Misgeld and Dietzel, 1989

Region	Species	Age	Substrate	Medium	Application	Reference
Cerebellum, Purkinje cells	Mouse	P1	PL + Thy-1	HS, then SF	Purkinje cells enriched	Messer et al., 1984
Cerebellum, granule cells	Mouse, Rat	P7	PL	FCS or HS	Granule cells enriched	Meier et al., 1984
Hippocampus	Rat	E18-19	PL	Hplac-5, CEE-2	Morphology	Banker and Cowen, 1977
	Mouse	E13-18	Collagen	FCS+HS-10		Peacock et al., 1979
	Rat	E18	PL	FCS then SF	Process development	Mattson, 1988
	Rat	E19-20	PL	FCS-10	Morpho-Physiol correlation	Buchhalter and Dichter, 1991
					GABA and Som Develop.	Legido et al., 1990
Neocortex	Rat	E15	Feeder layer	RS-10	Morphol Physiol	Dichter, 1978
			Collagen and PL		Neurotransmitters	Snodgrass et al., 1980; Borg et al., 1985

*Partial listing of references reporting cultures derived from various regions of mammalian CNS. E = embryonic day, P = postnatal day, FCS = fetal calf serum, HS = horse serum, RS = rat serum, CEE = chick embryo extract, SF = serum free, NS = NuSerum, PL = polylysine, and Coll = collagen.

In order to obtain an optimal density of neurons in dissociated cell culture, the neurons (and glia) must stick to the substrate and differentiate. One prognostic sign suggesting a healthy culture is that the neurons are separate from one another immediately after plating rather than being present in aggregated balls. The latter are present when the cells adhere to each other during dissociation or following plating, owing to failure to adhere to the substrate or to too high a plating density. It should be noted that neurons have a tendency to reaggregate even after attaching to a substrate, presumably by cell movement along the substrate. Although neuronal migration may be healthy under some circumstances, it is clearly detrimental for cultures intended for electrophysiological experiments to be made up of a series of reaggregate balls of cells with stringing processes between them.

A variety of substrates has been utilized to grow primary cultures of the CNS. Uncoated tissue culture plastic is adequate to support the attachment and growth of astrocytes. However, neurons require a more complex substrate. Collagen preparations were among the first to be used to coat tissue culture plastic or glass cover slip surfaces. More recently, several polyamine compounds (polylysine, polyornithine) have been added to the collagen coating or are used alone. The optimal coating has been traditionally determined in each laboratory for each preparation.

A substrate frequently used to support a low-density plating of neurons *(see below)* is a glial feeder layer *(see* Table 1). In this technique, cell suspensions derived from postnatal neocortex or the region from which the neurons will be derived are plated onto plastic or glass cover slips with the knowledge that mainly astrocytes will survive and form a confluent layer within 7–10 d, depending on the plating density. Once confluent, the astrocytes are inhibited from further cell division. Freshly dissociated neurons are then plated onto this glial substrate. In this manner, neurons make immediate contact with a glial surface similar to that which develops during the first week in culture in higher-density neuron and glia mixed cultures. Studies examining different types of substrates have provided interesting, if not idiosyncratic, results. For example, survival and development of mouse dorsal root ganglion (DRG) neurons was better with a fibronectin, rather than a polylysine substrate (Horie and Kim, 1984). Yet, in another study, the combination of laminin and fibronectin was considered to be most effective (Orr and Smith, 1988). However, laminin, fibronectin, collagen, collagen-polylysine, and polylysine produced the same survival of human DRG (Yong et al., 1988). In a study of hippocampal process outgrowth (Nakayma et al., 1989), basal lamina produced greater outgrowth than a substrate consisting of collagen plus laminin. Thus, choice of substrate has clear implications for survival and process outgrowth.

Another critical, although infrequently discussed, aspect of the substrate is the need for the culture substrate to be preincubated with serum containing media for minutes to hours prior to plating. In our experience, this is a required step for the adherence of embryonic rat hippocampal and neocortical cells. It is postulated that crucial proteins precipitate from the serum onto the substrate.

Thus, the nature of the surface onto which neurons and glia are plated is complex. In practical terms, this will be determined by the number of neurons and duration of survival desired, in addition to the need for having access to isolated neurons and processes. Serum present in the growth media probably interacts with the substrate.

2.4. Media

Almost everyone who has developed a cell culture system for a CNS region has developed a unique culture medium. In fact, some laboratories have used several culture media in an attempt to optimize their cultures. These consist of a salt solution supplemented by a variety of amino acids, vitamins, trace nutrients, and some form of serum preparation. In general, it would be preferable to grow cultures in the absence of biological fluids, such as serum, since the precise contents are impossible to control; but maintaining central neurons for extended periods (greater than 2 wk) in culture has been exceedingly difficult in the absence of some form of serum (*see* chapter by Bottenstein, this vol.). Given the variety of media that are utilized in different laboratories, it is difficult to say that any one is superior to another, except under individual circumstances. As it is impossible to assess all forms of neuronal function for a given group of cells in culture (e.g., membrane properties, neurotransmitter development, synapse formation, peptide production, and so on), it is not possible to know whether even the media recommended by individual laboratories is optimal for the given specific cell type. Laboratories will usually optimize conditions based on the experiments being performed with the cultures.

The use of serum remains a complex issue for all laboratories culturing central neurons. In general, fetal serum is preferable to adult serum, but this has not always been the case. In addition, neurons are grown in serum from a variety of species, commonly including calf (Wilkinson et al., 1974), horse (Dichter, 1978; Misgeld and Dietzel, 1989; De Deyn and MacDonald, 1987), rat (Smith et al., 1989), and human (Rothman, 1983; Banker and Cowan, 1979). It is also clear that some batches of serum, even those supposedly derived from the same kind of source, are toxic, whereas other batches are permissive. Attempts have been made to fractionate serum to find components that are particularly beneficial and others that are particularly harmful (Kaufman and Barrett, 1983), but these have met with only limited success and are not generally in use. Each laboratory experiments with sera from several sources and settles on a certain type and concentration.

Following the initial plating, the medium is replaced at predetermined intervals. The optimal amount and frequency of medium exchange has not been rigorously studied. The frequency of exchange varies from every 3 d (Dichter, 1978) to not at all (Banker and Cowan, 1979; Bartlett and Banker, 1984). Furthermore, the medium can be completely replaced or only a portion exchanged. This is an important consideration for at least two reasons. The first is that the medium that has been in the presence of the culture is conditioned by the presence of neuron trophic factors secreted by glia. Second, fresh medium may contain factors (e.g., amino acids) that are toxic to neurons and such toxicity may be influenced by the number of glia present and the natural development of some receptors, e.g., N-methyl-D-aspartic acid (NMDA) receptors, as the neurons mature

(Rosenberg and Aizenman, 1989). Although not usually considered as independent variables, the frequency and extent of medium replacement may have a significant effect on neuron survival.

2.5. Special Conditions

2.5.1. Glia

As alluded to above, the presence of glia (predominantly astrocytes) is a critical component in the growth of mammalian CNS neurons. Glia contribute to the nutritive microenvironment by producing trophic factors that are required for neuron maintenance and development. In cultures that contain large numbers of glia, no supplements are necessary. However, studies that require analysis of neuron process growth and morphology are facilitated by culture conditions that produce neurons growing in isolation without a glial layer, which can obscure neuronal processes contained therein. The number of glia in the dissociated cell culture can be reduced by low plating density (Banker and Cowan, 1977; Banker and Cowan, 1979; Mattson et al., 1989), addition of mitotic inhibitors (Dichter, 1978; Orr and Smith, 1988), or serum-free media (Bottenstein, 1983). The trophic factors produced by glia can be replaced in a variety of ways, including: placing an explant of tissue or cover slips with confluent glia in the dish, use of media that have been present in glia containing cultures (so-called "conditioned media") (Banker and Cowan, 1977; Banker and Cowan, 1979; Banker, 1980), or placing a cover slip containing glia face down, and making contact with the media covering the cover slip containing the neurons to be studied (Caceres et al., 1986; Dotti et al., 1988). Although all of these techniques have been used with success to support neuron growth, it should be noted that survival is significantly reduced when compared to mixed cultures containing large numbers of glia in addition to the desired neurons.

Another less appreciated function of glia in dissociated cell cultures is that of detoxification. It has recently been demonstrated that the excitotoxicity of glutamate via the NMDA receptor is reduced two log units by the presence of a confluent glial layer (Rosenberg and Aizenman, 1989). Thus, the percentage of serum and other sources of amino acids in the neuron growth media may need to be adjusted according to the glial content of the culture. The detoxification of other compounds by glia remains to be determined.

2.5.2. Inhibitors of Mitosis

Although glia are likely to perform important trophic and detoxification functions, it is necessary to inhibit glial growth once a confluent sheath has formed covering the substrate. If this is not done, the glia will overgrow and eliminate the neurons. A variety of agents have been used to inhibit glial division, including ARA-C, ARA-A, IUBR, uridine, fluorodeoxyuridine (FUdR), and radiation-induced inhibition of mitosis. The choice of inhibitor has been arbitrary given the lack of studies examining the effects of different inhibitors on any aspect of neuronal function. Recently, however, it has been demonstrated that ARA-C directly inhibits neuron survival by inhibition of deoxycytidine-dependent processes (Martin et al., 1990). Thus, it is conceivable that different inhibitors may have direct effects on neurons that could affect the experimental question under consideration.

2.5.3. Elevated Extracellular Potassium

The possibility that inhibition of neural activity reduces survival has received experimental support in tissue culture studies in which action potential generation was blocked with a sodium channel blocker (tetrodotoxin) (Brenneman and Eiden, 1986; Brenneman et al., 1983,1984). Another way to perform this type of experiment is to enhance neural activity by elevated potassium-induced depolarization (Mattson and Kater, 1988b). An extracellular potassium concentration of 20 mM produced enhanced survival compared to a standard concentration of 5 mM, whereas 50–100 mM significantly reduced survival. The mechanism is unknown but may be related to release of trophic substances (neurotransmitters, neuromodulators).

2.5.4. Extracellular Volume

Several studies utilizing hippocampal neurons have demonstrated that reduced extracellular volume enhances neuron survival (Banker, 1980; Brewer et al., 1987; Mattson and Kater,

1988b). A recent study directly compared survival in two different extracellular volumes and found a significantly greater survival in the lower volume (Mattson and Kater, 1988b). Another technique to reduce the immediate extracellular space is to place a cover slip over the cover slip containing the hippocampal neurons (Mattson and Kater, 1988b; Brewer et al., 1987). The mechanism is unknown but may be related to trophic factors produced by neurons or glia that are concentrated in the reduced extracellular space.

2.5.5. Plating Density

The number of cells plated is determined by the purpose of the experiment. Thus, a culture containing many neurons is required for biochemical studies, whereas the presence of robust isolated neurons on a glial layer is useful for cellular electrophysiology. Fine morphological analysis is facilitated by having no overlap between the processes of adjacent neurons, with all processes exposed. Early studies of neuronal tissue culture indicated that there was a range of plating densities optimal for survival. Densities above or below this range appeared to fail, although for probably very different reasons. It should be noted that when the number of cells in a freshly dissociated cell suspension from embryonic CNS is counted in a hemocytometer, most of the viable (i.e., trypan blue excluding) cells appear round, without processes. Thus, the ultimate neuron or glial identity is unknown. As mentioned above, low-density cultures require supplementation with glia-derived factors and are more sensitive to exogenous and endogenous excitatory amino acids. High-density cultures require more vigorous use of antimitotic agents. Thus, the technical considerations interact and cannot be addressed as independent issues.

3. Specific Techniques
for Obtaining Dissociated Neurons
from Rat Hippocampus in Long-Term Culture

Our laboratory has an extensive experience culturing neocortex and hippocampus (Dichter, 1978; Legido et al., 1990; Buchhalter and Dichter, 1991). We will describe the techniques currently employed for one of these regions in detail, in order to use this as one model for examining the techniques of CNS culture. The technique of dissociated hippocampal cultures was originally developed by Banker and Cowan (1977) and was subsequently modified for a variety of purposes, including the analysis of morphology (Banker and Cowan, 1979; Peacock et al., 1979b; Kawaguchi and Hama, 1987), development (Banker and Cowan, 1977; Banker and Cowan, 1979; Peacock and Walker, 1983; Dotti et al., 1988), the effects of trophic factors (Mattson et al., 1988a,1989; Mattson and Kater, 1988b), and electrophysiology (Peacock, 1979a; Peacock and Walker, 1983; Segal, 1983; Rothman and Samaie, 1985b; Rogawski, 1986; Yaari et al., 1987; Bekkers and Stevens, 1989; Buchhalter and Dichter, 1991). We have used this system to study neuronal growth and differentiation in culture, the physiology of the hippocampal neurons in culture, the properties of the synapses they form with one another, and the pharmacology of both excitatory and inhibitory neurotransmitters and of neuromodulatory agents. Cultures were prepared for these applications by the technique presented below.

Embryos are rapidly removed on embryonic days 19 or 20 following carbon dioxide-induced narcosis and cervical dislocation of the pregnant Sprague-Dawley rat (Charles River). The embryos are decapitated, brains are carefully removed, and placed in warm N-2-hydroxyethylpiperazine N-2-ethane sulfonic acid (HEPES)-buffered saline solution (HBS). Under a low-power dissection microcope, the brain is stabilized with forceps, and each hemisphere is peeled off the diencephalon, thereby exposing the hippocampus. The borders of the hippocampus are clearly defined by the overlying meningeal vessels and the ventricle. Fine iridectomy scissors are used to cut out the hippocampus. The meninges are removed to decrease the number of fibroblasts in the culture. The hippocampi are incubated with trypsin (0. 027%, Sigma) in a 35°C, 95% air, 5% CO_2 incubator for 40 min. The trypsin solution is removed by several washes with warm Ca^{2+} and Mg^{2+}-free HBS. The hippocampi are placed in warm growth media composed of DMEM (Whittaker) with 25 mM HEPES supplemented by 10% fetal calf serum (Hyclone), 10% Hams F-12 (Whittaker) with L-glutamine, and 50 µL/mL penicillin-streptomycin (Sigma), and triturated with a sterile 9" Pasteur pipet 15 times. Viability is then assessed by mixing a 0. 1 cc aliquot of the cell suspension with 0.3 cc trypan blue. The total number of cells excluding (viable) and not excluding trypan (not viable) is counted with a hemocytometer and the viability calculated. The viability is routinely greater than 85%. Cells are plated at a density of 600,000 viable cells in a total volume of 1.5 cc into a 35 mm Petri dish containing five, 12 mm cover slips that have been coated with poly-L-lysine (Peninsula Laboratories). The dishes were preincubated with growth media for 48 h prior to plating. Medium is partially exchanged 3 times each wk by aspiration of approximately 0.5 cc and replacement with that amount of fresh medium. At 7–10 d, the glial background becomes confluent, and subsequent mitosis is inhibited with cytosine arabinoside (5 µM) for 48 h. Cultures prepared in this manner provide neurons that survive in vitro for 4–8 wk on a confluent layer of glia.

4. Morphological Development and Physiological Properties of Hippocampal Neurons Maintained in Dissociated Cell Culture

In our laboratory, we have used dissociated cultures of hippocampal neurons to study active and passive membrane properties, agonist-induced changes in membrane channels, peptide production, and colocalization with amino acid neurotransmitters, synaptic transmission and the correlation of neuronal morphology with physiology.

4.1. Morphology

The ability to identify cellular morphologies as neuronal by phase contrast microscopy is important in studies that evaluate the neurotoxicity of agents added to culture, as well as in those electrophysiological and immunohistochemical studies in which neurons are subdivided into morphological subtypes. In order to address this issue, we have stained neurons with the relatively specific marker, tetanus toxin, at 24 and 72 h, and 1 wk in culture. By 72 h, most tetanus toxin-positive cells have assumed morphologies that are clearly neuronal and are easily identifiable with phase contrast microscopy. At the time of the initial plating, most cells are spherical with an occasional process noted. By 24 h, cells have begun to differentiate into morphologic subtypes. This process continues with increase in neuron size and complexity of processes (Fig. 1). The morphologies noted are typical of the more mature cultures and are similar to morphologies seen in vivo. The cultures contain pyramidal, stellate-multipolar, and bipolar neurons (Fig. 2). These results are concordant with those of other studies (Banker and Cowan, 1977; Banker and Cowan, 1979).

4.2. Electrophysiology

One of the most frequent uses for neurons in culture is for the analysis of electrophysiological properties of mammalian central neurons, especially those properties of the neurons for which it is essential to study cells in isolation (Fig. 3). For example, it has been reported that hippocampal neurons when studied in the slice preparation, have electrophysiological characteristics that vary with cellular morphology (Lacaille and Schwartzkroin, 1988a,b; Schwartzkroin and Mathers, 1978). We examined the passive and active properties of neurons of pyramidal and a specific nonpyramidal morphology (stellate) in order to determine if such correlations could be noted when the neurons are grown in culture. No significant differences were found in the resting membrane potential and input resistance, action potential amplitude, rate of rise and fall, duration at half maximal amplitude, and adaptation to steady state depolarization between the two cell types (Buchhalter and Dichter, 1991). Thus, whatever caused the two cell types to develop different electrophysiological characteristics *in situ* did not appear to be active in vitro. This type of study is facilitated by the direct visual identification of neurons in cell culture.

Biophysical characterization of membrane currents in hippocampal neurons has also been effectively performed utilizing neurons maintained in culture. Properties of a variety of potassium and calcium currents have been studied (Segal, 1983; Segal

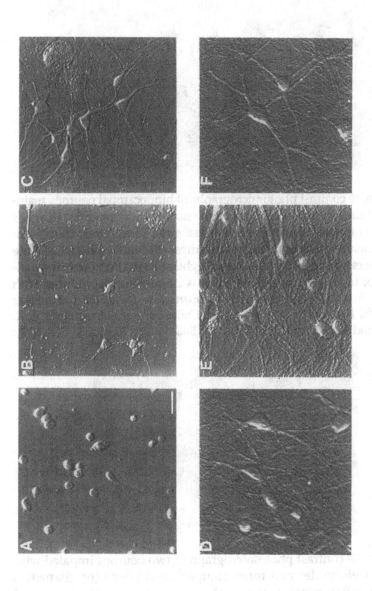

Fig. 1. Development of hippocampal neurons in culture. A-F: differential interference contrast photomicrographs of neurons at various times after plating 1 h, 24 h, 1 wk, 2 wk, 3 wk, and 7 wk, respectively. Calibration bar = 50 microns and applies to all photomicrographs.

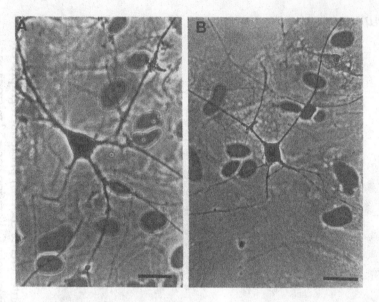

Fig. 2. Phase contrast photomicrographs of hippocampal neurons maintained in culture for 3 wk. The neurons were stained with a modified Wright's stain. (A) Pyramidal neuron—note the triangular soma, stout apical dendrite, and smaller diameter basilar dendrites. (B) Nonpyramidal, stellate neuron. Processes and background cells slightly blurred because of the shallow plane of focus. In A and B, scale bar is 20 microns. Reproduced with permission from *Electrophysiological Comparison of Pyramidal and Stellate Nonpyramidal Neurons in Dissociated Cell Culture of Rat Hippocampus*, J. R. Buchhalter and M. A. Dichter, *Brain Research Bulletin*, Vol. 26, pp. 333–338, 1991.

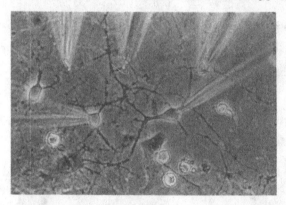

Fig. 3. Phase contrast photomicrograph of two neurons impaled with intracellular electrodes and three miniperfusion pipets (tip diameters approximately 5 microns) containing pharmacologic agents that are ejected by application of air pressure. Reproduced with permission from *Electrophysiological Comparison of Pyramidal and Stellate Nonpyramidal Neurons in Dissociated Cell Culture of Rat Hippocampus*, J. R. Buchhalter and M. A. Dichter, *Brain Research Bulletin*, Vol. 26, pp. 333–338, 1991.

and Barker, 1984; Rogawski, 1986; Yaari et al., 1987; Miller, 1987; Sah et al., 1988), and the effects of various neuromodulators and drugs on these currents have been determined (Trussell and Jackson, 1987; Wang et al., 1989,1990).

Hippocampal neurons in culture have also proven to be of extreme value for the analysis of excitatory and inhibitory transmission and the regulation of excitatory and inhibitory neurotransmitter receptors. Fast excitatory neurotransmission has been shown most likely to be caused by activation of a subclass of AMPA/quisqualate receptors by the neurotransmitter glutamate. This amino acid appears to activate a class of channels that open rapidly and with high probability after application of the agonist but then close (or inactivate) in the continued presence of the agonist. It is necessary for the agonist to be removed for these channels to be de-inactivated and ready for another set of openings (Tang et al., 1989). The behavior of AMPA/quisqualate-activated channels is distinctly different from those channels activated by NMDA. These channels open more slowly and with a lower probability after exposure to agonist. They continue to open throughout exposure to glutamate and do not appear to enter an inactivated state. These channels are also affected by extracellular Mg^{2+} in a voltage-dependent manner. At the normal resting potential, NMDA-induced currents are reduced or blocked by ambient concentrations of Mg^{2+} (Mayer and Westbrook, 1985). It has also been demonstrated that NMDA receptors on hippocampal neurons in culture have a binding site for glycine, occupation of which appears to decrease the desensitization in response to glutamate (Mayer et al., 1989). NMDA currents are also affected by pH. Increased extracellular H^+ concentrations suppress NMDA-induced current by decreasing the probability of channel opening (Tang et al., 1990). Basic conditions produce the opposite effects.

One of the most important and difficult issues to explore in the intact mammalian CNS is the cellular basis of synaptic transmission. The opportunity to stimulate and record from two synaptically connected neurons under direct visual inspection is one of the most powerful applications of dissociated hippocmpal cultures. Early studies from this laboratory indicated that inhibitory and excitatory postsynaptic potentials could be recorded with the intracellular technique beginning at approximately 1 wk (Frosch et al., 1983). By 3 wk in culture, monosynaptic connections were found between 60–90% of neurons pairs simultaneously impaled. Similar studies have been used to describe the kinetics and pharmacology of EPSPs of spinal cord neurons in culture (Nelson et al., 1986). In more recent studies, whole cell patch clamp recording has replaced the intracellular technique. This allows partial voltage control of the postsynaptic membrane and increased stability of recording neurons at younger ages. Simultaneous pre- and postsynaptic recording has been applied to hippocampal neurons to investigate presynaptic excitatory (Forsythe and Clements, 1990) and inhibitory (Harrison, 1990) agonist–receptor interactions. Finally, the synaptic basis of long-term potentiation has been investigated with this technique and initial work suggests a presynaptic locus (Bekkers and Stevens, 1990).

4.3. Peptides

Somatostatin is a neuropeptide that is believed to play a modulatory role in extra-hypothalamic CNS locations (Delfs and Dichter, 1983). In hippocampal cultures prepared as above, somatostatin could be demonstrated in hippocampal neurons as early as 3 d of age. Production increased from undetectable to over 4000 pg/mL in the media by 3–5 wk in vitro as measured by radioimmunoassay (Legido et al., 1990). Costaining studies that used immunohistochemical techniques, revealed that approx 10% of neurons in 3-wk-old cultures contained somatostatin and that 63% of neurons that were positive for somatostatin were also GABA immunoreactive (Fig. 4). No differences in colocalization were noted between neurons of pyrmidal, multipolar, and bipolar morphologies.

4.4. Toxic Factors

In addition to providing direct access to individual neurons, the dissociated cell culture technique allows complete control of the external milieu and thereby the opportunity to assess the effects and underlying mechanisms of

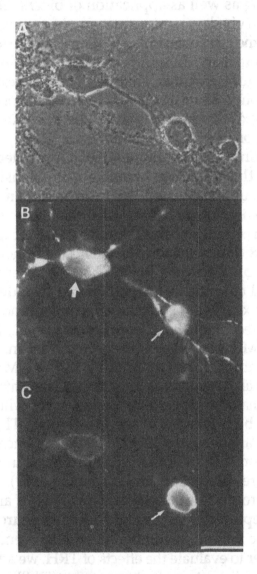

Fig. 4. Colocalization of SOM and GABA in individual neurons. (A) Phase contrast. (B) SOM-IR labelling of 2 neurons (thick and thin arrows). (C) GABA-IR is co-localized in 1 of the neurons (thin arrow); the other neuron does not stain for GABA. Calibration bar = 25 microns. Reproduced with permission from *Expression of Somatostatin and GABA Immunoreactivity in Cultures of Rat Hippocampus,* A. Legido, S. Reichlin, M. A. Dichter, and J. R. Buchhalter, *Peptides,* Vol. 11, pp. 103–109, 1990.

potential neurotoxins. The preparation allows application of cell counting as well as application of biochemical assays of neuronal survival.

One of the most outstanding examples can be found in studies of excitotoxicity. In addition to its important role as an excitatory neurotransmitter, glutamate has significant neurotoxic effects. Although the relationship between excitatory neurotransmitters and excitotoxic damage had been previously suggested, the application of excitatory amino acids (EAAs) to hippocampal neurons in culture (Rothman, 1985a) provided the first insight into the ionic mechanisms. An extensive series of experiments utilizing dissociated neocortical and hippocampal neurons in culture has revealed that the toxicity is owing to activation of both non-NMDA and NMDA receptors (Choi et al., 1987b,1988). An acute toxicity manifested by neuronal swelling caused by activation of non-NMDA receptors causing sodium and chloride influx is followed by a delayed, more profound toxicity that is calcium dependent and mediated by the NMDA receptor (Choi, 1987a; Rothman et al., 1987). The response to NMDA develops with age in culture and parallels the development of glutamate binding to the NMDA receptor (Peterson et al., 1989). In addition to these insights, the cell culture system has also suggested a mechanism by which glutamate levels may be regulated in the intact CNS. The toxicity of glutamate was attenuated 100-fold by the presence of abundant astrocytes compared to the toxicity produced in astrocyte-depleted cultures (Rosenberg and Aizenman, 1989).

Thyrotropin releasing hormone (TRH) is another hypothalamic peptide that is thought to have neurotrophic and antiepileptic actions, in addition to its role in endocrine function. In order to evaluate the effects of TRH, we added it to our cultures at various stages in development (Phillips et al., 1990). Much to our surprise, this peptide proved to be highly toxic to hippocampal neurons at early stages in vitro. Preliminary data suggest that this is not mediated by an NMDA receptor mechanism. Thus, the dissociated neuron model provides an excellent opportunity to explore toxicity studies during which the age of the cells and extracellular milieu can be exquisitely controlled and monitored by direct visual inspection.

4.5. Trophic Factors

The same advantages that facilitate toxicity studies are also applied to the study of factors that promote growth of neurons. This area is usually considered in determination of serum factors and has been greatly facilitated by use of serum-free media. This is considered in detail in the chapter by Bottenstein in this volume.

The ability to exmine the early events of process differentiation into axons and dendrites of single neurons through time represents another unique opportunity provided by dissociated neurons in culture. Using a low plating density of hippocampal neurons and videomicroscopy, it has been observed that processes all begin as short extensions from the soma (Dotti et al., 1988). Within 24 h, one begins to rapidly elongate and, in doing so, becomes the axon, the others becoming dendrites. The underlying mechanisms are unknown, but a system exists for further studies. Using the same type of system, it has been demonstrated that glutamate in nontoxic doses can selectively regulate the outgrowth of dendrites (Mattson et al., 1988a).

5. Conclusion

The above brief review demonstrates several points regarding the utility of neurons in dissociated cell culture. This technique allows unsurpassed access and visual identification of individual soma and processes of neurons that are not currently available by other techniques. Critical applications include membrane biophysics, agonist–receptor interactions, synaptic physiology, and process development. The other class of studies that is best served by this methodology is one that requires absolute control of the extracellular environment. These studies include the effects of toxic as well as trophic factors. The principal downside of dissociated tissue culture is the loss of the rich network of regional connections that characterizes the mammalian nervous system. Furthermore, all results must be interpreted with caveats accounting for growth in a two-dimensional environment and an extracellular milieu that are not identical to those found *in situ*. Given these limitations, neurons grown in dissociated cell culture have provided a wealth of information about the development and functioning of the mammalian CNS.

References

Ahmed Z., Walker P. S., and Fellows R. E. (1983) Properties of neurons from dissociated fetal rat brain serum-free culture. *J. Neurosci.* **3**, 2448–2462.

Banker G. A. (1980) Trophic interactions between astroglial cells and hippocampal neurons in culture. *Science* **209**, 809,810.

Banker G. A. and Cowan M. (1977) Rat hippocampal neurons in dispersed cell culture. *Brain Res.* **126**, 397–425.

Banker G. A. and Cowan W. M. (1979) Further observations of hippocampal neurons in dispersed cell culture. *J. Comp. Neurol.* **187**, 469–494.

Bartlett W. P. and Banker G. A. (1984) An electron microscopic study of the development of axons and dendrites by hippocampal neurons in culture. *J. Neurosci.* **4**, 1944–1953.

Bekkers J. M. and Stevens C. F. (1989) NMDA and non-NMDA receptors are co-localized at individual excitatory synapses in cultured rat hippocampus. *Nature* **341**, 230–233.

Bekkers J. M. and Stevens C. F. (1990) Presynaptic mechanism for long-term potentiation in the hippocampus. *Nature* **346**, 724–729.

Borg J., Spitz B., Hamel G., and Mark J. (1985) Selective culture of neurons from rat cerebral cortex: morphological characterization, glutmate uptake, and related enzymes during maturation in various culture media. *Dev. Br. Res.* **18**, 37–49.

Bottenstein J. E. (1983) Defined media for dissociated neural cultures, in *Current Methods in Cellular Neurobiology* (Barker J. L. and McKelvy J. F., eds.), Wiley, New York, pp. 107–130.

Brenneman D. E., Neale E. A., Habig W. H., Bowers L. M., and Nelson P. G. (1983) Developmental and neurochemical specificity of neuronal deficits produced by electrical impulse blockade in dissociated spinal cord cultures. *Dev. Br. Res.* **9**, 13–27.

Brenneman D. E., Fitzgerald S., and Nelson P. G. (1984) Interactions between trophic and electrical activity in spinal cord cultures. *Dev. Br. Res.* **15**, 211–217.

Brenneman D. E. and Eiden L. E. (1986) Vasoactive intestinal peptide and electrical activity influence neuronal survival. *Proc. Natl. Acad. Sci.* **83**, 1159–1162.

Brewer G. J., Peterson C., and Cotman C. W. (1987) Long term survival of rat hippocampal neurons at low density: advantages of low oxygen. *Soc. Neurosci. Abst.* **13**, 256.

Buchhalter J. R. and Dichter M. A. (1991) Electrophysiological comparison of pyramidal and stellate nonpyramidal neurons in dissociated cell culture of rat hippocampus. *Brain Res. Bull.* **26**, 333–338.

Buse E. (1985) A method for the collection of defined areas from the embryonic rat brain for cell and tissue culture. *J. Neurosci. Methods* **14**, 177–186.

Caceres A., Banker G. A., and Binder L. (1986) Immunocytochemical localization of tubulin and microtubule-associated protein 2 during the development of hippocampal neurons in culture. *J. Neurosci.* **6**, 714–722.

Choi D. W. (1987a) Ionic dependence of glutamate neurotoxicity. *J. Neurosci.* 7, 369–379.

Choi D. W., Maulucci-Gedde M., and Krigstein A. R. (1987b) Glutamate neurotoxicity in cortical cell culture. *J. Neurosci.* 7, 357–368.

Choi D. W., Koh J., and Peters S. (1988) Pharmacology of glutamate neurotoxicity in cortical cell culture: attenuation by NMDA antagonists. *J. Neurosci.* 8, 185–196.

Dal Toso R., Giorgi O., Soranzo C., Kirschner G., Ferrari G., Favaron M., Benvegnu D., Presti D., Vicini S., Toffano G., Azzone G. F., and Leon A. (1988) Development and survival of neurons in dissociated fetal mesencephalic serum-free culture: Effects of cell density and of an adult mammalian striatal-derived neuronotrophic factor (SDNF). *J. Neurosci.* 8(3), 733–745.

De Deyn P. P. and MacDonald R. L. (1987) CGS 9896 and ZK 91296, but not CGS 8216 and RO 15-1788, are pure benzodiazepine receptor antagonists on mouse neurons in culture. *J. Pharm. Exp. Therap.* 2421, 48–55.

Delfs J. and Dichter M. (1983) Effects of somatostatin on cortical neurons in culture. *J. Neurosci.* 3, 1176–188.

Dichter M. A. (1978) Rat cortical neurons in cell culture: culture methods, cell morphology, electrophysiology, and synapse formation. *Brain Res.* 149, 279–293.

Dotti C. G., Sullivan C. A., and Banker G. A. (1988) The establishment of polarity by hippocampal neurons in culture. *J. Neurosci.* 8, 1454–1468.

Dubinsky J. M. (1987) Development of inhibitory synapses among striatal neurons in vitro. *J. Neurosci.* 9, 3955–3965.

Forsythe I. D. and Clements J. D. (1990) Presynaptic glutamate receptors depress excitatory monosynaptic transmission between mouse hippocampal neurones. *J. Physiol.* 429, 1–16.

Frosch M., Barnes D., and Dichter M. (1983) Synapses between hippocampal neurons in culture. *Soc. Neurosci. Abst.* 9, 518.

Harrison N. L. (1990) On the presynaptic action of baclofen at inhibitory synapses between cultures rat hippocampal neurones. *J. Physiol.* 422, 433–446.

Horie S. and Kim S. U. (1984) Improved survival and differentiation of newborn and adult mouse neurons in F12–defined medium by fibronectin. *Brain Res.* 294, 178–181.

Kaufman L. and Barrett J. (1983) Serum factor supporting long-term survival of rat central neurons in culture. *Science* 220, 1394–1396.

Kawaguchi Y. S. and Hama K. (1987) Two types of nonpyramidal cells in the rat hippocampal formation identified by intracellular recording and HRP injection. *Brain Res.* 411, 190–195.

Kay A. R. and Wong R. K. S. (1986) Isolation of neurons suitable for patch clamping from adult mammalian central nervous system. *J. Neurosci. Methods* 16, 227–238.

Kriegstein A. and Dichter M. (1983) Morphological classification of rat cortical neurons in cell culture. *J. Neurosci.* 3, 1634–1647.

Lacaille J. C. and Schwartzkroin P. A. (1988a) Stratum lacunosum-moleculare interneurons of hippocampal CA1 region I: Intracellular response characteristics, synaptic responses, and morphology. *J. Neurosci.* **8**, 1400–1410.

Lacaille J. C. and Schwartzkroin P. A. (1988b) Stratum lacunosum-moleculare interneurons of hippocampal CAI region II: Intrasomatic and intradendritic recordings of local circuit synaptic interactions. *J. Neurosci.* **8**, 1411–1424.

Legido A., Reichlin S., Dichter M., and Buchhalter J. R. (1990) Expression of somatostatin and GABA immunoreactivity in cultures of rat hippocampus. *Peptides* **11**, 103–109.

Martin D. P., Wallace T. L., and Johnson E. M. (1990) Cytosine arabinoside kills postmitotic neurons in a fashion resembling trophic factor deprivation: Evidence that a deoxycytidine-dependent process may be required for nerve growth factor signal transduction. *J. Neurosci.* **10**, 184–193.

Masuko S., Nakajima Y., Nakajima S., and Yamaguchi K. (1986) Noradrenergic neurons from the locus ceruleus in dissociated cell culture: culture methods, morphology, and electrophysiology. *J. Neurosci.* **6**, 3229–3241.

Mattson M. P., Dou P., and Kater S. B. (1988a) Outgrowth-regulating actions of glutamate in isolated hippocampal pyramidal neurons. *J. Neurosci.* **8**, 2087–2100.

Mattson M. P. and Kater S. B. (1988b) Isolated hippocampal neurons in cyropreserved long-term cultures: Development of neuroarchitechture and sensitivity to NMDA. *Int. J. Dev. Neurosci.* **6**, 439–452.

Mattson M. P., Murrain M., Guthrie P. B., and Kater S. B. (1989) Fibroblast growth factor and glutamate: Opposing roles in the generation and degeneration of hippocampal neuroarchitechture. *J. Neurosci.* **9**, 3728–3733.

Mayer M. and Westbrook G. (1985) The action of N-methyl-D-aspartic acid on mouse spinal neurones in culture. *J. Physiol.* **361**, 65–90.

Mayer M. L., Vyklicky L., and Clements J. (1989) Regulation of NMDA receptor desensitization in mouse hippocampal neurons by glycine. *Nature* **338**, 425–427.

Meier E., Drejer J., and Schousboe A. (1984) GABA induces functionally active low-affinitive GABA receptors on cultured cerebellar granule cells. *J. Neurochem.* **43**, 1737–1744.

Messer A., Snodgrass G. L., and Maskin P. (1984) Enhanced survival of cultured cerebellar Purkinje cells by plating on antibody to *Thy-l. Cell. Mol. Neurobiol.* **4**, 285–290.

Miller R. J. (1987) Multiple calcium channels and neuronal function. *Science* **235**, 46–52.

Misgeld U. and Dietzel I. (1989) Synaptic potentials in the rat neostriatum in dissociated embryonic cell culture. *Brain Res.* **492**, 149–157.

Nakayama T., Sugiyama H., and Furuya S. (1989) Basal lamina enhances hippocampal neurite outgrowth in vitro. *Dev. Br. Res.* **49**, 145–149.

Nelson P. G., Pun R. Y. K., and Westbrook G. L. (1986) Synaptic excitation in cultures of mouse spinal cord neurons: Receptor pharmacology and behavior of synaptic currents. *J. Physiol.* **372,** 169–190.

Orr D. J. and Smith R. A. (1988) Neuronal maintenance and neurite extension of adult mouse neurones in nonneuronal cell-reduced cultures is dependent on substratum coating. *J. Cell Sci.* **91,** 555–561.

Peacock J. H. (1979a) Electrophysiology of dissociated hippocmpal cultures from fetal mice. *Brain Res.* **169,** 247–260.

Peacock J. H., Rush D. F., and Mathers L. H. (1979b) Morphology of dissociated hippocampal cultures from fetal mice. *Brain Res.* **169,** 231–246.

Peacock J. H. and Walker C. R. (1983) Development of calcium action potentials in mouse hippocampal cell cultures. *Dev. Brain Res.* **8,** 39–52.

Peterson C., Neal J. H., and Cotman C. W. (1989) Development of N-methyl-D-aspartate excitotoxicity in cultured hippocampal neurons. *Dev. Br. Res.* **48,** 187–195.

Phillips J., Buchhalter J. R., and Winokour A. (1990) Neurotoxic effects of thyrotropin releasing hormone on fetal rat hippocampal neurons. *Soc. Neurosci. Absts.* **16,** 518.

Rogawski M. A. (1986) Single voltage-dependent potassium channels in cultured rat hippocampal neurons. *J. Neurophys.* **56,** 481–493.

Rosenberg P. A. and Aizenman E. (1989) Hundred-fold increase in neuronal vulnerability to glutamate toxicity in astrocyte-poor cultures of rat cerebral cortex. *Neurosci. Lett.* **103,** 162–168.

Rothman S. M. (1983) Synaptic activity mediates death of hypoxic neurons. *Science* **220,** 536–537.

Rothman S. M. (1985a) The neurotoxicity of excitatory amino acids is produced by passive chloride influx. *J. Neurosci.* **5,** 1483–1489.

Rothman S. M. and Samaie M. (1985b) Physiology of excitatory synaptic transmission in cultures of dissociated rat hippocampus. *J. Neurophys.* **54,** 701–713.

Rothman S. M., Thurston J. H., and Hauhart R. E. (1987) Delayed neurotoxicity of excitatory amino acids in vitro. *Neuroscience* **22,** 471–480.

Sah P., Gibb A. J., and Gage P. W. (1988) Potassium current activated by depolarization of dissociated neurons from adult guinea pig hippocampus. *J. Gen. Physiol.* **92,** 263–278.

Schwartzkroin P. A. and Mathers L. H. (1978) Physiological and morphological identification of a nonpyramidal hippocampal cell type. *Brain Res.* **157,** 1–10.

Segal M. (1983) Rat hippocampal neurons in culture: Responses to electrical and chemical stimuli. *J. Neurophys.* **50–56,** 1249–1264.

Segal M. and Barker J. (1984) Rat hippocmpal neurons in culture: Potassium conductances. *J. Neurophys.* **51,** 1409–1433.

Shahar A., de Vellis J., Vernadakis A., and Haber B. (eds.) (1989) *A Dissection and Tissue Culture Manual of the Neruous System.* Liss, New York.

Smith S. M., Zorec R., and McBurney R. N. (1989) Conductance states activated by glycine and GABA in rat cultured spinal neurones. *J. Memb. Biol.* **108,** 45–52.

Snodgrass S. R., White W. F., and Dichter M. (1980) Biochemical correlates of GABA function in rat cortical neurons in culture. *Brain Res.* **190,** 123–138.

Tang C.-M., Dichter M., and Morad M. (1989) Quisqualate activates a rapidly inactivating high-conductance ionic channel in hippocampal neurons. *Science* **243,** 1474–1477.

Tang C.-M., Dichter M., and Morad M. (1990) Modulation of the N-methyl-D-aspartate channel by extracellular H+. *Proc. Natl. Acad. Sci.* **87,** 6445–6449.

Trussell L. O. and Jackson M. B. (1987) Dependence of an adenosine-activated potassium current on a GTP-binding protein in mammalian central neurons. *J. Neurosci.* **7,** 3306–3316.

Wang H.-L., Bogen C., Reisine T., and Dichter M. (1989) Somatostatin-14 and somatostatin-28 induce opposite effects on potassium currents in rat neocortical neurons. *Proc. Natl. Acad. Sci.* **86,** 9616–9620.

Wang H.-L., Dichter M. and Reisine T. (1990) Lack of cross desensitization of somatostatin-14 and somatostatin-28 receptors coupled to potassium channels in rat neocortical neurons. *Molec. Pharm.* **38,** 357–361.

Whately S. A., Hall C., and Lim L. (1981) Hypothalamic neurons in dissociated cell culture: The mechanism of increased survival times in the presence of nonneuronal cells. *J. Neurochem.* **36,** 2052–2056.

Wilkinson M., Gibson C. J., Bressler B. H., and Inman D. R. (1974) Hypothalamic neurons in dissociated cell culture. *Brain Res.* **82,** 129–138.

Yaari Y., Hamon B., and Lux H. D. (1987) Development of two types of calcium channels in cultured mammalian hippocampal neurons. *Science* **235,** 680–682.

Yamamoto M., Steinbusch H. W. M., and Jessell T. M. (1981) Differentiated properties of identified serotonin neurons in dissociated cultures of embryonic rat brain stem. *J. Cell. Biol.* **91,** 142–152.

Yong V. W., Horie H., and Kim S. U. (1988) Comparison of six different substrata on the plating efficiency, differentiation, and survival of human dorsal root ganglion neurones in culture. *Dev. Neurosci.* **10,** 222–230.

Astrocytes

Bernhard H. J. Juurlink and Leif Hertz

1. Why Culture Astrocytes?

1.1. What Is an Astrocyte?

The central nervous system (CNS) is characterized by an immense structural complexity with intimate associations of its constituent cell types: neurons and different types of glial cells, mainly oligodendrocytes and astrocytes. A century ago Andriezen (1893) described what are now called protoplasmic astrocytes found in grey matter and fibrous astrocytes found in white matter. As their name implies, astrocytes are generally process-bearing cells. A characteristic of all astrocytes, but particularly those in grey matter, is the enormous surface area owing to the extensive branching of the cellular processes; this process formation results in astrocytes being structurally highly complex. The cellular processes of fibrous astrocytes are generally fingerlike, whereas those of the protoplasmic astrocytes are variable in shape and closely follow the contours of adjacent cells. The protoplasmic astrocytic processes often end as thin sheets that surround synapses (*see* Narlieva, 1988). The protoplasmic astrocyte can be divided into a number of distinct morphological forms (Palay and Chan-Palay, 1974; Kosaka and Hama, 1986). In addition, the cellular processes of astrocytes impinge on small blood vessels as perivascular endfeet and astrocytic foot processes also terminate on the pia mater forming the *glia limitans* (Peters et al., 1976). Such intimate morphological relationships

From: *Neuromethods, Vol. 23: Practical Cell Culture Techniques*
Eds: A. Boulton, G. Baker, and W. Walz ©1992 The Humana Press Inc.

between astrocytes and the other cell types of the CNS have caused neurobiologists to speculate that there are also close metabolic and functional relationships among these cells. Hypotheses about glial functions abounded already at the turn of the century. It is remarkable how much of today's research on astrocytes is directed toward answering questions that were first asked at that time. Much of what seemed mysterious then has become clear, not because our generation is more intelligent than the ones before us, but because of the techniques and instruments available to us now (Somjen, 1988). Until the development of techniques to study astrocytes independently of other neural cell types, it was only possible to assess astrocyte functions indirectly by, e.g.,

1. Examining the morphology of the cells;
2. Using cytochemical approaches;
3. Using immunocytochemical approaches;
4. Using radioisotopic tracers to delineate the spatiotemporal formation of astrocytes and relating this to CNS total functional development; and
5. Using radioisotopic tracers to delimit metabolic compartments (whose cellular locations were unknown) *in situ* or in brain slices.

Because of the complex and intimate relationships of astrocytes to the other cellular elements, it was difficult, even with these techniques, to say much more about the functions of astrocytes than Penfield's (1932) "they support, they seem to nourish, they take part in repair following destructive lesions."

1.2. Why Culture Astrocytes?

Ramon y Cajal wrote in 1909 that even now (i.e., 1909) the functions of glial cells are unknown, and what is worse, will remain so for a long time to come because no methods exist to study these functions. This statement held true for exactly 50 years until Hydén (1959) developed techniques to microdissect samples of glial cells as well as individual neurons. Such preparations could be and were used successfully to investigate the functions of living glial cells and neurons. Since then, the ability to study living glial cells and neurons progressed rapidly with the development of gradient centrifugation methods to isolate various neural cell types from dissociated nervous tissue by Norton and coworkers (Norton and Poduslo, 1970) and development of cultures highly enriched in either neurons or astrocytes by Booher and Sensenbrenner (1972). This latter methodology has been extremely useful for the study of astrocytic functions since, with the development of tissue culture techniques, investigators had tools to dynamically investigate the properties of astrocytes and build on this knowledge to postulate likely functions of astrocytes *in situ*. Furthermore, knowledge gained about functions of astrocytes in vitro is being used to devise more profitable experiments *in situ*, resulting in a much better understanding of astrocyte functions in vivo.

1.3. Astrocytes in Culture

1.3.1. Brief History

Although Shein (1965) and Varon and Raiborn (1969) had already obtained cultures highly enriched in glial cells, the majority of astrocyte culture methodologies are variations of the techniques first described by Booher and Sensenbrenner (1972). These investigators were primarily interested in growing neurons, and dissociated human, rat, and chick brains obtained from several different stages of development using either a mechanical sieving technique or an enzymatic (trypsin) technique. The dissociated cells were planted into plastic tissue culture flasks. They observed that both neuronal and nonneuronal cells developed in the cultures established from embryonic brains, whereas cultures established from older brains contained only nonneuronal cells. Thus, a fundamental principle of neural cell culture was exposed: that cells with simple morphologies are most likely to survive the tissue dissociation procedures and hence, to survive and differentiate in culture. Virtually all of today's techniques for the preparation of homogenous astrocytic cultures are more or less modifications of the Booher-Sensenbrenner procedure. The history of the dispersion of this technique has been reviewed previously (Hertz et al., 1982).

To develop enriched cultures of astrocytes, it becomes a simple matter of choosing a time in the development of a particular brain region when neuronogenesis is completed and neurons are attaining a complex morphology, and also a time when there are large numbers of immature astroglial cells proliferating. This is discussed more extensively in Hertz et al. (1985a). For the rat or mouse neopallium, the appropriate time is at the newborn stage of development. The neopallium, which includes the neocortex plus underlying tissue, is often improperly referred to as neocortex; in the newborn mouse or rat, the neopallium consists of neocortex plus the underlying intermediate and subventricular zones (*see* Boulder Committee [1970] for terminology). The subventricular zone at this stage of development consists mainly of glial precursor cells that migrate into the overlying intermediate zone and cortex; the astrocytes that develop in cultures established from the neopallium arise mainly from these glial precursor cells of the subventricular zone (Juurlink et al., 1981a). The intermediate zone at the newborn stage of development is occupied by migrating glial precursor cells and axons; these axons will be myelinated and form the white matter of the neopallium.

1.3.2. What Types of Astrocytes Are Present in Culture?

It has been recognized for about a decade that primary astroglial cultures prepared from perinatal rodent brain contain two neuroglial populations (Fig. 1). One population consists of large, flat cells that exhibit glial fibrillary acidic protein (GFAP) (Hertz et al., 1982). Within the CNS, GFAP is an unequivocal marker for astrocytes (Eng et al., 1971; Bignami and Dahl, 1977; Schachner et al., 1977). These large, flat astrocytes can be induced to form processes by the addition of cAMP analogs to the culture medium (Moonen et al., 1975; Manthorpe et al., 1979; Sensenbrenner et al., 1980; Juurlink et al., 1981a). The second population consists of smaller cells that spontaneously form processes; in some laboratories these cells were found to be GFAP-positive (e.g., Manthorpe et al., 1979; Fedoroff et al., 1983; Juurlink and Hertz, 1985), and hence, astrocytic in nature, whereas in other

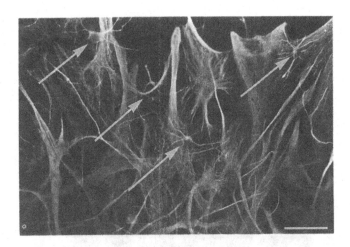

Fig. 1. Micrograph of 4-wk old rat astrocytes in primary culture established from the neopallia of newborn rat pups. Immunocytochemical localization of GFAP is demonstrated. Note that the astrocytes consist of smaller process-bearing type-2 astrocytes (arrows) and larger flatter type-1 astrocytes. Bar = 40 µm.

laboratories, they were found to be galactocerebroside- or 2', 3'-cyclic nucleotide 3'-phosphohydrolase-positive (e.g., McCarthy and de Vellis, 1980), and hence, oligodendrocytic in nature. These contradictory findings are now readily explained by the findings of Raff and coworkers (Raff et al., 1983a,b; Raff and Miller, 1984), who observed that in dissociated immature rat optic nerve, there are at least two populations of proliferative neuroglial cells: ganglioside-negative large, flat cells that give rise to large ganglioside-negative, GFAP-positive astrocytes, termed type-1 astrocytes, and a population of smaller ganglioside-positive process-bearing cells, termed O2A progenitor cells (e.g., Fig. 2). The GQ1c ganglioside present on the plasmalemma of rat O2A progenitor cells is recognized by the A2B5 monoclonal antibody (Eisenbarth et al., 1979). In addition, the O2A progenitor cells are characterized by the presence of chondroitin sulfate proteoglycan on the plasmalemma (Gallo et al., 1987; Stallcup and Beasley, 1987). Depending on culture environment, O2A progenitors in single-cell microculture can be induced to differentiate into either GQ1c ganglioside-positive, chondroitin sulfate

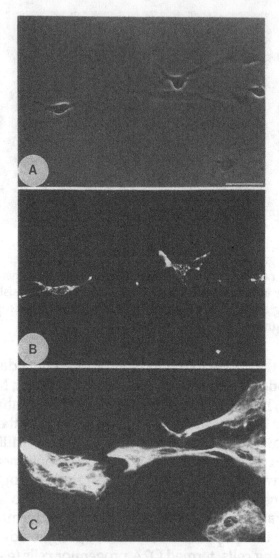

Fig. 2. Micrographs of 5-d old rat primary culture established from the neopallium of 4-d old rat pups. Bar = 40 µm. **(A)** Smaller process-bearing O2A progenitor cells can be seen associated with larger flat type-1 astrocytes. Phase contrast microscopy. **(B)** Same field as in A demonstrating that the localization of A2B5 binding is restricted to the small process-bearing cells (i.e., O2A progenitor cells). **(C)** Same field demonstrating that the localization of GFAP is restricted to the large flat cells (i.e., type-1 astrocytes) at this early stage of culture.

proteoglycan-positive, and GFAP-positive type-2 astrocytes, or into oligodendrocytes that express none of the aforementioned markers (Temple and Raff, 1985). Oligodendrocytes are recognized by the presence of galactocerebroside on the plasmalemma (Raff et al., 1978). Figure 3 illustrates the lineage relationships of these three glial cell types.

One factor that has been demonstrated to induce O2A progenitors to differentiate into type-2 astrocytes is ciliary neurotrophic factor; however, only 25–35% of the progenitors do so (Hughes et al., 1988). For optic nerve-derived O2A lineage cells, the combination of ciliary neurotrophic factor and extracellular matrix-associated molecules induce the majority of progenitors to differentiate into type-2 astrocytes (Lillien et al., 1990). O2A progenitors are not unique to the optic nerve cultures and have also been demonstrated in cultures established from cerebellar cortex (Levi et al., 1986; Trotter and Schachner, 1989) and cerebral cortex (Aloisi et al., 1988; Behar et al., 1988; Ingraham and McCarthy, 1989; Juurlink and Hertz, 1991). Although initially recognized in tissue culture, O2A progenitor cells as well as type-1 and type-2 astrocytes do not appear to be artifacts of the culture system since they have been demonstrated *in situ* (Miller and Raff, 1984; Miller et al., 1985).

1.3.3. Properties of Astrocytes

1.3.3.1. TYPE-1 ASTROCYTES

Some physiological functions of type-1 astrocytes seem to be quite generally accepted today (Hertz, 1989a,b,1990a,b,1991), maybe first and foremost

1. Involvement in potassium homeostasis at the cellular level;
2. Accumulation of glutamate and some other amino acid transmitters and utilization of part of the accumulated amino acids as metabolic fuel;
3. Expression of receptors for a multitude of neurotransmitters; and
4. The recently established role in calcium signaling.

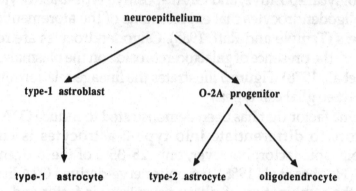

Fig. 3. Illustration of the development of two astroglial cell lineages. The lineage relationships between the O2A progenitor cells and type-1 astroblasts is not known.

It may be indicative of today's respectability of research on type-1 astrocytes that it lasted decades before their role in potassium homeostasis and transmitter function was accepted, whereas their contribution to calcium signaling in brain already is becoming well-established after less than 5 yr. Still more type-1 astrocytic functions may be unravelled during the coming decades, but at the same time, research on astrocytes seems to be entering a new phase in which more emphasis can be placed on the role of these cells in the function of the intact CNS, mainly through interactions with neurons, but to a minor extent also through interactions with other cells in the CNS.

Functional interactions between type-1 astrocytes and neurons obviously require that the two cell types can converse. The talk from neurons to astrocytes occurs to a large extent in the same manner as that between neurons and other neurons or neurons and effector cells (e.g., smooth muscle), i.e., by release of neurotransmitters, often from "varicosities" rather than from genuine synapses (Beaudet and Descarries, 1984). This mechanism of impulse transmission is much more diffuse than that of classical synaptic transmission. The extracellular potassium concentration, which is determined partly by neuronal release and partly by astrocytic redistribution, constitutes another language in which neurons can communicate with astrocytes. The question of communication in the opposite direction, i.e., from astrocytes to neurons, is less established and more controversial. This may occur by way of the extracellular potassium concentration, since even minor deviations from the physiological level of the potassium concentration in the extracellular fluid exert pronounced effects on neuronal excitability (Sykova, 1983; Walz and Hertz, 1983; Walz, 1989). Genuine transmitter release, i.e., a depolarization-induced release of a transmitter requiring the presence of calcium, does not appear to occur from type-1 astrocytes. However, these cells do exhibit a potassium-induced, noncalcium-dependent release of glutamate that can be expected to exert an action on neuronal glutamate receptors because of the close apposition between neurons and astrocytes (Schousboe et al., 1988); no corresponding release of γ-aminobutyric acid (GABA) occurs. Also, owing to the high potassium conductance of astrocytes that has been unequivocally demonstrated in vivo (Somjen, 1979), potassium ions move rapidly across astrocytic membranes in both directions as determined by the electrochemical potential. This, together with active transport of potassium (Walz and Hertz, 1983; Hertz, 1990a) and a wave of alterations in free intracellular calcium concentration evoked by local stimulation and spreading through adjacent astrocytes (Cornell-Bell et al., 1990), may provide type-1 astrocytes with the capability of altering neuronal excitability not only locally, but also over longer distances.

Type-1 astrocytes also give metabolic support to neurons. Neurons are metabolically handicapped cells (or metabolically highly specialized cells) since they exhibit little or no activity of two key enzymes: glutamine synthetase (Norenberg and Martinez-Hernandez, 1979) and pyruvate carboxylase (Yu et al., 1983; Shank et al., 1985). The first of these catalyzes synthesis of glutamine from glutamate, and the second, formation of oxaloacetate, a constituent of the tricarboxylic acid (Krebs') cycle, from glucose. These enzymes appear to be essential for *de novo* synthesis of precursors for glutamate and GABA in neurons, a process that is essential because of the astrocytic uptake of glutamate and GABA. Type-1 astrocytes have relatively high activities of both enzymes, and the inevitable conclusion is that

in order to maintain normal glutamatergic and GABAergic transmitter function, glutamine and/or α-ketoglutarate must travel
from astrocytes, where they are formed, to neurons, where they
are utilized (Hertz, 1989a,b,1991).

1.3.3.2. TYPE-2 ASTROCYTES

Until recently, it was only possible to study type-2 astrocytes in mixed glial cultures, and hence, only limited investigations could be performed on these cells. These studies have
demonstrated that type-2 astrocytes have a number of properties that traditionally have been considered to be neuronal and
not astrocytic in nature. For example, type-2 astrocytes avidly
take up GABA (Levi et al., 1983; Johnstone et al., 1986; Levi et al.,
1986; Reynolds and Herschkowitz, 1987; Barres et al., 1990), but
in contrast to type-1 astrocytes (Schousboe et al., 1977; Bardakjian
et al., 1979), they do not metabolize it readily; rather, type-2
astrocytes store GABA and release it upon stimulation with glutamate or glutamate analogs (Gallo et al., 1989). Similar to
GABAergic neurons and unlike type-1 astrocytes, type-2 astrocytes synthesize GABA, albeit possibly by a novel pathway from
putrescine (Barres et al., 1990). These observations suggest that
type-2 astrocytes may mainly affect neurons in an inhibitory fashion, whereas type-1 astrocytes may mainly affect neurons in an
excitatory fashion. Type-2 astrocytes not only display a "neuronal" channel phenotype (Barres et al., 1988,1990), but also a much
greater density of sodium channels than type-1 astrocytes
(Minturn et al., 1990). Furthermore, the majority of the sodium
channels in the type-2 astrocytes appear to be of the neuronal
variety rather than of the glial variety (Barres et al., 1988). The
significance of the presence of neuron-like channels in type-2
astrocytes is presently not clear.

Why is it that these cells are classified as astrocytes? The
primary reason is that their intermediate filaments consist of
GFAP (Raff et al., 1983a,b). The enzyme glutamine synthetase is
preferentially localized in the CNS to astrocytes (Norenberg and
Martinez-Hernandez, 1979; Mearow et al., 1989) and is also considered to constitute a good molecular marker of the astroglial
family of cells, although several recent reports have suggested
that oligodendrocytes also express some glutamine synthetase

activity (Warringa et al., 1988; Cammer, 1990; D'Amelio et al., 1990; Tansey et al., 1991). Similar to their in vivo counterparts, type-1 astrocytes in vitro exhibit high glutamine synthetase activity that is inducible by cortisol (Hallermayer et al., 1981; Juurlink et al., 1981b). Type-2 astrocytes in vitro also have an intense basal glutamine synthetase activity that is sevenfold greater than that of type-1 astrocytes and is also inducible by cortisol (Juurlink and Hertz, 1991). Thus, type-2 astrocytes appear to be *bona fide* astrocytes as determined by the above two quintessential astrocytic markers. The reason for this very high glutamine synthetase activity in type-2 astrocytes is not readily evident, but since glutamine serves as a nontoxic store of easily transferable nitrogen and of glutamate (Cooper et al., 1983), it suggests that type-2 astrocytes play a major role in nitrogen homeostasis and/or neurotransmitter inactivation.

1.3.3.3. WHAT ARE TYPE-1 AND TYPE-2 ASTROCYTES?

Type-1 astrocytes appear to be the most common type of astrocyte in the CNS and are also the astrocytes that undergo reactive gliosis during CNS injury (Miller et al., 1986). The majority of studies on astroglia in culture have been carried out on type-1 astrocytes, and the demonstrated properties, such as neurotransmitter inactivation (Drejer et al., 1982; Holopainen and Kontro, 1986), extracellular ion regulation (Walz, 1989; Hertz, 1990a), induction of blood-brain barrier properties in endothelium (Janzer and Raff, 1987), and the formation of *glia limitans*-like structures (Devon and Juurlink, 1988) suggest that these cells are involved in isolation and barrier functions. There is some evidence that type-2 astrocytes correspond to the perinodal astrocyte (ffrench-Constant and Raff, 1986a; Raff, 1989). The evidence is that both perinodal astrocytic processes and type-2 astrocytes express the J1 cell adhesion glycoprotein recognized by the HNK-1 and NSP-4 monoclonal antibodies. In addition, both the perinodal astrocyte and the type-2 astrocyte have high densities of Na^+ channels in their membranes (Minturn et al., 1990). How do the type-1 and type-2 cultured astrocytes relate to the classical grey matter protoplasmic and white matter fibrous astrocytes? Not that well, but it is likely that the traditional classification is an arbitrary one based solely on morphology;

it has been amply demonstrated, particularly in vitro, that the morphology of astrocytes is very responsive to external signals. Perhaps a better classification of astrocytes is in an astrocyte category having broad functions, including isolation and barrier functions, as well as participation in cellular interactions whose major consequence is ensuring the ability of neurons to efficiently receive and integrate information (i.e., type-1 astrocytes), and in a category (type-2 astrocytes) having the more restricted function of cooperating with oligodendrocytes to ensure that transmission of information occurs efficiently, and thus, is mainly concerned with the functioning of the nodal apparatus.

Astrocytes form approx 30% of the cellular vol of the cerebral cortex (Pope, 1978) and one would therefore expect that astrocytes in culture should have more than a threefold higher activity of astrocyte-specific enzymes than cerebral homogenates. The maximum hydrocortisone-induced activity of glutamine synthetase in type-1 astrocytes in culture is only about twice the activity of cerebral tissue (Juurlink et al., 1981b; Fages et al., 1988), whereas the maximal activity of type-2 astrocytes is twice that of type-1 astrocytes (Juurlink and Hertz, 1991). If the relative activities of type-1 and type-2 astrocytes are the same in vitro as in vivo, it would suggest that approx 25% of the astrocytes in the whole cerebrum are type-2 astrocytes.

2. Methodologies of Culturing Astrocytes

2.1. Introduction

We will first describe in detail highly purified mouse type-1 astrocyte cultures (i.e., >95% homogeneity). We will then describe procedures to obtain:

1. Rat astrocyte cultures;
2. Highly purified rat type-1 astrocyte cultures; and finally,
3. Highly purified rat type-2 astrocyte cultures.

Mouse type-1 astrocyte cultures are primary cultures, whereas rat astrocyte cultures of type-1 usually are and type-2 always are secondary cultures. For details of tissue culture terminology, see Fedoroff (1977). Cells taken directly from the organism and maintained for more than 24 h in vitro are considered to be primary cultures whereas cells taken from primary cultures and subcultured are referred to as secondary cultures, tertiary cultures, and so on, according to the number of times the cells have been subcultured.

2.2. Mouse Astrocytes

2.2.1. Standard Mouse Type-1 Astrocyte Cultures

2.2.1.1. INTRODUCTION

The method used to obtain cultures of mouse cerebral astrocytes in our laboratories is modified (Hertz et al., 1982; 1985a,b; 1989a) from the procedure first described by Booher and Sensenbrenner (1972). The main strategies used are:

1. Selection of nervous tissue where neuronogenesis is completed and that contains immature astroglial cells as the major proliferative cell population;
2. The use of vigorous mechanical tissue dissociation procedures that destroy the majority of neurons but allow the survival of small undifferentiated astroglial precursor cells;
3. The employment of a filtration step that sieves out meningeal remnants and blood vessels; and
4. The use of appropriate culture conditions resulting in the preferential survival and proliferation of astroglial cells.

2.2.1.2. PROCEDURE

1. Kill newborn mouse pups either by cervical dislocation or by an ether overdose. Dip heads in 2% iodine in 70% ethanol followed by 70% ethanol. This reduces the chances of contaminating the brain by microorganisms present on the skin during the subsequent dissection. Cut off heads and pin heads ventral side down onto a styrofoam or wax surface.
2. Isolate the cerebral hemispheres aseptically from newborn mouse pups. This is most conveniently done using three pairs of forceps. The first pair is used to remove the skin from the head, the second pair is used to separate and lift the flat bones from the surface of the brain, and the third pair is used to separate the cerebral hemispheres from the olfactory bulbs in front and the midbrain behind and finally to remove the hemispheres from the cranial cavity (Fig. 4).

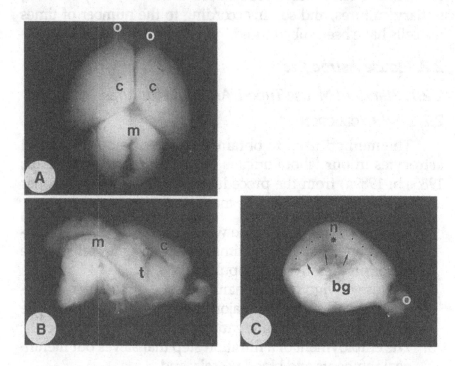

Fig. 4. Micrographs demonstrating the isolation of the neopallium. (A)
Dorsal view of isolated cerebral hemispheres (c) with olfactory bulbs (o)
and midbrain (m) still attached. (B) Medial view of left part of brain after
the two hemispheres are separated. Much of the view of the left cerebral
hemisphere (c) is obscured by the thalamus (t) and midbrain (m). (C)
Medial view of left hemisphere after removal of midbrain and thalamus;
note olfactory bulb (o) and basal ganglia (bg). The hippocampal sulcus is
readily seen (arrows). In the next step of the dissection, use forceps to
nibble through the junction between the entorhinal area (*) and the
neopallium (n); this junction is indicated approximately by the dotted
line. (D) *(opposite page)* Medial view of cerebral hemisphere after hippocampal formation is removed. The thin shell of tissue above the lateral
ventricle (v) is the neopallium (n); the floor of the ventricle is formed by
the basal ganglia (bg). Anteriorly, the olfactory bulb (o) is still attached.
Using forceps as shears, the neopallium is separated from the basal ganglia.
(E) *(opposite page)* Medial view of the neopallium (n) separated from the
basal ganglia (bg), except for its most posterior part. To complete the dissection, the neopallium is completely separated from the basal ganglia and
meninges removed.

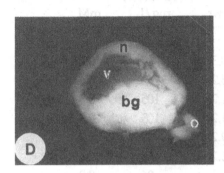
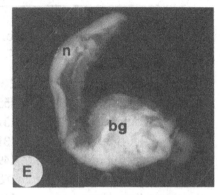

Place the hemispheres in medium containing 20% horse serum (5–10 mL in a 60-mm Petri dish) and keep in this medium for the subsequent steps. The medium is described in Table 1 and is a modification of Eagle's Minimum Essential Medium (MEM), as discussed in Section 3.3.1.

3. Remove the meninges (i.e., the connective tissue coverings of the brain), olfactory bulbs, hippocampi, and basal ganglia from the hemispheres, thus leaving the neopallia, i.e., the cortical tissue dorsal and lateral to the lateral ventricles (*see* Fig. 4). Neuronogenesis is completed in the neopallium during the late fetal stages of development; at the newborn stage there are large numbers of astrocyte precursors present.

4. Cut neopallia into approx 1-mm cubes and vortex tissue (Bullaro and Brookman, 1976) at maximum speed (Deluxe Mixer, Scientific Products, Baxter, Mississauga, ON, Canada) for 1 min. This procedure destroys the majority of neurons. There are other ways of mechanically dissociating the tissue, one of which is to gently push the tissue through a nylon mesh with a pore size of 80 μm, but *see* Section 4.1.1. To obtain cultures free of vascular elements and meningeal cells, it is important not to use enzymes during the tissue dissociation.

5. Pass the cell suspension sequentially through Nitex™ meshes (from L. and S. H. Thompson and Co., Ltd., Montréal, Québec, Canada) of pore sizes 80 μm and 10 μm. This procedure removes blood vessels, remaining meningeal sheets, and cellular aggregates.

Table 1
Composition of Eagle's Modified MEM

	Mol wt	Amount, mg/L	mM
Amino acids			
L-cystine	240.3	48	0.2
L-histidine	155.2	62	0.4
L-isoleucine	131.2	105	0.8
L-leucine	131.2	105	0.8
L-methionine	149.2	30	0.2
L-phenylalanine	165.2	66	0.4
L-threonine	119.1	96	0.8
L-tryptophan	204.2	20	0.1
L-tyrosine	181.2	72	0.4
L-valine	117.2	94	0.8
L-arginine·HCl	210.7	253	1.2
L-lysine·HCl	182.7	146	0.8
L-glutamine	146.1	365	2.5
Vitamins			
folic acid	–	4	–
riboflavin	–	0.4	–
calcium pantothenate	–	4	–
choline chloride	–	4	–
i-inositol	–	8	–
niacinamide	–	4	–
pyridoxal·HCl	–	4	–
thiamine·HCl	–	4	–
Salts and glucose			
$MgSO_4 \cdot 7H_2O$	246.5	200	0.8
KCl	74.6	400	5.4
$NaHCO_3$	84.0	2200	26.2
NaCl	58.4	6800	116
$NaH_2PO_4 \cdot H_2O$	138	140	1.0
$CaCl_2 \cdot 2H_2O$	147.0	264	1.8
D-glucose	180.1	1350	7.5
pH indicator			
phenol red	–	20	–

6. Seed into Falcon or NUNC tissue culture Petri dishes (e.g., 3–4 60-mm Petri dishes/brain, which corresponds to ~3 × 10⁴ viable cells/cm²). With such a low seeding density very few oligodendrocytes or type-2 astrocytes develop in these cultures. With Falcon Primaria dishes, the seeding density can be lowered (~20 35-mm dishes/brain, i.e., 1.5 × 10⁴ viable cells/cm²) since initial cell attachment is improved.

7. Incubate at 37°C in a 95/5% (v/v) mixture of atmospheric air and CO_2 with a relative humidity of 90%.

8. After 3 d (this time period is important), remove medium and feed cultures with fresh medium containing 20% horse serum. From this point on, feed cultures 2–3 times weekly with medium containing either 5–10% horse serum or 10% fetal bovine serum. After 2 wk of culture, when cells are confluent, dibutyryl cyclic adenosine 3′,5′-monophosphate (dBcAMP) may be added to the medium at a final concentration of 0.25 mM. This treatment causes pronounced biochemical and morphological differentiation of the cells (*see* Section 3.2.2.1.).

2.2.1.3. DESCRIPTION OF CULTURES

Initially, individuals who are used to culturing either glial cell lines or neurons are horrified upon examining neopallial astrocyte cultures within the first 3 d of culture. There is so much cellular debris present that one thinks the cultures are contaminated. In addition, it is difficult to find cells since very few of the cells in the suspension attach to the plastic substratum. In fact, only about 1–2% of the viable cells attach and proliferate (Juurlink et al., 1981a). Upon feeding, after 3 d of culture, sparsely distributed small islands of cells can be observed (Fig. 5A) These cells proliferate vividly (Fig. 5B) such that by 10–14 d, a confluent layer of cells is seen (Fig. 6A). Although seemingly a monolayer of cells, these confluent cultures are formed by multiple layers of cytoplasmic sheets (Fig. 7) resembling a *glia limitans* (Devon and Juurlink, 1988,1989).

In these cultures, 95% of the cells are GFAP-positive, and hence, astrocytic in nature (Hertz et al., 1982,1985a,b; Juurlink and Hertz, 1985). Almost all of these cells can be classified as type-1 astrocytes, with fewer than 1% being type-2 astrocytes.

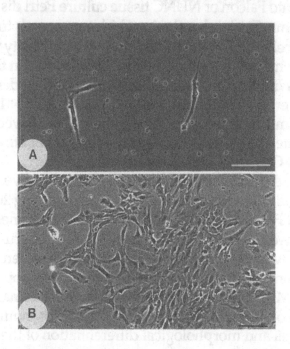

Fig. 5. Phase contrast micrograph of living cells in cultures established from newborn mouse neopallia. Bars = 100 μm. **(A)** Three-d-old culture demonstrating the paucity of type-1 astrocyte precursors that attach to the substratum. **(B)** One-wk-old culture demonstrating the very rapid proliferation of type-1 astrocytes in these cultures.

The remaining cells are macrophages/microglia (~3%) and fibroblast-like cells (<1%). Irregular feeding of the cultures promotes macrophage/microglia proliferation, likely because of the secretion of the macrophage/microglia mitogen colony-stimulating factor-1 (CSF-1) by type-1 astrocytes (Hao et al., 1990); hence, feeding the cultures every 2–3 d reduces the concentration of CSF-1 in the medium and thus keeps the numbers of macrophages/microglia low. In addition, the incorporation of dBcAMP also keeps macrophage/microglia numbers low, the reason for this is not known but it perhaps acts by reducing the secretion of CSF-1 by the astrocytes. Fibroblast-like cells comprise a very small proportion of the cells in our cultures. The reasons for this are: The meninges are carefully

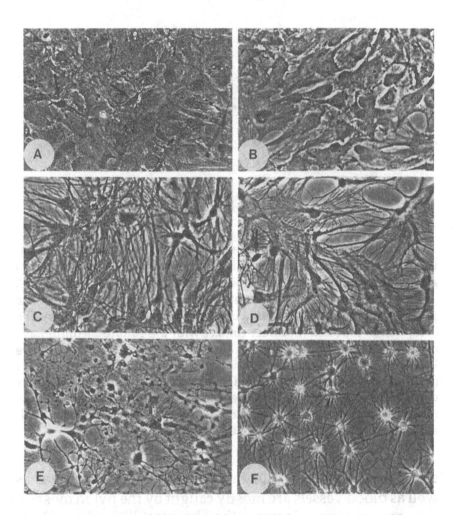

Fig. 6. Phase contrast micrographs of living mouse and rat type-1 astrocytes four weeks after the establishment of the primary cultures. All micrographs are at the same magnification. Arrows depict macrophages/microglia. Bar = 50 μm. **(A)** Mouse primary culture. **(B)** Rat secondary culture. **(C)** Mouse astrocytes after two weeks exposure to 0.25 m*M* dBcAMP. Note process-bearing appearance of astrocytes. **(D)** Rat secondary astrocyte cultures after 2 d of exposure to 0.25 m*M* dBcAMP. Cellular processes are already well-developed by this time. **(E)** Mouse primary culture after 20 h exposure to 100 n*M* staurosporine. Note considerable cell necrosis. **(F)** Rat secondary culture after 20 h exposure to 100 n*M* staurosporine. Note stellation of rat type-1 astrocytes.

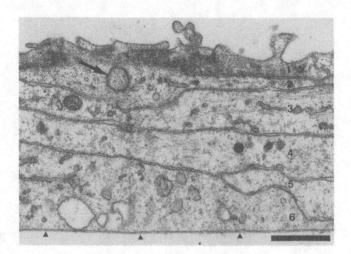

Fig. 7. Electron micrograph of a 4 wk old primary astrocyte culture established from newborn mouse neopallium (from unpubished work of Devon and Juurlink). This is a dense culture that has not been treated with dBcAMP. Small arrowheads depict the culture substratum. Six overlapping sheet-like processes (numbered) as well as a finger-like process (arrow) are seen. The cytoplasmic processes in direct contact with the culture medium have numerous projecting villi. These cultures resemble a *glia limitans*, and are ideally suited for examining cell-to-cell movement of macromolecules (Juurlink and Devon, 1990). That dense astrocyte cultures have multiple overlapping processes with a considerable amount of true intercellular space must be taken into account when examining such phenomena as drug binding kinetics or ion fluxes. Bar = 0.5 μm.

cleaned from the neopallia, and the remaining meninges as well as blood vessels are mostly caught by the nylon meshes. One can, of course, increase the fibroblast-like cells in these cultures either by initiating cultures from brains of older pups (*see* Trimmer et al., 1984) or by using enzymatic means of tissue dissociation. The older the pup, the more difficult it is to remove the meninges completely. Enzymatic means of tissue dissociation not only dissociate neural tissue, but also result in the dissociation of the blood vessels; hence, one can no longer sieve out the blood vessels from the cell suspension. Both the use of brain tissue from pups older than 1 d and the use of trypsin for tissue dissociation probably explains the very high (24–28%) contamination with fibronectin-positive fibroblast-like cells in the cultures described by Stieg et al. (1980).

In conclusion, to obtain mouse cultures that consist of more than 95% cortical astrocytes, we consider it critical to:

1. Use newborn pups, i.e., no older than 24 h postnatum;
2. Select only the neopallium;
3. Use only mechanical means to dissociate the tissue; and
4. Use nylon meshes to sieve out aggregates, remnants of meninges, and blood vessels.

No subsequent separation of type-1 and type-2 astrocytes is required since very few type-2 astrocytes develop in these cultures.

2.3. Rat Astrocytes

2.3.1. Enriched Cultures of Rat Astrocytes

2.3.1.1. INTRODUCTION

The following is our technique for the preparation of rat astrocyte cultures; these cultures consist of a mixture of type-1 astrocytes, O2A progenitor cells, and type-2 astrocytes. This procedure is a minor variation of that described by McCarthy and de Vellis (1980) in which the O2A lineage cells are subsequently separated from the type-1 astrocytes.

2.3.1.2. PROCEDURE

1. Kill newborn rat pups either by cervical dislocation or by an ether overdose. Dip heads in 2% iodine in 70% ethanol followed by 70% ethanol. This reduces the chances of contaminating the brain by microorganisms present on the skin during the subsequent dissection. Cut off heads and pin heads ventral side down onto a styrofoam or wax surface.
2. Isolate the cerebral hemispheres aseptically from newborn rat pups as previously described for the mouse. Place the hemispheres in modified MEM medium (Table 1) containing 20% horse serum (5–10 mL in a 60-mm Petri dish) and keep in this medium for the subsequent steps.

3. Remove the meninges, olfactory bulbs, hippocampi, and basal ganglia from the hemispheres, thus leaving the neopallia, i.e., the cortical tissue dorsal and lateral to the lateral ventricles.

4. Cut neopallia into approx 1-mm cubes and vortex tissue at maximum speed (Deluxe Mixer, Scientific Products, Baxter) for 1 min.

5. Pass the cell suspension sequentially through Nitex™ meshes (from L. and S.H. Thompson and Co., Ltd., Montréal, Québec, Canada) of pore sizes 80 μm and 10 μm. This procedure removes blood vessels, remaining meningeal sheets, and cellular aggregates.

6. Seed into Falcon or NUNC tissue culture Petri dishes (e.g., 6–8 60-mm Petri dishes/brain, which corresponds to ~1.5 × 10^4 viable cells/cm^2) or at half the density into Falcon primaria dishes.

7. Incubate at 37°C in a 95/5% (v/v) mixture of atmospheric air and CO_2 with a relative humidity of 90%.

8. After 3 d, remove medium and feed cultures with fresh medium containing 20% horse serum. From this point on, feed cultures 2–3 times weekly with medium containing either 5–10% horse serum or 10% fetal bovine serum.

2.3.1.3. DESCRIPTION OF CULTURES

This procedure is essentially identical to that used for preparing mouse astrocytes. Nevertheless, rat cortical astroglial cultures differ from those of mouse in that large numbers of O2A lineage cells develop in these cultures (Fig. 8). The O2A lineage cells are often described as being situated on top of a monolayer of type-1 astrocytes. Although this is true for some of the O2A lineage cells, the cytoplasmic processes of many of the O2A lineage cells are directly adherent to the substratum or sandwiched between the processes of the multiple layers of type-1 astrocytes. The number of O2A lineage cells can be reduced by decreasing the initial planting density. This, of course, also reduces the density of type-1 astrocytes. Researchers interested in studying features such as the metabolism of purified rat type-1 astrocytes generally use the shaking method of the

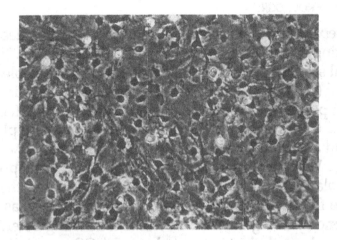

Fig. 8. Phase contrast micrograph of living cells in 10-d old cultures established from the neopallia of 2-d old rat pups. Numerous small phase dark process-bearing O2A lineage cells can be observed, as well as a confluent carpet consisting of type-1 astrocytes. This is a typical field and illustrates the large numbers of O2A lineage cells that are present throughout these cultures. Because these cultures are very dense, more macrophages/ microglia (small phase bright cells) are present than in primary mouse neopallial cultures. Bar = 50 μm.

McCarthy and de Vellis (1980) to remove the O2A lineage cells from the type-1 astrocytes.

2.3.2. Standard Rat Type-1 Astrocyte Cultures

2.3.2.1. INTRODUCTION

The most common method used for the preparation of rat cortical type-1 astrocyte cultures is that described by McCarthy and de Vellis (1980), occasionally with minor modifications. This technique is described in greater detail by Cole and de Vellis (1989). For the preparation of rat cerebral astrocyte cultures, most researchers isolate the cerebral hemispheres from 1 or 2 day old rat pups and use entire hemispheres for the generation of the cultures. We, however, only use the neopallium. The following is the procedure we use (Juurlink and Hertz, 1991); it is essentially only a minor variation of the procedure described by McCarthy and de Vellis (1980).

2.3.2.2. PROCEDURE

1. Prepare a mixed rat astroglial culture as previously described using 75 cm² flasks (Falcon) as the culture vessels. The initial seeding density should be about 1.5×10^4 viable cells/cm² (i.e., plant the cells obtained from the neopallia of one rat pup into three flasks) and total medium vol/flask is 8 mL. Cultures are fed 2–3 times a wk. At the time of planting and at each feeding, the flasks are gassed with 5% CO_2 in air. When cultures are fed, the 5% CO_2 atmosphere is replaced with straight air, and gassing displaces the air in the flask with an atmosphere having 5% CO_2. Gassing is done by tilting the flask on end and inserting a sterile cotton plugged pipet connected to a 5% CO_2 supply into the flask and gassing for about 10 s. This procedure ensures that deviations from a pH of 7.4 are not prolonged.

2. After 10–14 d, wash cultures with a balanced salt solution and replace with a growth medium consisting of modified MEM supplemented with 1 mM NaHCO₃ instead of the usual 26 mM, 15 mM β-glycerophosphate and 5% horse serum. The flasks are then capped tightly and placed onto a gyratory shaker situated in a walk-in incubator maintained at 37°C and shaken at 200 gyrations/min. Since the capped flasks are not completely impermeable to gases and the walk-in incubator that we use contains an air atmosphere, the use of β-glycerophosphate together with the reduction of the NaHCO₃ concentration ensures that physiological pH can be maintained in the absence of CO_2 in the environment.

3. After shaking overnight, remove medium and replace with fresh β-glycerophosphate-buffered growth medium and shake for an additional 24 h. After this period, very few O2A lineage cells should remain adherent to the type-1 astrocytes. Remove the medium and wash the type-1 astrocytes with three changes of a Ca^{2+}-Mg^{2+}-free balanced salt solution.

4. The original cultures can now be fed with growth medium and used as primary cultures, but it is generally more convenient to subculture the cells into Petri dishes; these cells

now form secondary cultures. No systematic studies have been performed whether this affects functional characteristics, but primary brain culture cells do change for at least a period of several days after they have been subcultured (Hertz et al., 1985a). To subculture, add 6 mL 0.025% trypsin in Ca^{2+}-Mg^{2+}-free balanced salt solution to each flask and incubate for 15 min at room temperature. (During the dissociation of brain tissue to establish the astrocyte cultures, we strongly recommend avoiding enzymatic means of tissue dissociation since this results in considerable contamination of the cultures with fibroblasts and endothelial cells; however, this logic does not apply to removing cells from the tissue culture dish.) After this, gently pipet cells off the substratum, remove the solution and centrifuge for 10 min at 180g. Decant the supernatant and resuspend the cell pellet in a small vol of $NaHCO_3$-buffered modified MEM containing 5–10% horse serum.

5. Count the cells using a hemocytometer and plant at approx $2.5–5 \times 10^4$ cells/cm^2 in Petri dishes or onto coverslips. Within a wk, these cultures will be confluent.

2.3.2.3. Description of Cultures

The secondary rat cultures described consist of >95% type-1 astrocytes (Fig. 6B) and thus, are very comparable in composition to the primary mouse type-1 astrocyte cultures described earlier.

2.3.3. Enriched Rat Type-2 Astrocyte Cultures

2.3.3.1. Introduction

In the previously described procedure, the majority of the cells removed from the primary mixed astrocytic cultures are O2A lineage cells, but some macrophages/microglia are also present. By removing the macrophages/microglia from this population, one can obtain highly enriched type-2 astrocyte cultures as described below. In this procedure, several details are different from the ones described in Section 2.3.1.1. These modifications were introduced to maximize the yield of O2A precursor cells. If one wants to obtain both type-1 and type-2 astrocytes from the same mixed primary culture, it is possible with either method; however, in order to obtain a reasonable yield of type-2 astrocytes, we recommend using this present version.

We use the following procedure (Juurlink and Hertz, 1991), which is essentially the procedure described by McCarthy and de Vellis (1980) to obtain O2A lineage-derived cells. Under the conditions described, this procedure will yield cultures that are >90% type-2 astrocytes. We recommend preparing the cultures from the neopallium only, since in this part of the cerebrum, neuronogenesis is completed. The GQ1c ganglioside recognized by the A2B5 monoclonal antibody and used as a marker of rat O2A progenitor cells as well as rat type-2 astrocytes is also present on the cell membranes of neurons (Eisenbarth et al., 1979); hence, the presence of neurons in one's cultures complicates the identification of O2A progenitor cells as well as type-2 astrocytes. Neuronogenesis still occurs in the hippocampus and olfactory bulbs for several weeks postnatum (Angevine, 1965; Hinds, 1966).

2.3.3.2. PROCEDURE

1. We have found that P4 (i.e., from 4-d old rat pups) neopallium yields the largest numbers of O2A lineage cells. P4 rat pups are killed and neopallia isolated as previously described. Once the brains are isolated, all subsequent procedures are performed in growth medium consisting of modified MEM supplemented with 20% horse serum. The neopallia are diced into cubes of ~1 mm and gently forced through a Nitex™ mesh of a pore size of 80 μm. Note that no subsequent sieving through a 10 μm mesh should be done. The Nitex™ mesh is taped such that it covers the mouth of a 50 or 100 mL beaker; the Nitexed beaker is wrapped in aluminum foil and autoclaved prior to use. We use the curved side of a curved Pasteur pipet to force the tissue through the mesh. This is done very gently to minimize the amount of meninges and blood vessels that pass through the mesh.

2. The cells are counted and planted into 75 cm^2 tissue culture flasks at a density of 3×10^4 cells/cm^2 in a total vol of 8 mL/ flask. The flasks are then gassed and placed into a humidi-

fied incubator containing a 5% CO_2 atmosphere and maintained at 37°C. The cultures are fed with fresh growth medium containing 20% horse serum after 3 d and then twice weekly, at each feeding, the flasks are gassed with 5% CO_2 before the flasks are returned to the incubator.

3. After 10–14 d of culture, the O2A lineage cells are isolated using the procedure summarized in Fig. 9. The cultures are first shaken on a gyratory platform at 200 gyrations/min for 30 min to detach cells that are loosely adherent to the confluent layer of type-1 astrocytes—these are mostly macrophages/microglia. The gyratory apparatus is situated in a large walk-in incubator maintained at 37°C.

4. Medium is removed from flasks and cultures are washed several times with a Ca^{2+}-Mg^{2+}-free balanced salt solution. Growth medium consisting of modified MEM supplemented with 1 mM $NaHCO_3$ (instead of the usual 26 mM), 15 mM β-glycerophosphate and 5% horse serum is added to the flasks. As previously mentioned, the β-glycerophosphate acts as the buffer and ensures pH stability during the next step. A small amount of $NaHCO_3$ is incorporated in the medium to enable the astrocytes to fix CO_2, a critical feature of astrocyte metabolism (Yu et al., 1983; Shank et al., 1985).

5. The flasks are then placed on a gyratory platform situated in the walk-in incubator and shaken at 200 gyrations/min overnight (~18 h). The medium with cells in suspension is then removed and placed into 100-mm tissue culture Petri dishes (the medium from one flask is placed into one Petri dish) and incubated for 3 h in a humidified incubator with a 100% air atmosphere and maintained at 37°C. This step allows the type-1 astrocytes and macrophages/microglia to become firmly adherent to the substratum, whereas the O2A lineage cells become only lightly adherent. The tissue culture flasks containing the type-1 astrocytes are fed with a $NaHCO_3$-buffered MEM supplemented with 5% horse serum, gassed with 5% CO_2, and placed in a CO_2 incubator. We generally use these cells to prepare type-1 astrocyte conditioned medium (*vide infra*).

PRIMARY ASTROCYTE CULTURES SHAKEN ON ROTARY SHAKER AT 37° C
FOR 30 MIN
↓
REMOVE MEDIUM WITH SUSPENDED CELLS
↓
RINSE CULTURES AND ADD FRESH MEDIUM BUFFERED WITH β–
GLYCEROPHOSPHATE. SHAKE OVERNIGHT (~ 18 HR)
↓
REMOVE SUSPENDED CELLS AND PLACE IN 100 mm PETRI DISHES FOR A
3 HR INCUBATION AT 37° C IN A HUMIDIFIED AIR ATMOSPHERE
↓
REMOVE MEDIUM AND GENTLY WASH PETRI DISHES TO REMOVE DEBRIS
↓
REMOVE THE LOOSELY ADHERENT O-2A LINEAGE CELLS BY GENTLY
PIPETTING THE SURFACE OF THE DISHES WITH A BICARBONATE
BUFFERED GROWTH MEDIUM CONTAINING 0.02% DNase
↓
CENTRIFUGE CELLS AT 180g FOR 10 MIN AND RESUSPEND IN GROWTH
MEDIUM CONTAINING 0.02% DNase
↓
CENTRIFUGE CELLS AT 150g FOR 15 MIN THROUGH A 3.5% BSA CUSHION
↓
RESUSPEND CELLS IN GROWTH MEDIUM CONTAINING 0.02% DNase AND
FILTER SEQUENTIALLY THROUGH NYLON MESHES OF PORE SIZES 35μm
AND 15μm
↓
PLANT CELLS IN GROWTH MEDIUM CONTAINING 0.02% DNase ONTO
POLYORNITHINE COATED SURFACES
↓
AFTER 15 MIN REMOVE MEDIUM AND ADD DNase-FREE GROWTH MEDIUM
CONTAINING 20% TYPE-1 ASTROCYTE CONDITIONED MEDIUM

Fig. 9. Flow sheet illustrating the isolation of O2A lineage cell from rat primary neopallial cultures. For details, *see* Section 2.3.3.2.

6. The medium from the 100-mm Petri dishes is removed and the dishes are gently washed with a balanced salt solution; this tends to remove much of the cellular debris. Thereafter, a NaHCO$_3$-buffered modified MEM supplemented with 5% horse serum and containing 0.02% DNase (Sigma, St. Louis, MO, type 1) is gently pipeted over the surface of the Petri dish to dislodge the loosely adherent O2A lineage cells from the substratum, but leaving the type-1 astrocytes and macroglia/microglia attached to the substratum. DNA acts as a glue, and thus, the presence of DNase reduces the aggregation of O2A lineage cells that would be caused by DNA released from damaged cells.

7. The cell suspensions are removed from the Petri dishes and centrifuged at 180g for 10 min, resuspended in DNase-containing growth medium. The cell suspension is layered onto 3.5% bovine serum albumin in modified MEM and cells are centrifuged through this cushion for 15 min at 150g. Much of the debris in the cell suspension will stay in the albumin cushion.

8. The cell pellet is resuspended in growth medium containing 0.02% DNase and filtered sequentially through nylon meshes of pore sizes 35 μm and 15 μm. Despite the use of DNase, some cell aggregates still form and these are filtered out of the cell suspension by the nylon meshes. This is an important step since the aggregates are composed of type-1 astrocytes as well as O2A lineage cells. With removal of these cell aggregates, the cells remaining in suspension consist almost exclusively of O2A lineage cells.

9. A cell count is performed and cells are planted onto either polyornithine or polylysine-coated glass or plastic surfaces. We usually plant the cells at a density of 2.5×10^4 cells/cm². After 15 min, the medium is removed and replaced by growth medium (modified MEM supplemented with 5% horse serum) containing 20% (v/v) medium conditioned by type-1 astrocytes. In the absence of conditioned medium, O2A lineage cells degenerate within 5 d. The conditioned medium is prepared as follows: purified type-1 astrocyte cultures are fed with modified MEM supplemented with 5% horse serum and after 2 d, the medium is collected, centrifuged at 180g to remove cells, and stored at −20°C until ready to be used.

2.3.3.3. DESCRIPTION OF CULTURES

With the above procedure one can obtain $0.5–1.0 \times 10^6$ O2A lineage cells from cultures established from the neopallia obtained from one rat pup. The cultures that are established with these cells are composed of more than 95% O2A lineage cells (Fig. 10A). When first established, the cultures consist of ~30% type-2 astrocytes and ~70% O2A progenitor cells, but by 1 wk in the presence of conditioned medium obtained from type-1 astro-

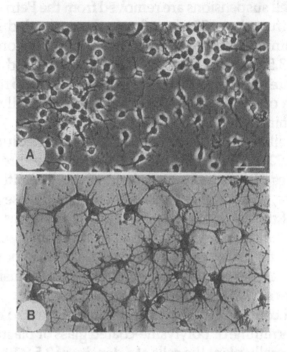

Fig. 10. O2A lineage cells isolated from rat primary neopallial cultures.
Bar = 50 μm. **(A)** Phase contrast micrograph of living cells 1 d after isolation;
approx 70% of these cells are O2A progenitor cells and 30% type-2 astrocytes. **(B)** Three wk old cultures demonstrating GFAP localization using a
biotin-avidin-alkaline phosphatase (nonfluorescent) immunocytochemical
technique. At this time in culture, 90% of the cells are type-2 astrocytes.

cyte cultures, the type-2 astrocytes comprise ~90% of the cells
(*see* Fig. 10B) in the cultures, whereas the O2A progenitors comprise ~6% of the cells (Juurlink and Hertz, 1991). Thus, contrary
to the findings of Lillen et al. (1990) with optic nerve-derived
O2A progenitors (where the presence of extracellular matrix-
associated molecules are necessary for the majority of the
precursor cells to differentiate into type-1 astrocytes), under our
culture conditions, neopallial-derived O2A progenitors do not
require the presence of extracellular matrix-associated molecules
for their differentiation into type-2 astrocytes. The cultures can
be maintained for at least 4 wk with little change in cellular composition; e.g., at 3 wk, the cultures consist of 91.0 ± 4.6% type-2
astrocytes, 5.5 ± 2.8% O2A progenitor, 1.5 ± 1.5% type-1 astrocytes, and 2.1 ± 1.5% unidentified cells.

3. Agents that Affect the Differentiation of Type-1 Astrocytes

3.1. Introduction

Both type-1 and type-2 astrocytes in culture develop from undifferentiated precursor cells. Therefore, in most laboratories, cultures of astrocytes are generally used for experimentation after remaining in culture for an appropriate time (e.g., 3 wk in the case of mouse type-1 astrocyte cultures) for differentiation to occur. The differentiation of these cells in culture is to a remarkable extent "on schedule" when compared to their in vivo development (Hertz et al., 1978; Abney et al., 1981). In many laboratories, including our own, minor alterations are made in the composition of the growth medium after several weeks in an attempt to enhance or modify morphological and functional differentiation of the cells. Presently, this is done routinely for type-1 astrocytes.

3.2. Morphogens and Astrocyte Differentiation

3.2.1. Morphogens and Astrocyte Differentiation In Vivo

It has now become very well-established that compounds which in the mature organism function as neurotransmitters may play a major developmental role as "morphogens" (Turing, 1952) during development (Lauder, 1988). There is good reason to believe that the same agents also have a differentiating effect on functional (biochemical and biophysical) parameters, i.e., also act as "functiogens." The monoamine transmitters dopamine, serotonin, and noradrenaline, as well as some polypeptide transmitters, appear to act as morphogens (promoting morphological differentiation) and functiogens (promoting functional differentiation) for astrocytes (Brenneman et al., 1987; Hatten, 1985; Hansson and Rönnbäck, 1988; Hertz, 1990b; Meier et al., 1991).

Neocortical noradrenergic innervation has been observed in rodents as early as embryonic day 16, but the development of the neocortical fibers continues through the first 2 postnatal wk (Foote et al., 1983). Astrocytes proliferate and differentiate at a late, mainly postnatal stage in both mouse and rat (Privat, 1975). Thus, the morphogenic signals they receive must operate postnatally. Both β-adrenergic and α-adrenergic receptors have been demonstrated recently on mature astrocytes in vivo (Aoki et al., 1987; Stone and Ariano, 1989). This adrenergic innervation of astrocytes is relevant for astrocytic functions by modulating, e.g., adenylate cyclase activity, free intracellular calcium concentration (Salm and McCarthy, 1990), glucose metabolism, protein phosphorylation, and probably also channel-mediated K^+ transport (for refs., see Hertz, 1989a,1991).

3.2.2. Use of Adrenergic Agonists and Second Messenger Analogs In Vitro

3.2.2.1. COMPOUNDS AFFECTING cAMP

Primary cultures of mouse and rat astrocytes are prepared routinely from newborn animals and contain no neurons; therefore, they do not receive noradrenergic signals that may be essential for their normal differentiation. This may mean that these cells morphologically and functionally remain undifferentiated, unless noradrenergic signals normally received in vivo are replaced by modulation of the external medium, which replaces the microenvironment that the cells have in vivo.

Cultured astrocytes express β-adrenergic (McCarthy et al., 1988) as well as α-adrenergic (Ebersolt et al., 1981; Bockaert and Ebersolt, 1988) and serotonergic (Whitaker-Azmitia, 1988) receptors. Since exposure to β-adrenergic agonists leads to an increased intracellular concentration of cyclic AMP (cAMP), it might be possible to mimic this effect by including a cAMP analog such as dibutyryl cyclic adenosine monophosphate (dBcAMP) in the culture medium and thus increase the intracellular level of cAMP (e.g., Facci et al., 1987).

The functional effects of dBcAMP supplementation of the medium have been recently reviewed (Hertz, 1990b; Meier et al., 1991). Among the most pronounced alterations in astrocytes in primary cultures after exposure to dBcAMP for 1 wk are process formation (Figs. 6B and C), a doubling of carbonic anhydrase activity and of Na^+, K^+-ATPase activity, a tripling of monoamine oxidase B activity, an alteration of a cotransport mechanism for GABA with one sodium ion to cotransport with two sodium ions (enabling a more concentrative uptake) and, perhaps most dramatically, the appearance of a voltage-sensitive L channel for calcium, indicated by calcium channel activity in astrocytes (MacVicar 1984; Barres et al., 1989), a high affinity binding of nitrendipine to astrocytes (Hertz, 1990b), a potassium-stimulated uptake of calcium that is potently inhibited by nimodipine (Hertz et al., 1989b), and the ability of an elevated concentration of potassium to stimulate glycogenolysis (K. V. Subbarao and L. Hertz, unpublished results). These alterations might be essential for astrocytic function in the adult CNS.

Although dBcAMP has profound morphogenic and functiogenic effects on astrocytes in primary cultures, it would probably be naive to believe that dBcAMP treatment completely substitutes for all normal transmitter effects on astrocytic development. In contrast to the continued and uniform exposure to dBcAMP in culture, transmitter signals in vivo will be limited in time, different at different locations, and able to create concentration gradients. Intracellular messengers other than dBcAMP may also be involved. Nevertheless, we do believe that supplementation of the medium with dBcAMP does lead to an enhanced functional differentiation of type-1 astrocytes. This conclusion does not mean that astrocytes in primary cultures display no abnormal features (Juurlink and Hertz, 1985). Thus, their content of GFAP (measured by rocket immunoelectrophoresis) is higher than that of the brain in vivo (Bock et al., 1975; Hertz et al., 1978), and the ratio between vimentin and GFAP is also higher than should be expected in mature astrocytes (Goldman and Chiu, 1984). However, this applies to primary cultures of type-1 astrocytes regardless of whether they have been treated with dBcAMP, and the increase in GFAP after addition of dBcAMP is rather slow. We do not believe that dBcAMP-treated astrocytes are the homologs of reactive astrocytes found in vivo as has been suggested by Fedoroff et al. (1984), although this is not to say that type-1 astrocytes in vitro do not exhibit certain features that reactive astrocytes also exhibit. Direct comparisons between reactive astrocytes in vivo and dBcAMP-treated astrocytes in vitro have been few in number. In a preliminary study, Wandosell et al. (1990) reported that dBcAMP-treated astrocytes do not exhibit the same surface markers as reactive astrocytes in vivo; however, the cells had only been treated for 2 d with dBcAMP and longer treatment would probably be required to unequivocally conclude that these cells are not reactive astrocytes.

3.2.2.2. COMPOUNDS AFFECTING PROTEIN KINASE C ACTIVITY

Since astrocytes also express α-1 adrenergic receptors and serotonergic receptors that operate via the phosphoinositol-protein kinase C second messenger system, it should be noted that Mobley et al. (1986) observed a morphogenic effect after addition of phorbolesters, compounds known to stimulate protein kinase C. At present, no information seems to be available about possible functiogenic effects.

3.2.3. Medium Potassium Concentration

An elevated concentration of potassium has a profound differentiating effect on certain neurons that require a partially depolarizing potassium concentration (probably substituting for presynaptic activation received in vivo), e.g., for development of release of neurotransmitter glutamate (for refs. see Meier et al., 1991; Peng et al., 1991). A similar requirement is not found in astrocytes, although culturing at an elevated potassium concentration does affect some developmental aspects of both rabbit (Reichelt et al., 1989) and mouse (Peng et al., 1991) astroglial cells. As a further indication of the importance of the composition of the culturing medium, it should be noted that as little as 3 d of exposure to an elevated potassium concentration causes a massive increase in astrocytic Na^+,K^+,-ATPase activity (K. V. Subbarao and L. Hertz, unpublished experiments), whereas exposure to a decreased potassium concentration has the opposite effect.

3.2.4. Amino Acid Transmitters

Very little evidence has been obtained that transmitter amino acids, which so dramatically affect neurons, have any morphogenic effects on astrocytes (Meier et al., 1991).

3.3. Medium Composition

3.3.1. Chemically Defined Components of the Medium

It was demonstrated early that slight alterations in the composition of the medium can have profound effects on the properties of astrocytes. Moonen et al. (1975) had observed that rat astrocytes grown in the presence of fetal bovine serum respond very differently to dBcAMP, depending on whether Eagle's BME or Eagle's MEM was used in the growth medium. When grown in MEM, the cells responded to dBcAMP by process formation that took several days to occur and was poorly reversible. In contrast, when cells were grown with BME as the medium, dBcAMP addition did not result in process formation except when serum was removed, in which case there was a rapid but transient process formation. Although there are quantitative differences in the components of the two media, qualitatively the only differences are the presence of biotin in BME and its absence in MEM and the presence of valine in MEM and its absence in BME.

The modifications in our medium from MEM consist of doubling all of the amino acid concentrations except for glutamine (kept at 2.5 mM), quadrupling the vitamin concentrations and raising glucose to 7.5 mM, whereas all the other components remain unchanged. Although this medium is quite similar to Dulbecco's medium, there are qualitative and quantitative differences. The qualitative differences are as follows: Dulbecco's medium contains 0.25 μM Fe(NO$_3$)$_3$, 0.4 mM glycine, and 0.4 mM L-serine, whereas modified MEM does not contain these components. The quantitative differences from modified MEM (Table 1) are that Dulbecco's medium contains more NaHCO$_3$ (44 mM), arginine (0.4 mM), and glutamine (4.0 mM) and less NaH$_2$PO$_4$ (0.088 mM), NaCl (110 mM), glucose (5.6 mM), histidine (0.2 mM), and tryptophan (0.08 mM).

There are many examples of various laboratories obtaining different results when using the same cell preparations; many of these differences are possibly attributable to slight differences in the composition of the chemically defined component of the medium.

3.3.2. Serum Supplementation

Most growth media are supplemented with serum. The function of the serum is to supply hormones and growth factors that are required by the cells. Serum is generally used because these requirements of the cells are poorly understood. The two commonly used sera are fetal bovine serum and horse serum. The choice of serum may influence the experimental results obtained (Juurlink and Hertz, 1985). In addition to species differences, there may be batch-to-batch differences within the serum of any given species. Therefore, it is recommended not to purchase any serum batch without testing it on the cultures of interest, and once a satisfactory batch has been identified, one should acquire an amount sufficient to allow the use of the serum of that batch for an extended period. For an inexperienced investigator establishing a new tissue culture laboratory, it would be wise to obtain initially a small amount of serum and medium of good quality from an established investigator in the area. This allows the new investigator to gain experience with the cultures of interest as well to check other aspects of the culture procedure being established.

For many purposes, it is essential to use serum-free media. Media designed specifically for the culture of astrocytes are now available (*see* chapter by Bottenstein, this vol.). These media sustain morphologically satisfactory development, but without further investigation it cannot be assumed that all aspects of functional development are identical to those in serum-supplemented medium.

4. Variability of Astrocytic Cultures

4.1. Differences in the Behavior Between Rat and Mouse Astrocytes in Culture

4.1.1. Astroglial Precursor Cell Proliferation

Using modified Eagle's MEM (Table 1) plus horse serum as the supplement we have observed a major difference between rat and mouse neopallial astrocyte cultures in the numbers of O2A progenitors that develop in the culture. Large numbers appear in rat cultures but few in mouse cultures. In both cultures, there is avid proliferation of type-1 astrocyte precursor cells. Using the vortex procedure to dissociate the neopallium followed by sieving the cells through Nitex™ meshes of pore sizes 80 μm and 10 μm, usually very few O2A lineage cells are detected in cultures established from mouse neopallium. However, if one dissociates the neopallium by gently forcing the tissue through a mesh with a 80 μm pore size and seeds this cell suspension into culture with no subsequent filtration through a smaller mesh, then small cellular aggregates are retained in the cell suspension, and O2A lineage cells are associated with such cell aggregates. These cells can be seen once the type-1 astroglial cells spread over the plastic substratum (Fig. 11A). The number of such aggregates containing O2A lineage cells appears to be maximal in cultures established from P4 mouse neopallium. By 2 wk in such cultures, the O2A lineage cells can be seen in discrete islands throughout the culture (Fig. 11B); however, generally, they do not appear to form more than ~10% of the total culture population. This is in marked contrast to the situation seen in rat neopallial cultures where the numbers of O2A lineage cells is similar shortly after the cultures are initiated but the cells proliferate so rapidly that they soon form a major part of the cell population in the cultures (Fig. 8). The basis of this difference in the proliferative ability between rat and mouse O2A progenitor cells is not known, but it explains why highly enriched mouse type-1 astrocyte cultures are readily obtained in primary cultures, whereas highly enriched rat type-1 astrocyte cultures can presently be obtained only by removing the O2A lineage cells using the procedures previously outlined.

4.1.2. Functional Differences

4.1.2.1. INTRODUCTION

For any comparison between type-1 astrocytes in mouse and rat cultures or between different rat astrocyte preparations, it is essential to remember that rat primary (i.e., unseparated) cerebral astrocytic cultures consist of a mixture of type-1 astrocytes and O2A lineage cells, and that homogenous rat type-1 astrocyte cultures are only obtained after mechanical separation of the two cell populations, whereas the standard type-1 mouse

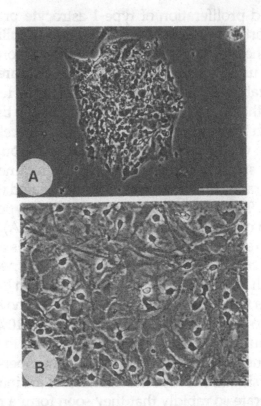

Fig. 11. Phase contrast micrograph of living cells from a culture established from the neopallia of 4-d-old mouse pups. Rather than using the vortex method for tissue dissociation, the neopallium was dissociated by gently pushing the tissue through an 80 µm nylon mesh and cells planted immediately in culture. A few small tissue aggregates remain intact and attach to the substratum. Within such aggregates are O2A progenitor cells. Bars = 100 µm. (A) Three day culture demonstrating one such aggregate that has attached to the substratum; note small O2A progenitor cells as well as larger type-1 astrocyte precursor cells. (B) Two-wk-old culture demonstrating a portion of a small island of O2A lineage cells; such islands occupy approx 10% of the culture surface area. Compare this with comparable rat primary cultures (Fig. 8). Bar = 50 µm.

astrocyte culture is homogeous and requires no separation. Comparisons between unseparated mouse and rat astrocyte cultures therefore, may not be valid since one is comparing mouse type-1 astrocyte characteristics with a mixture of rat type-2 astrocytes and O2A progenitors and their derivatives (type-2 astrocytes and oligodendrocytes).

4.1.2.2. POTASSIUM PERMEABILITY

Potassium fluxes in primary cultures of rat type-1 astrocytes are about one order of magnitude lower than in corresponding cultures of mouse type-1 astrocytes (Walz and Kimelberg, 1985; Walz and Mozaffari, 1987). The reason for this is unclear, but since potassium fluxes in rat type-1 astrocytes increase with lowered serum concentration and with increasing culture age (Juurlink and Hertz, 1985), it seems possible that the culture conditions currently used for rat type-1 astrocytes are not optimal for the maturation of the cells. Lowering the serum concentration promotes differentiation of astrocytes rather than the proliferation of the astrocytes (Fedoroff and Hall, 1979; Juurlink et al., 1981b). With a decreased cell density there is also a decreased stratification of the astroglial cultures (Devon and Juurlink, 1988), and hence, a greater proportion of the astrocytic cell membranes is exposed directly to the culture medium. It has been demonstrated for mouse type-1 astrocytes that a lower cell density results in an increase of potassium flux when measured per unit protein (Juurlink and Hertz, 1985; Walz and Mozaffari, 1987). Thus, it seems possible that the increase in potassium fluxes observed in rat type-1 astrocytes when medium serum concentration is lowered is related to lower cell density. Variations in cell density, however, cannot completely explain the differences between mouse and rat cultures since very low potassium fluxes were observed in highly enriched low density secondary cultures of rat type-1 astrocytes (B. H. J. Juurlink and L. Hertz, unpublished experiments).

In agreement with the much lower potassium permeability in rat type-1 astrocytes than in mouse type-1 astrocytes, potassium conductance is almost 10 times higher in the cultured mouse cells (Nowak et al., 1987) compared to cultured rat cells (Quandt and MacVicar, 1986). This high conductance seen in cultured mouse type-1 astrocytes corresponds to in vivo findings (Somjen, 1979). Since there is no reason to believe that mouse and rat astrocytes differ fundamentally in vivo, it is likely that the differences seen in vitro are a result of subtle differences in culturing requirements.

4.1.3. Modulation of Protein Kinase Activity

Protein kinase C activation by phorbolesters causes a dramatic decrease in potassium permeability in mouse type-1 astrocytes (maybe mimicking a transmitter response) (Hertz, 1990a). No corresponding effect is seen with rat type-1 astrocytes (B. H. J. Juurlink and L. Hertz, unpublished experiments), but this could be because of the low potassium permeability in these cultures under culture conditions. Therefore, studies on the effects of neurotransmitters and other agents on potassium channel activity in rat type-1 astrocytes in primary cultures should be interpreted with great caution.

In contrast to the morphological response to phorbolesters by rat type-1 astrocytes observed by Mobley et al. (1986), phorbolester administration had no effect on mouse type-1 astrocytes in our experiments. However, we also did not observe any morphological effect in rat type-1 astrocyte cultures. These differences from the results of Mobley et al. (1986) probably reflect differences in the composition of the growth medium used. In contrast, staurosporine, a protein kinase inhibitor causes a rapid and dramatic change in the morphology of rat type-1 astrocytes (Fig. 6F) at concentrations ranging from 50–150 nM, but is toxic to mouse astrocytes at these concentrations (Fig. 6E). Staurosporine is a potent protein kinase C inhibitor (Tamaoki et al., 1986) but also affects other protein kinases (e.g., Yanagihara et al., 1991).

4.2. Regional and Temporal Differences

4.2.1. Type-1 Astrocytes

The type-1 astrocytes in culture that have been investigated most extensively have been obtained from the cerebral cortex. Type-1 astrocytic cultures can also be obtained from other brain regions such as the hippocampus, striatum, mesencephalon, cerebellum, and brain stem (e.g., Hansson and Rönnbäck, 1989; Rousselet et al., 1990). Preparations from different regions may exhibit morphologies somewhat different from cortical type-1 astrocytes (note Fig. 12). They may also exhibit different functional properties such as different rates of glutamate uptake (Schousboe and Divac, 1979), different responses to dopamine

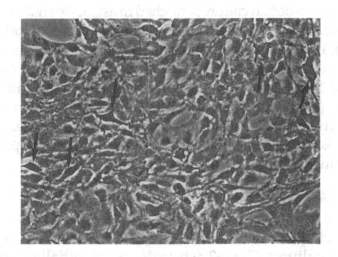

Fig. 12. Phase contrast micrograph of living type-1 astrocytes in cultures established from cerebella of 6-d-old mouse pups. Relatively large numbers of macrophages/microglia (arrows) are present in these cultures. In cerebellar cultures established from mouse pups of this age, many granule cell neurons may be present during the early part of the culture period. With regular feeding, these neurons degenerate over the first few weeks. The degenerating neurons appear to stimulate the proliferation of the macrophages/microglia. Bar = 50 μm.

(Hansson et al., 1984), and differences in how they affect the development of neuronal morphology (Rousselet et al., 1990). When preparing astrocytes from any particular brain region, it is important to pay attention to the stage of development of that particular brain region at the time the culture is prepared. Some of the differences in astrocyte characteristics previously noted may be related to the developmental stage of the particular brain region from which the cultures have been prepared. As an example, Doering et al. (1983) have observed that when astrocytes obtained from cultures initiated from prenatal mouse neopallium were transplanted into mouse brain, their nuclear diameters were indistinguishable from astrocytes that had developed *in situ*. In contrast, when astrocytes obtained from cultures initiated from newborn mouse neopallium were transplanted into mouse brain, their nuclear diameters were significantly larger than astrocytes developed *in situ*.

One can obtain cultures enriched in astrocytes not only from embryonic and fetal CNS, but also from adult brain (Norton and Farooq, 1989). However, there is generally no advantage to starting cultures from adult animals since the cultured cells are derived from glial cell precursors remaining in the adult brain anyway (Norton and Farooq, 1989). One situation when it might be advantageous to be able to obtain cultures from adult tissue is the possibility of developing astrocytic cultures from postmortem human brains (Pontén and MacIntyre, 1968).

4.2.2. Type-2 Astrocytes

The procedure for obtaining type-2 astrocytes described in this chapter deals with astrocytes isolated from cerebral cortical cultures. Type-2 astrocytes were initially described in cultures of optic nerve (Raff et al., 1983a,b), but it is not possible to isolate large numbers of type-2 astrocytes from such cultures. Type-2 astrocytes have also been described in cultures established from postnatal rat cerebellum, and it is possible to develop cultures enriched in type-2 astrocytes from the cerebellum (Levi et al., 1986; Trotter and Schachner, 1989); however, to date, no cerebellar type-2 astrocytic cultures have been reported with the same enrichment and stability over time as the cerebral type-2 cultures described in Section 2.3.3. We have also observed O2A progenitor cells in cultures established from E15 mouse spinal cord (B. H. J. Juurlink, unpublished observations) and in principle, it should be possible to develop cultures highly enriched in type-2 astrocytes from the spinal cord.

O2A progenitors have been observed not only in cultures established from embryonic and perinatal rodent CNS, but also in cultures established from adult rat optic nerve (ffrench-Constant and Raff, 1986b; Wolswijk and Noble, 1989). This suggests that one should be able to isolate O2A progenitor cells from other regions of the adult CNS and develop highly enriched type-2 astrocyte cultures from these precursors. It is unlikely, however, that one can obtain large numbers of type-2 astrocytes from adult CNS.

5. Concluding Remarks

During the scarcely 20 years since the publication of the technique for culturing astrocytes by Booher and Sensenbrenner (1972), primary cultures of astrocytes have become powerful tools for investigating the differentiation and functions of astrocytes. In fact, to a large extent, our current understanding of astrocyte function in vivo is based on data collected using cultured astrocytes. This obliges us, those who use cultures, to take precautions that the methodology is not used carelessly. One should ensure that the cultures used are comprised mainly of astrocytes. Contamination of the cultures to a great extent by another cell population may result in erroneous interpretations of astrocyte functions. As an example, it appears to be commonly agreed that astrocytes secrete a variety of cytokines and act as auxiliary cells for immune responses in the CNS (Hertz et al., 1990); however, certain of these cytokines are secreted in much larger quantities by macrophages/microglia. As pointed out by Giulian (1988), contamination of "pure" astrocytic cultures with macrophages/microglia might provide incorrect information on astrocyte function. Another example is that the likely contamination of astrocyte preparations by endothelial cells when cultures are established from enzyme dissociated brain tissue may lead to erroneous interpretations of astrocytic functions (Paterson and Hertz, 1989).

Investigators should always keep in mind that primary cultures of mouse and rat astrocytes differ in cellular composition. Mouse primary astrocyte cultures consist almost exclusively of type-1 astrocytes, whereas rat primary cultures consist of a mixture of type-1 astrocytes and O2A lineage cells (i.e., O2A progenitor cells, type-2 astrocytes and/or oligodendrocytes). We have emphasized that purification of rat astrocyte cultures as described in McCarthy and de Vellis (1980) is necessary to obtain cultures consisting almost exclusively of rat type-1 astrocytes. Failure to do so might affect interpretation of some parameters to great extent (e.g., the very high glutamine synthetase activity in type-2 astrocytes) but be of little concern with regard to interpretation of other parameters. Moreover, even when one obtains purified type-1 astrocyte cultures, there are still some unexplained differences between the type-1 astrocytes in culture of the two species, e.g., differences in potassium permeability.

If the investigator keeps such caveats in mind and remembers to be cautious in interpretation if disagreement appears to exist between characteristics observed in vivo and in the cultured cells, then cultures of astrocytes become exquisite instruments by which one may winkle out astrocyte functions.

References

Abney E. R., Bartlett P. P., and Raff M. C. (1981) Astrocytes, ependymal cells, and oligodendrocytes develop on schedule in dissociated cell cultures of embryonic rat brain. *Dev. Biol.* **83**, 301–310.

Aloisi F., Agresti C., and Levi G. (1988) Establishment, characterization and evolution of cultures enriched in type-2 astrocytes. *J. Neurosci. Res.*, **21**, 188–198.

Andriezen W. L. (1893) The neuroglia elements of the brain. *Br. Med. J.* **2**, 227–230.

Angevine J. B. (1965) Time of neuron origin in the hippocampal region. An autoradiographic study in the mouse. *Exp. Neurol. Suppl.* **1**, 1–39.

Aoki C., Joh T. H., and Pickel V. M. (1987) Ultrastructural localization of immunoreactivity for beta-adrenergic receptors in cortex and neostriatum of rat brain. *Brain Res.* **437**, 264–282.

Bardakjian J., Tardy M., Pimoule C., and Gonnard P. (1979) GABA metabolism in cultured glial cells. *Neurochem. Res.* **4**, 517–527.

Barres B. A., Chun L. L. Y., and Corey D. P. (1988) Ion channel expression by white matter glia: I. Type 2 astrocytes and oligodendrocytes. *Glia* **1**, 10–30.

Barres B. A., Chun L. L. Y., and Corey D. P. (1989) A calcium current in cortical astrocytes: induction by cAMP and neurotransmitters, and permissive effects of serum factors. *J. Neurosci.* **9**, 3169–3175.

Barres B. A., Koroshetz W. J., Swartz K. J., Chun L. L. Y., and Corey D. P. (1990) Ion channel expression by white matter glia: the O2A progenitor cell. *Neuron* **4**, 507–524.

Beaudet A. and Descarries L. (1984) Fine structure of monoamine axon terminals in cerebral cortex, in *Monoamine Innervation of Cerebral Cortex* (Descarries L., Reader T. R., and Jasper H. H., eds.), Liss, New York, pp. 77–93.

Behar T., McMorris F. A., Novotny E. A., Barker J. L., and Dubois-Dalq M. (1988) Growth and differentiation properties of O-2A progenitors purified from rat cerebral hemispheres. *J. Neurosci. Res.* **21**, 168–180.

Bignami A. and Dahl D. (1977) Specificity of glial fibrillary acidic protein for astroglia. *J. Histochem. Cytochem.*, **25,** 466–469.

Bock E., Jorgensen O. S., Dittman L., and Eng L. F. (1975) Determination of brain-specific antigens in short-term cultivated rat astroglia cells and in rat synaptosomes. *J. Neurochem.* **25,** 867–870.

Bockaert J. and Ebersolt C. (1988) a-Adrenergic receptors on glial cells, in *Glial Cell Receptors* (Kimelberg H. K., ed.), Raven Press, New York, pp. 35–51.

Booher J. and Sensenbrenner M. (1972) Growth and cultivation of dissociated neurons and glial cells from embryonic chick, rat and human cells in flask cultures. *Neurobiology* **2,** 97–105.

Boulder Committee (1970) Embryonic vertebrate central nervous system: revised terminology. *Anat. Rec.* **166,** 257–266.

Brenneman D. E., Neale E. A., Foster G. A., d'Autremont S. W., and Westbrook G. L. (1987) Non-neuronal cells mediate neurotrophic action of vasoactive intestinal peptide, *J. Cell Biol.* **104,** 1603–1610.

Bullaro J. C. and Brookman D. H. (1976) Comparison of skeletal muscle monolayer cultures initiated with cells dissociated by the vortex and trypsin methods. *In Vitro* **12,** 564–570.

Cammer W. (1990) Glutamine synthetase in the central nervous system is not confined to astrocytes. *J. Neuroimmunol.* **26,** 173–178.

Cole R. and de Vellis J. (1989) Primary cultures of astrocyte and oligodendrocyte cultures from primary rat glial cultures, in *A Dissection and Tissue Culture Manual for the Nervous System* (Shahar A., de Vellis J., Vernadakis A., and Haber B. eds.), Liss, New York, pp. 121–133.

Cooper A. J. L., Vergara F., and Duffy T. E. (1983) Cerebral glutamine synthetase, in *Glutamine, Glutamate, and GABA in the Central Nervous System* (Hertz L., Kvamme E., McGeer E. G., and Schousboe A. eds.), Liss, New York, pp. 77–93.

Cornell-Bell A. H., Finkbeiner S. M., Cooper M. S., and Smith S. J. (1990) Glutamate induces calcium waves in cultured astrocytes: long-range glial signaling. *Science* **247,** 470–473.

D'Amelio F., Eng L. F., and Gibbs M. A. (1990) Glutamine synthetase immunoreactivity is present in oligodendroglia of various regions of the central nervous system. *Glia* **3,** 335–341.

Devon R. M. and Juurlink B. H. J. (1988) Structural complexity of primary cultures of astrocytes as revealed by transverse sections. *Glia* **1,** 151–155.

Devon R. M. and Juurlink B. H. J. (1989) Dynamic morphological responses of mouse astrocytes in primary cultures following medium changes. *Glia* **2,** 266–272.

Doering L. C., Fedoroff S., and Devon R. M. (1983) Fibrous astrocytes and reactive astrocyte-like cells in transplants of cultured astrocyte precursor cells. *Dev. Brain Res.* **6,** 183–189.

Drejer J., Larsson O. M., and Schousboe A. (1982) Characterization of L-glutamate uptake into and release from astrocytes and neurons cultured from different brain regions. *Exp. Brain Res.* **47,** 259–269.

Ebersolt C., Perez M., and Bockaert J. (1981) α_1 and α_2 adrenergic receptors in mouse brain astrocytes from primary cultures. *J. Neurosci. Res.* **6**, 643–652.

Eisenbarth G. S., Walsh F. S., and Niremberg M. (1979) Monoclonal antibody to a plasma membrane antigen of neurons. *Proc. Natl. Acad. Sci.* **76**, 4913–4917.

Eng L. F., Vanderhaeghen J. J., Bignami A., and Gerstl B. (1971) An acidic protein isolated from fibrous astrocytes. *Brain Res.* **28**, 351–354.

Facci L., Skaper S. D., Levin D. L., and Varon S. (1987) Dissociation of the stellate morphology from intracellular cyclic AMP levels in cultured rat brain astroglial cells: Effects of ganglioside GM_1 and lysophosphatidylserine. *J. Neurochem.* **48**, 566–573.

Fages C., Khelil M., Rolland B., Bridoux A. M., and Tardy M. (1988) Glutamine synthetase: a marker of an astroglial subpopulation in primary cultures of defined brain areas. *Dev. Neurosci.* **10**, 47–56.

Fedoroff S. (1977) Primary cultures, cell lines, and cell strains, in *Cell, Tissue and Organ Cultures in Neurobiology* (Fedoroff S. and Hertz L., eds.), Academic Press, New York, pp. 265–286.

Fedoroff S. and Hall C. H. (1979) Effect of horse serum on neural cell differentiation in tissue culture. *In Vitro* **15**, 641–648.

Fedoroff S., White R., Neal J., Subramanyan L., and Kalnins V. I. (1983) Astrocyte cell lineage. II. Mouse fibrous astrocytes and reactive astrocytes in cultures have vimentin- and GFP-containing intermediate filaments. *Dev. Brain Res.* **7**, 303–315.

Fedoroff S., McAuley W. A. J., Houle J. D., and Devon R. M. (1984) Astrocyte cell lineage. V. Similarity of astrocytes that form in the presence of dBcAMP in cultures to reactive astrocytes in vivo. *J. Neurosci. Res.* **12**, 15–27.

ffrench-Constant C. and Raff M. C. (1986a) The oligodendrocyte-type-2 astrocyte lineage is specialized for myelination. *Nature* **323**, 335–338.

ffrench-Constant C. and Raff M. C. (1986b) Proliferating bipotential glial progenitor cells in adult rat optic nerve. *Nature* **319**, 499–502.

Foote S. L., Bloom F. E., and Aston-Jones G. (1983) Nucleus locus ceruleus: new evidence of anatomical and physiological specificity. *Physiol. Rev.* **63**, 844–914.

Gallo V., Bertolotto A., and Levi G. (1987) The proteoglycan chondroitin sulfate is present in a subpopulation of cultured astrocytes and in their precursors. *Dev. Biol.* **123**, 282–285.

Gallo V., Giovannini C., Suergiu R., and Levi G. (1989) Expression of excitatory amino acid receptors by cerebellar cells of the type-2 astrocyte lineage. *J. Neurochem.* **52**, 1–9.

Giulian D. (1988) The immune response and astrogliosis: control of astroglial growth by secretion of microglial peptides, in *The Biochemical Pathology of Astrocytes* (Norenberg M. D., Hertz L., and Schousboe A., eds.), Liss, New York, pp. 91–105.

Goldman J. E. and Chiu F. C. (1984) Dibutyryl cyclic AMP causes intermediate filament accumulation and actin reorganization in astrocytes. *Brain Res.* **306**, 85–95.

Hallermayer K., Harmening C., and Hamprecht B. (1981) Cellular localization and regulation of glutamine synthetase in primary cultures of brain cells from newborn mice. *J. Neurochem.*, **37**, 43–52.

Hansson E. and Rönnbäck L. (1988) Neurons from substantia nigra increase the efficacy and potency of second messenger arising from striatal astroglia dopamine receptor. *Glial* **1**, 393–397.

Hansson E., Rönnbäck L., and Sellström Å. (1984) Is there a 'dopaminergic glial cell'? *Neurochem. Res.* **9**, 679–689.

Hansson E. and Rönnbäck L. (1989) Primary cultures of astroglia and neurons from different brain regions, in *A Dissection and Tissue Culture Manual for the Nervous System* (Shahar A., de Vellis J., Vernadakis A., and Haber B., eds.), Liss, New York, pp. 92–104.

Hao C., Guilbert L. J., and Fedoroff S. (1990) Production of colony-stimulating factor-1 (CSF-1) by mouse astroglia in vitro. *J. Neurosci. Res.* **27**, 314–323.

Hatten M. E. (1985) Neuronal regulation of astroglial morphology and proliferation *in vitro. J. Cell Biol.* **100**, 384–396.

Hertz L. (1989a) Is Alzheimer's disease an anterograde degeneration, originating in the brainstem, and disrupting metabolic and functional interactions between neurons and glial cells. *Brain Res. Rev.* **14**, 335–353.

Hertz L. (1989b) Functional interactions between neurons and glial cells, in *Regulatory Mechanisms of Neuron to Vessel Communication in Brain* (Govoni S., Battani F., Magnoni M. S., and Trabluchi M., eds.), Springer, Heidelberg, pp. 271–303.

Hertz L. (1990a). Regulation of potassium homeostasis by glial cells, in *Differentiation and Functions of Glial Cells* (Levi G., ed.), Liss, New York, pp. 225–234.

Hertz L. (1990b) Dibutyryl cyclic AMP treatment of astrocytes in cultures as a substitute for normal morphogenic and 'functiogenic' transmitter signals, in *Molecular Aspects of Development and Aging of the Nervous System* (Lauder J., Privat A., Giacobini E., Timiras P., and Vernadakis A., eds.), Plenum, New York, pp. 227–243.

Hertz L. (1991) Neuronal-astrocytic interactions in brain development, brain function and brain disease, in *Plasticity and Regeneration of the Nervous System* (Timiras P. S., Privat A., Giacobini E., Lauder J., and Vernadakis A., eds.), Plenum, New York, pp. 143–159.

Hertz L., Bock E., and Schousboe A. (1978) GFA content, glutamate uptake and activity of glutamate metabolizng enzymes in differentiating mouse astrocytes in primary cultures. *Dev. Neurosci.* **1**, 226–238.

Hertz L., Juurlink B. H. J., Fosmark H., and Schousboe A. (1982) Astrocytes in primary culture, in *Neuroscience Approached Through Cell Culture* (Pfeiffer S. E., ed.), CRC Press, Boca Raton, FL, pp. 157–174.

Hertz L., Juurlink B. H. J., and Szuchet S. (1985a) Cell cultures, in *Handbook of Neurochemistry* (Lajtha A., ed.), Plenum, New York, pp. 603–661.

Hertz L., Juurlink B. H. J., Szuchet S., and Walz W. (1985b) Cell and tissue cultures, in *Neuromethods, vol. 1* (Boulton A. and Baker G. B., eds.), Humana Press, Clifton, NJ, pp. 117–167.

Hertz L., Juurlink B. H. J., Hertz E., Fosmark H., and Schousboe A. (1989a) Preparation of primary cultures of mouse (rat) astrocytes, in *A Dissection and Tissue Culture Manual for the Nervous System* (Shahar A., de Vellis J., Vernadakis A., and Haber B., eds.), Liss, New York, pp. 104–108.

Hertz L., Bender A. S., Woodbury D. M., and White H. S. (1989b) Potassium–stimulated calcium uptake in astrocytes and its potent inhibition by nimodipine. *J. Neurosci. Res.* 22, 209–215.

Hertz L., McFarlin D. E., and Waksman B. H. (1990) Astrocytes: auxillary cells for immune response in the central nervous system? *Immunol. Today* 11, 265–268.

Hinds J. W. (1966) Autoradiographic study in the mouse olfactory bulb. I. Time of origin of neurons and neuroglia. *J. Comp. Neurol.* 134, 287–304.

Holopainen I. and Kontro P. (1986) High–affinity uptake of taurine and b-alanine in primary cultures of rat astrocytes. *Neurochem. Res.* 11, 211–215.

Hughes S. M., Lillien L. E., Raff M. C., Rohrer H., and Sendtner M. (1988) Ciliary neurotrophic factor induces type-2 astrocyte differentiation in culture. *Nature* 335, 70–73.

Hydén H. (1959) Quantitative assay of compounds in isolated, fresh nerve cells and glial cells from control and stimulated animals. *Nature* 184, 433–435.

Ingraham C. A. and McCarthy K. D. (1989) Plasticity of process-bearing glial cell cultures from neonatal rat cerebral cortical tissue. *J. Neurosci.* 9, 63–69.

Janzer R. C. and Raff M. C. (1987) Astrocytes induce blood-brain properties in endothelial cells. *Nature* 325, 253–257.

Johnstone S. R., Levi G., Wilkin G. P., Schneider A., and Ciotti M. T. (1986) Subpopulations of rat cerebellar cells in primary culture: morphology, cell surface antigens and [^{3}H]GABA transport. *Dev. Brain Res.* 24, 63–75.

Juurlink B. H. J. and Devon R. M. (1990) Macromolecular translocation—a possible function of astrocytes. *Brain Res.* 533, 73–77.

Juurlink B. H. J. and Hertz L. (1985) Plasticity of astrocytes in primary cultures: an experimental tool and a reason for methodological caution. *Dev. Neurosci.* 7, 263–267.

Juurlink B. H. J. and Hertz L. (1991) Establishment of highly enriched type-2 astrocyte cultures and quantitative determination of intense glutamine synthetase activity in these cells. *J. Neurosci. Res.* 30, 531–539.

Juurlink B. H. J., Fedoroff S., Hall C., and Nathaniel E. J. H. (1981a) Astrocyte cell lineage. I. Astrocyte progenitor cells in mouse neopallium. *J. Comp. Neurol.* 200, 375–391.

Juurlink B. H. J., Schousboe A., Jørgensen O. S., and Hertz L. (1981b) Induction by hydrocortisone of glutamine synthetase in mouse primary astrocyte cultures. *J. Neurochem.* 36, 136–148.

Kosaka T. and Hama K. (1986) Three-dimensional structure of astrocytes in the rat dentate gyrus. *J. Comp. Neurol.* **249**, 242–260.

Lauder J. M. (1988) Neurotransmitters as morphogens. *Progr. Brain Res.* **73**, 365–387.

Levi G., Wilkin G. P., Ciotti M. T., and Johnstone S. (1983) Enrichment of differentiated stellate astrocytes in cerebellar interneuron cultures as studied by GFAP immunofluorescence and autoradiographic uptake patterns with [³H] D-aspartate and [³H]GABA. *Dev. Brain Res.* **10**, 227–241.

Levi G., Gallo V., and Ciotto M. T. (1986) Bipotential precursors of putative type 2 astrocytes and oligodendrocytes in rat cerebellar cultures express distinct surface features and 'neuron-like' g-amino acid transport. *Proc. Natl. Acad. Sci.* **83**, 1504–1508.

Lillien L. E., Sendtner M., and Raff M. C. (1990) Extracellular matrix-associated molecules collaborate with ciliary neuronotrophic factor to induce type-2 astrocyte development. *J. Cell Biol.* **111**, 635–644.

MacVicar B. A. (1984) Voltage-dependent calcium channels in glial cells. *Science* **226**, 1345–1347.

Manthorpe M., Adler R., and Varon S. (1979) Development, reactivity and GFA immunofluorescence of astroglial containing monolayer cultures from rat cerebrum. *J. Neurocytol.* **8**, 605–621.

McCarthy K. D. and de Vellis J. (1980) Preparation of separate astroglial and oligodendroglial cell cultures from rat cerebral tissue. *J. Cell Biol.* **85**, 890–902.

McCarthy K. D., Salm A., and Lerea L. S. (1988) Astroglial receptors and their regulation of intermediate filament protein phosphorylation, in *Glial Cell Receptors* (Kimelberg H. K., ed.), Raven Press, New York, pp. 1–22.

Mearow K. M., Mill J. F., and Vitkovic L. (1989) The ontogeny and localization of glutamine synthetase gene expression in rat brain. *Mol. Brain Res.* **6**, 223–232.

Meier E., Hertz L., and Schousboe A. (1991) Neurotransmitters as developmental signals. *Neurochem. Int.* **19**, 1–15.

Miller R. H. and Raff M. C. (1984) Fibrous and protoplasmic astrocytes are biochemically and developmentally distinct. *J. Neurosci.* **4**, 585–592.

Miller R. H., David S., Patel R., Abney E. A., and Raff M. C. (1985) A quantitative immunohistochemical study of macroglial cell development in the rat optic nerve: *in vivo* evidence for two distinct astrocytic lineages. *Dev. Biol.* **11**, 35–41.

Miller R. H., Abney E. R., David S., ffrench-Constant C., Lindsay R., Patel R., Stone J., and Raff M. C. (1986) Is reactive gliosis a property of a distinct subpopulation of astrocytes? *J. Neurosci.* **6**, 22–29.

Minturn J. E., Black J. A., Angelides K. J., and Waxman S. G. (1990) Sodium channel expression detected with antibody 7493 in A2B5+ and A2B5⁻ astrocytes from rat optic nerve in vitro. *Glia* **3**, 358–367.

Mobley P. L., Scott S. L., and Cruz E. G. (1986) Protein kinase C in astrocytes: a determinant of cell morphology. *Brain Res.* **398**, 366–369.

Moonen G., Cam Y., Sensenbrenner M., and Mandel P. (1975) Variability of the effects of serum-free medium, dibutyryl-cyclic AMP on newborn rat astroblasts. *Cell Tissue Res.* **163**, 365–372.

Narlieva N. (1988) Multilamellar glial envelopes of synapses in the pontine nuclei of the cat. *Acta Anat.* **131**, 227–230.

Norenberg M. D. and Martinez-Hernandez A. (1979) Fine structural localization of glutamine synthetase in astrocytes of rat brain. *Brain Res.* **161**, 303–310.

Norton W. T. and Poduslo S. E. (1970) Neuronal soma and whole neuroglia of rat brain: a new isolation technique. *Science* **167**, 1144–1146.

Norton W. T. and Farooq M. (1989) Astrocytes cultured from mature brain derive from glial precursor cells. *J. Neurosci.* **9**, 769–775.

Nowak L., Ascher P., and Berwald-Netter Y. (1987) Ionic channels in mouse astrocytes in culture. *J. Neurosci.* **7**, 101–109.

Palay S. L. and Chan-Palay V. (1974) *Cerebellar Cortex. Cytology and Organization.* Springer-Verlag, New York.

Paterson I. A. and Hertz L. (1989) Sodium-independent transport of noradrenaline in mouse and rat astrocytes in primary culture. *J. Neurosci. Res.* **23**, 71–77.

Penfield W. (1932) *Cytology & Cellular Pathology of the Nervous System. vol. 2.* Paul B. Hoeber Inc., New York.

Peng L., Juurlink B. H. J., and Hertz L. (1991) Development of cerebellar granule cells in the presence and absence of excess extracellular potassium—do the two culture systems provide a means of distinguishing between events in transmitter-related and non-transmitter-related glutamate pools. *Dev. Brain Res.* **63**, 1–12.

Peters A., Palay S. L., and Webster H. deF. (1976) *The Fine Structure of the Nervous System: the Neurons and Supporting Cells.* W. B. Saunders Co., Philadelphia.

Pontén J. and MacIntyre E. H. (1968) Long term culture of normal and neoplastic human glia. *Act. Pathol. Microbiol. Scand.* **74**, 465–486.

Pope A. (1978) Neuroglia: quantitative aspects, in *Dynamic Properties of Glia Cells* (Schoffeniels E., Franck G., Hertz L., and Tower D. B., eds.), Pergamon Press, Oxford, pp. 13–20.

Privat A. (1975) Postnatal gliogenesis in the mammalian brain. *Int. Rev. Cytol.* **40**, 281–323.

Quandt F. N. and MacVicar B. A. (1986) Calcium activated potassium channels in cultured astrocytes. *Neuroscience* **19**, 29–41.

Raff M. C. (1989) Glial cell diversification in the rat optic nerve. *Science* **243**, 1450–1455.

Raff M. C. and Miller R. H. (1984) Glial cell development in the rat optic nerve. *Trends Neurosci.* **7**, 469–472.

Raff M. C., Miller R. H., and Noble M. (1983a) A glial progenitor cell that develops *in vitro* into an astrocyte or an oligodendrocyte depending upon culture medium. *Nature* **303**, 390–396.

Raff M. C., Abney E. R., Cohen J., Lindsay R., and Noble M. (1983b) Two types of astrocytes in cultures of developing rat white matter: differences in morphology, surface gangliosides, and growth characteristics. *J. Neurosci.* 3, 1289–1300.

Raff M. C., Mirsky R., Fields K. L., Lisak R. P., Dorfman S. H., Silberberg D. H., Gregson N. A., Liebowitz S., and Kennedy M. C. (1978) Galactocerebroside is a specific cell-surface antigenic marker for oligodendrocytes in culture. *Nature* 274, 813–816.

Ramon y Cajal S. (1909) *Histologie du Systeme Nerveux de l'Homme et des Vertébrés*. Masson & Cie., Paris.

Reichelt W., Dettmer D., Brückner G., Brust P., Eberhardt W., and Reichenbach A. (1989) Potassium as a signal for both proliferation and differentiation of rabbit retina (Müller) glia growing in culture. *Cell Signal.* 1, 187–194.

Reynolds R. and Herschkowitz H. (1987) Oligodendroglial and astroglial heterogeneity in mouse primary central nervous system culture as demonstrated by differences in GABA and D-aspartate transport and immunocytochemistry. *Dev. Brain Res.* 36, 13–25.

Rousselet A., Autillo-Touati A., Araud D., and Prochiantz A. (1990) In vitro regulation of neuronal morphogenesis and polarity by astrocyte-derived factors. *Dev. Biol.* 137, 33–45.

Salm A. K. and McCarthy K. D. (1990) Norepinephrine-evoked calcium transients in cultured type 1 astroglia. *Glia* 3, 529–538.

Schachner M., Hedley-White E. T., Hsu D. W., Schoonmaker G., and Bignami A. (1977) Ultrastructural localization of glial fibrillary acidic protein in mouse cerebellum by immunoperoxidase staining. *J. Cell Biol.* 75, 67–73.

Schousboe A., Hertz L., and Svenneby G. (1977) Uptake and metabolism of GABA in astrocytes cultured from dissociated mouse brain hemispheres. *Neurochem. Res.* 2, 217–229.

Schousboe A. and Divac I. (1979) Differences in glutamate uptake in astrocytes cultured from different brain regions. *Brain Res.* 177, 407–409.

Schousboe A., Drejer J., and Hertz L. (1988) Uptake and release of glutamate and glutamine in neurons and astrocytes in primary cultures, in *Glutamine and Glutamate in Mammals* (Kvamme E., ed.), CRC Press, Boca Raton, FL, pp. 21–38.

Sensenbrenner M., Devillers G., Bock E., and Porte A. (1980) Biochemical and ultrastructural studies of cultured rat astroglial cells. Effect of brain extract and dibutyryl cyclic AMP on glial fibrillary acidic protein and glial filaments. *Differentiation* 17, 51–61.

Shank R. P., Bennett G. S., Freytag S. O., and Campbell G. LeM. (1985) Pyruvate carboxylase: an astrocyte-specific enzyme implicated in the replenishment of amino acid neurotransmitter pools. *Brain Res.* 329, 364–367.

Shein H. M. (1965) Propagation of human fetal spongioblasts and astrocytes in dispersed cell cultures. *Exp. Cell Res.* **40,** 554–569.

Somjen G. G. (1979) Extracellular potassium in the mammalian central nervous system. *Ann. Rev. Physiol.* **41,** 159–177.

Somjen G. G. (1988) Nervenkitt: Notes on the history of the concept of neuroglia. *Glia* **1,** 2–9.

Stieg P. E., Kimelberg H. K., Mazurkiewicz J. E., and Banker G. A. (1980) Distribution of glial fibrillary acidic protein and fibronectin in primary astroglial cultures from rat brain. *Brain Res.* **199,** 493–500.

Stallcup W. B. and Beasley L. (1987) Bipotential glial precursor cells of the optic nerve express NG2 proteoglycan. *J. Neurosci.* **7,** 2737–2744.

Stone E. A. and Ariano M. A. (1989) Are glial cells targets of the central noradrenergic system? *Brain Res. Rev.* **14,** 297–309.

Sykova E. (1983) Extracellular K^+ accumulation in the central nervous system. *Progr. Biophys. Mol. Biol.* **42,** 135–189.

Tamaoki T., Nomoto H., Takahashi I., Kato Y., Morimoto M., and Tomita F. (1986) Staurosporine, a potent inhibitor of phospholipid/Ca^{2+} dependent protein kinase. *Biochem. Biophys. Res. Commun.* **135,** 397–402.

Tansey F. A., Farooq M., and Cammer W. (1991) Glutamine synthetase in oligodendrocytes and astrocytes: new biochemical and immunocytochemical evidence. *J. Neurochem.* **56,** 266–272.

Temple S. and Raff M. C. (1985) Differentiation of a bipotential glial progenitor cell in single cell culture. *Nature* **313,** 223–225.

Trimmer P. A., Evans T., Smith M. M., Harden T. K., and McCarthy K. D. (1984) Combination of immunocytochemistry and radioligand receptor assay to identify β-adrenergic receptor subtypes on astroglia in vitro. *J. Neurosci.* **4,** 1598–1606.

Trotter J. and Schachner M. (1989) Cells positive for the O4 surface antigen isolated by cell sorting are able to differentiate into astrocytes or oligodendrocytes. *Dev. Brain Res.* **46,** 115–122.

Turing A. M. (1952) The chemical basis of morphogenesis. *Trans. Roy. Soc. Lond. Ser. B* **237,** 37–72.

Varon S. and Raiborn C. W. (1969) Dissociation, fractionation, and culture of embryonic brain cells. *Brain Res.* **12,** 180–199.

Walz W. (1989) Role of glial cells in the regulation of the brain ion microenvironment. *Progr. Neurobiol.* **33,** 309–333.

Walz W. and Hertz L. (1983) Functional interactions between neurons and astrocytes. II. Potassium homeostasis at the cellular level. *Progr. Neurobiol.* **20,** 133–183.

Walz W. and Kimelberg H. K. (1985) Differences in cation transport properties of primary astrocyte cultures from mouse and rat brain. *Brain Res.* **340,** 333–340.

Walz W. and Mozaffari B. (1987) Culture environment and channel-mediated potassium fluxes in astrocytes. *Brain Res.* **412,** 405–408.

Wandosell F., Bovolenta P., and Nieto-Sampedro M. (1990) Reactive astrocytes and DiBcAMP-treated astrocytes have different surface markers. *Soc. Neurosci. Abst.* **16,** 351.

Warringa R. A. J., van Berlo M. F., Klein W., and Lopes-Cardozo M. (1988) Cellular localization of glutamine synthetase and lactate dehydrogenase in oligodendrocyte-enriched cultures from rat brain. *Neurochemistry* **50,** 1461–1468.

Whitaker-Azmitia P. M. (1988) Astroglial serotonin receptors, in *Glial Cell Receptors* (Kimelberg H. K., ed.), Raven Press, New York, pp. 107–120.

Wolswijk G. and Noble M. (1989) Identification of an adult-specific glial progenitor cell. *Development* **105,** 387–400.

Yanagihara N., Tachikawa E., Izumi F., Yasugawa S., Yamamoto H., and Miyamoto E. (1991) Staurosporine: an effective inhibitor for Ca^{2+}/calmodulin–dependent protein kinase II. *J. Neurochem.* **56,** 294–298.

Yu A. C. H., Drejer J., Hertz L., and Schousboe A. (1983) Pyruvate carboxylase activity in primary cultures of astrocytes and neurons. *J. Neurochem.* **41,** 1484–1487.

Oligodendrocytes

Jean de Vellis
and Araceli Espinosa de los Monteros

1. Introduction

Oligodendrocytes were first described by Robertson in 1899 as small cells with a few processes of variable length. In 1921, del Rio Hortega published a detailed histological description of these cells and coined the descriptive word "oligodendrocyte." The localization of these cells along axonal tracts, around neuronal cell bodies, and in proximity to capillaries served as the basis for del Rio Hortega's classification of oligodendrocytes into three subgroups: interfascicular, perineural satellite, and perivascular. The heterogeneity of oligodendrocytes has also been observed at the ultrastructural level and by the use of immunocytochemical markers (for review and ref. *see* Espinosa de los Monteros and de Vellis, 1988; Goldman and Vaysse, 1991). However, there is no evidence for more than one oligodendrocyte lineage (Skoff and Knapp, 1991).

The main function ascribed to oligodendrocytes is the elaboration of extensive membranes that they wrap around axons to form the compact multilayered myelin sheath (Bunge et al., 1962). One oligodendrocyte can myelinate up to 50 axons. Once myelin is elaborated, the oligodendrocyte remains vital for its maintenance and turnover. Myelin itself is necessary for saltatory conduction of the nerve impulse. Absence of myelin is lethal in humans and animals that express a mutation in myelin proteins, such as proteolipid protein (PLP) or myelin basic protein (MBP).

From: *Neuromethods, Vol. 23: Practical Cell Culture Techniques*
Eds: A. Boulton, G. Baker, and W. Walz ©1992 The Humana Press Inc.

Demyelinating disorders in adults, such as multiple sclerosis, also lead to neurological dysfunctions. Besides myelin, oligodendrocytes seem to play a major role in iron homeostasis and storage in the central nervous system (CNS). Oligodendrocytes are the major source of the iron transport glycoprotein (for review, *see* Espinosa et al., 1989).

Oligodendrocyte development can be divided into two distinct but chronologically overlapping phases; the first includes the generation and proliferation of progenitor cells, whereas the second is characterized by the sequential activation of genes involved in myelinogenesis and iron metabolism. Although the biochemical composition and molecular organization of the myelin membrane are well understood, our knowledge of the regulation of key myelin and oligodendroglial genes is still rudimentary. The phenotypic and molecular characterization of oligodendrocyte progenitor proliferation and differentiation has been largely derived from cell culture studies (for review and ref., *see* Norton, 1984; Kumar and de Vellis, 1987; Noble, 1991). Myelination in vitro was first demonstrated by Peterson and Murray in 1955 in organotypic (slice, explant) cultures (Bornstein and Murray, 1958). In this type of culture, the three-dimensional organization of nervous tissue is not disrupted, thus preserving the normal relationships between the various cell types. Furthermore, developmental and maturation events follow a time-course that reproduces the in vivo situation (Breen and de Vellis, 1975). Reaggregating cultures established from dissociated perinatal brain cells also lead to myelination, but to a lesser extent than explant cultures (Matthieu et al., 1978; Lu et al., 1980). In dissociated brain cell cultures, myelin-like structures have been observed both in the presence and absence of neurons (for review, *see* Rome et al., 1986; Espinosa de los Monteros, 1987; Knapp et al., 1987).

Several investigators have studied the mechanisms involved in myelinogenesis and relevant cell-to-cell interactions by using oligodendrocytes at premyelinating and mature stages (Espinosa de los Monteros et al., 1986; Szuchet et al., 1986). The preparation of cultures established from newborn dissociated rat brain cells gives rise to mixed glial cultures, containing both oligodendrocytes and astrocytes (Labourdette et al., 1979; McCarthy et al., 1978). This type of primary culture is the most commonly used. It is grown either in the presence of fetal calf serum (McCarthy et al., 1978; McCarthy and de Vellis, 1980) or calf serum (Labourdette et al., 1979; Rome et al., 1986). The preparation of this culture system will be detailed in Section 3.2. The microscopic observation of these cultures provides a view of the progressive evolution of cell morphology and cell types. Five to six days after plating, a layer of flat polygonal astrocytes (Type 1 astrocytes; *see* Noble, 1991) has formed. Within the next 48 h, phase-dark cells appear on top of the astrocytic layer. Their number increases rapidly, forming little rows of cells, single cells, or clusters (Espinosa de los Monteros et al., 1985; Holmes et al., 1988). By day nine, proliferation has slowed down considerably, and the cell bodies become round and birefringent, extending thin long cell processes. Cell clumps increase in size and number (Holmes et al., 1988). These birefringent cells derive from the flat phase-dark cells that first appeared at 5–6 d (Espinosa de los Monteros et al., 1985). The cultures remain heterogenous up to several weeks, containing progenitor cells as well as mature oligodendrocytes (Holmes et al., 1988; Espinosa de los Monteros et al., 1989). In 1980, McCarthy and de Vellis developed a simple method to obtain pure oligodendrocytes; the process is based on two properties of the mixed glial cultures.

First, the oligodendrocytes are located on top of astrocytes. Second, the adhesion strength of oligodendrocytes to astrocytes at early stages (6–10 d of culture) is weaker than the strength of adhesion of astrocytes to the culture dish plastic. The current status of this technique is described in Section 3.3. This technique allows preparation of pure oligodendroglial cells at relatively early ages in vitro, providing progenitor cells and preoligodendrocytes (*see* Section 2. for definitions, and Gard and Pfeiffer, 1990; Saneto and de Vellis, 1985).

Oligodendrocytes at 2–3 wk in primary culture have increased their adhesion; consequently, the cells can no longer be easily detached by the shake-off procedure of McCarthy and de Vellis. In addition to high initial plating cell density, Labourdette et al. (1979,1980) substituted calf serum for fetal calf serum. Calf serum stimulates oligodendrocyte maturation. These primary cultures contain the maximum of oligodendrocytes around 15 d in vitro. These cells form clusters from which branched processes emerge. Enzymatic and mechanical procedures failed to detach the oligodendrocyte from the underlying astrocytes. An oligodendrocyte selective medium (OSM) was developed by Espinosa de los Monteros et al. (1986) that solely permitted the survival of oligodendrocytes, whereas astrocytes died. Since OSM did not allow the long-term survival of oligodendrocytes, an oligodendrocyte defined medium (OLDEM) was optimized (Espinosa de los Monteros et al., 1988) for long-term survival (*see* Section 3.6.2.).

2. Lineage and Developmental Cell Markers

Glial cell development, in particular, the development of oligodendrocyte in vitro, is the best characterized system of CNS development (Arenander and de Vellis, 1989; Noble, 1991). The current success in cell identification, using antibodies against glial-specific antigens and molecular probes for detecting cell type-specific mRNAs, has led to a rapid increase in our under-standing of the key signaling agents that regulate the rate and direction of glial cell differentiation along specific lineage path-ways (Cameron and Rakic, 1991; Noble, 1991; Skoff and Knapp, 1991). The mature CNS is composed of three main classes of glial cells: oligodendrocytes, astrocytes, and microglia. Starting from early stem cells, Type 1 astrocytes develop along a separate lin-eage from oligodendrocytes and Type 2 astrocytes. Oligoden-drocytes are generated postnatally and pass through a series of cell phenotypes from undifferentiated stem cells to mature myelin-forming cells. This sequential process of maturation of oligodendrocytes can be reproduced in culture. In culture, a pro-genitor cell, the oligodendrocyte-Type 2 astrocyte (O-2A) pro-genitor, is a bipotential cell that has been well characterized in the optic nerve by Raff and his collaborators (Noble, 1991). How-ever, in vivo evidence for the Type 2 astrocyte remains elusive (Noble, 1991; Skoff and Knapp, 1991). In this chapter, we will focus solely on the differentiation of the so-called O-2A progeni-tor into oligodendrocytes.

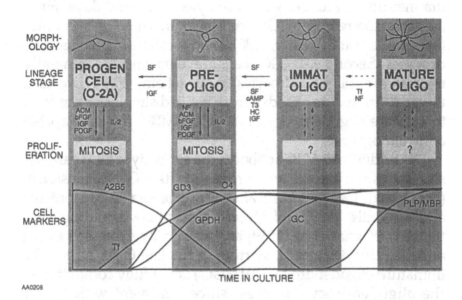

Fig. 1. The O-2A lineage. The characteristics that accompany the sequential differentiation of O-2A progenitor cells to mature oligodendrocytes in vitro. Various growth factors are listed that can regulate cell proliferation and/or differentiation. Dashed arrows indicate a particular direction may be possible, but without experimental support. The time-course of expression of cell type-specific markers is shown at the bottom. Note that each main stage of lineage development possesses a unique antigen profile. Growth factors: SF (serum factors), IGF (insulin-like growth factor), ACM (astrocyte-conditioned medium), bFGF (basic fibroblast growth factor), PDGF (platelet-derived growth factor), IL-2 (interleukin-2), NF (neuronal factor), T3 (triiodothyronine), HC (hydrocortisone), and Tf (transferrin). Cell markers: VIM (vimentin), GPDH (glycerophosphate dehydrogenase), GC (galactocerebroside), PLP (proteolipid protein), and MBP (myelin basic protein). Modified from Arenander and de Vellis.

In the rat, four main stages of development in vitro have been delineated (Fig. 1). The O-2A progenitor is a proliferating bipolar cell whose dominant phenotypic markers are glycolipids stained by the A2B5 monoclonal antibody (MAb). Cultures of O-2A progenitors, which are quite homogenous, can be easily prepared by a panning method (McKinnon et al., 1990). These cells, when cultured in defined medium containing 0.5% serum to enhance survival, rapidly differentiate in a relatively synchronous manner into mature oligodendrocytes in several days. First, 04^+/galactocerebroside (GC-) cells appear. These cells can proliferate (Gard and Pfeiffer, 1990), and we call them preoligodendrocytes. Second, 04^+/GC^+ cells develop and are generally referred to as oligodendrocytes. In our scheme (Fig. 1), we used the term "immature oligodendrocytes" to distinguish them from the mature oligodendrocytes 04^+/GC^+/MBP^+ that are capable of myelinogenesis.

In addition to A2B5 antibody, the antibody to GD3 ganglioside stains the second type progenitor (04^+/GC-). Transferrin (Espinosa de los Monteros et al., 1988a; Espinosa de los Monteros and de Vellis, 1939) and glycerol phosphate dehydrogenase (GPDH) (Kumar et al., 1989) are specifically expressed in the oligodendrocyte lineage, starting at the late O-2A stage. The immature oligodendrocyte (04^+/GC-) is not fully committed to the oligodendrocyte lineage, since treatment with ciliary neurotrophic factor (CNTF) or 10% fetal calf serum induces glial fibrillary acidic protein (GFAP) expression, the so-called Type-2 astrocyte (Noble, 1991). The plasticity of O-2A lineage is a very interesting phenomenon that indicates the importance of the environment in oligodendrocyte differentiation. This property of O-2A cells emphasizes the importance of the cell culture approach in identifying key regulatory agents. The oligodendrocyte (GC^+) also expresses 2', 3' cyclic nucleotide 5' phosphohydrolase, a myelin protein. The mature oligodendrocyte phenotype is identified by the sequential appearance of additional myelin-specific antigens (Norton, 1984): PLP, MBP, and myelin-associated glycoprotein (MAG). This classification scheme can be used to monitor the rate, direction, and extent of cell differentiation.

Proliferation of the O-2A progenitor is regulated by at least two growth factors: basic fibroblast growth factor (bFGF) (Saneto and de Vellis, 1985) and platelet-derived growth factor. These two factors together can maintain O-2A cells in the proliferative mode (McKinnon et al., 1990), whereas interleukin-2 inhibits cell division of proliferating progenitors (Saneto et al., 1986). Thyroid hormones (Matthieu et al., 1990), hydrocortisone (Kumar and de Vellis, 1987; Kumar et al., 1989) agents that increase intracellular cAMP (Breen et al., 1978; Kumar and de

Vellis, 1981), neuronal released factor(s) (Bologa et al., 1986), and insulin-like growth factor (IGF) (McMorris and Dubois-Dalcq, 1988) enhance oligodendrocyte differentiation.

3. Preparation of Oligodendrocyte Cultures from Newborn Rat Brain

3.1. Introduction

In 1974, Breen and de Vellis developed a method of dissociated rat cerebral cell culture that was optimized for the expression of the glucocorticoid-inducible glycerol phosphate dehydrogenase (GPDH) enzyme. Since GPDH is an oligodendroglial-specific marker in rats (Leveille et al., 1980), we had in fact optimized the culture conditions for oligodendrocytes. Breen and de Vellis (1974) further observed that GPDH was expressed only in cultures initially plated at high cell density (about $3 \times 10^5/cm^2$) and not at a low cell density. High cell density cultures were heterogenous. They contained process-bearing cells on top of a layer of "flat cells." At low cell-plating densities, only flat cells were present. High levels of GPDH induction were observed in all regions of the CNS, but only cultures from newborn cerebellum, cerebrum, and brain stem were characterized.

In subsequent studies from our laboratory, we chose to use newborn rat cerebrum as a starting material because of the abundance of this tissue compared to other brain areas and the purity of the oligodendrocyte preparation obtained from primary cultures. The type of cells in primary cerebral cultures was found to be dependent on the age of the starting tissue used to initiate the cultures and the length of time the cells were maintained in culture (McCarthy et al., 1978). The perinatal period, 0–2 d postnatally, was found optimum for growth of oligodendrocytes, identified on the basis of morphological, ultrastructural, and biochemical criteria (McCarthy et al., 1978; McCarthy and de Vellis, 1980). Neurons were absent from the cultures after 1 wk. The phase-dark process-bearing cells formed the top layer (Fig. 2A) and displayed typical ultrastructural features of oligodendroglial cells at various stages of differentiation. Nuclei were round and eccentrically located. Heterochromatin increased as

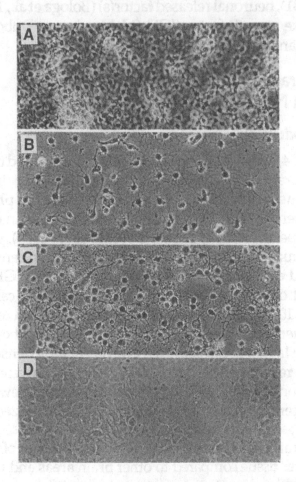

Fig. 2. **(A)** Phase contrast micrograph of primary mixed glial culture established from newborn rat brain and maintained for 12 d in DMEM/F12 medium supplemented with 10% fetal calf serum. Note the process-bearing cells, single or in clusters, over the phase-gray layer of astrocytes. **(B)** Phase contrast micrograph of oligodendrocytes purified from the primary culture (Fig. 2A) by the shaking procedure (*see* section 3.3.) and maintained in DMEM/F12 medium supplemented with 10% fetal calf serum for 5 d. Note the numerous thin cell processes radiating from the cell body. **(C)** Phase contrast micrograph of oligodendrocytes purified as in Fig. 2B but maintained in the serum-free medium OLDEM, described in Section 3.6.2., for 5 d. Note that the processes are thicker and extensively ramified. Cells are beginning to form myelin-like membranes and are more mature than in Fig. 2B. **(D)** Phase contrast micrograph of astrocytes 8 d after the oligodendrocytes were removed by shaking. The cells were maintained in DMEM/F12 supplemented with 5% fetal calf serum under continuous slow shake.

cells matured. In the cytoplasm, the chief features were: dilated cisternae of rough endoplasmic reticulum, microtubules, electron dense cytoplasm, numerous mitochondria, and the absence of glial filaments. The basal layer contained large flat cells, overlapping each other to make a multilayer of astrocytes. These cells displayed abundant bundles of intermediate filaments and other cytological features of astrocytes, previously described in brain tissue sections (Peters et al., 1975).

The stratification of oligodendrocytes and astrocytes in primary glial cultures suggested to us that it should be possible to detach the process-bearing cells selectively, leaving the astrocytes attached to the culture surface. Several chemical and physical methods were tried, but the agitation of flasks on a rotary shaker proved to be the simplest, least expensive, and most effective way to harvest the process-bearing cells selectively (McCarthy and de Vellis, 1980). We will now describe our current procedures for the preparation of primary cultures used to obtain highly enriched populations of oligodendrocytes (Fig. 2B, C) and astrocytes (Fig. 2D) from the same tissue source.

3.2. Procedure for the Preparation of Primary Glial Cultures from Newborn Rat Brain Cerebrum

We use 0- to 2-d-old rat cerebrum as a starting material, but other areas of the CNS can be used as well. However, a thorough immunocytochemical characterization with appropriate cell markers is necessary to characterize the cellular composition of the primary culture and to verify the purity of the secondary culture of isolated oligodendrocytes.

3.2.1. Dissection

The gross dissection is carried out on a table or counter near the tissue culture hood. The table is washed with 95% ethanol. The following items are arranged on the table:

1. Two Falcon #4014 containers with 100 mL of 95% ethanol each, and a wash bottle of 95% ethanol
2. Two 12-cc sterile syringes filled with Dulbecco modified Eagle's medium/Ham's F12 (1:1) supplemented with 10%

fetal calf serum (DMEM/F12 medium) and fitted with 18-gage needles
3. Next to one alcohol container place a pair of microdissecting scissors on a sterile gauze pad. Near the other container, a pair of sterile microdissecting scissors, one curved microdissecting forceps, and a microstainless steel spatula are placed on a pad of sterile gauze; and
4. Place 12–15 pieces of sterile gauze and two 60-mm Petri dishes just in front of you. Place a plastic disposal bag nearby and 1- to 2-d-old rat pups in a clean container.

Holding one pup over the plastic bag, rinse its head with alcohol from the wash bottle. Decapitate the head of the pup with scissors over a gauze pad, and place the body in the disposal bag. Using the same scissors, cut the skin at the midline of the head, cutting from the base of the head to the mideye area. Fold back the skin by pulling the loose skin forward and hold open using the thumb and forefinger, being careful not to touch the exposed skull. Dip all instruments in alcohol after each use, taking care to keep the scissors separate from the sterile instruments, and place them on a sterile gauze pad. Using the second pair of scissors, cut the skull at the midline fissure. Avoid cutting into brain tissue by lifting up the scissors while cutting. Holding the head and applying slight pressure with the thumb and index finger, remove the raised portion of the skull with the curved forceps. Excise the brain by carefully running a sterile spatula along the bottom and sides of the brain calvarium from the olfactory lobes to the medulla. Squirt media from the syringe onto the excised brain to remove blood and keep the tissue moist. Place all the excised brains, up to 15, in one 60-mm Petri dish. It is important to perform the dissection rapidly and aseptically, making sure that alcohol has drained from the instruments before they are used in the procedure.

Before dissecting the desired brain area, place the following items in the culture hood:

1. A container of 95% alcohol and two sterile gauze pads
2. One tissue forceps, two microdissecting forceps angled with fine sharp points, and one microdissecting forceps with slightly curved blades and
3. Two 35-mm and two 60-mm Petri dishes

Place the Petri dish containing the dissected rat brains into the hood. Follow aseptic procedures for all steps. Gently transfer one brain to an inverted lid of a 35-mm Petri dish, using micro-dissecting forceps. Place the lid on a sterile gauze pad to view brain areas more clearly. Steadying the brain with microdissect-ing forceps, gently pull away one cerebral hemisphere from the rest of the brain at the median fissure with the second pair of forceps. Transfer the cortex to a 60-mm Petri dish containing DMEM/F12 medium. Remove other brains in a similar manner. Then place a cortex on an inverted 35-mm Petri dish and peel off the meninges and blood vessels using two microdissecting tweezers. The tissue is placed into a fresh 60-mm Petri dish containing DMEM/F12 medium. Blot the removed meninges on a sterile gauze pad.

3.2.2. Tissue Dissociation

Dissociation of cerebral tissue can be accomplished mechani-cally (Lu et al., 1980) or enzymatically (McCarthy and de Vellis, 1980). We now prefer mechanical methods because of higher cell yield.

3.2.2.1. THE NITEX BAG METHOD

For the Nitex bag method (Lu et al., 1980) place the follow-ing items into the hood:

1. One 100-mm sterile Petri dish
2. Two sterile Cellector tissue sieves (one #60, 230 μm and one #100, 140 μm) into sterile Falcon #4104 containers. Cover with the opened sterilizing bags until ready to use
3. Six to eight 17 × 100-mm disposable sterile tubes with snap caps
4. A sterile glass rod and
5. A sterile monofilament mesh bag (Nitex™, 210 μm).

Use sterile forceps to place the Nitex™ bag in an open 100-mm Petri dish containing 15–20 mL of DMEM/F12. Holding the bag open with forceps, pour into it the tissue and media from the 60 mm Petri dish. Close the bag with forceps, immerse it into the culture media, and gently tease the tissue through the mesh by stroking very gently with the glass rod. When about half of the tissue has been released into the Petri dish, pour the cell suspension into the #60 Cellector sieve that has been placed into a sterile #4014 Falcon container. The stroking procedure with the bag is repeated in 20 mL of fresh media until only stromal tissues remain in the bag. The cell suspension is poured into the #60 Cellector sieve. The cells on the outside of the bag are washed off by using the syringe filled with media and poured into the sieve. Then, pour the filtrate from the #60 sieve into the #100 Cellector sieve and container. Simply allow the cells to filter by gravity. Wash the #100 sieve with 10 mL of fetal bovine serum to collect any adhering cells and to protect the cells' integrity during centrifugation. Centrifuge the cell suspension, using the 17×100-mm tubes, at 800 rpm in an IEC clinical centrifuge for 5 min at 150g. Pour off the supernatant, and resuspend the cell pellets in DMEM/F-12 medium, approx 100 mL for 20 cortices. Dilute an aliquot of the cell suspension 1/100 and count the cells in a hemocytometer or cell counter. Plate cells at $15 \times 10^6/cm^2$ in 75 cm^2 tissue culture flasks with 11 mL of media supplemented with 10% fetal calf serum. Two cerebral cortices from one rat brain generally suffice for preparing one flask. Flasks are placed in an incubator at 37°C in a moist 5% CO_2, 95% air atmosphere for 48–72 h, allowing cells to adhere and begin proliferating. The medium is changed every 48–72 h until the cultures are used. If the primary cultures are not grown for the oligodendrocyte shaking procedure, they can be grown, of course, in Petri dishes or other types of culture ware.

3.2.2.2. Mechanical Dissociation with the Stomacher Blender

This method has the advantage that the mechanical force applied can be controlled (Holmes et al., 1988). The following items should be placed in the hood:

1. Six to eight, 17×100-mm tubes with snap caps
2. Sterile lab-blender bags (80 mL)
3. Two sterile Cellector tissue sieves (one #60, 230 μm and one #100,140 μm) placed in #4014 containers
4. One 12-mL syringe fitted with an 18-gage needle and filled with 10 mL of fetal calf serum; and
5. A second 12-mL syringe filled with 10 mL of DMEM/ F12 medium.

The cortices/media suspension is poured into a sterile blender bag. Additional media is added to bring the total vol to 50–60 mL. The bag is then placed in the stomacher blender, leaving 2 cm of the bag visible above the closed door. The transformer that controls the electrical input to the blender is set at 80 V to obtain an optimal speed for the blender paddles. After a 2.5-min blending period, the bag is removed and placed in the hood. The cell suspension is poured into the #60 sieve, allowing gravity filtering. The sieve is rinsed with the syringe containing 10 mL of DMEM/F12. The filtrate is then poured into the #100 sieve. The cells are allowed to flow by gravity, and wash through the remaining cells with the syringe filled with 10 mL of fetal calf serum.

The cell pellets are then processed as described in the Nitex bag method (3.2.2.1.). Flasks (75 cm²), 60-mm Petri dishes, 35-mm Petri dishes, and 24-well plates are seeded with 16×10^6 cells in 11 mL of media, 1 to 1.5×10^6 cells in 3 mL of media, 1×10^6 cells in 2 mL of media, or 0.5×10^6 cells/well in 1 mL of media, respectively.

3.2.2.3. TISSUE DISSOCIATION BY TRYPSINIZATION

The cortices free of meninges are gently aspirated through a 10-mL pipet two or three times and then transferred to a sterile 50-mL screwcapped Erlenmeyer flask. Note the vol for calculating the amount of trypsin solution to add later. Shake the 50-mL tube for 15 min on a rotary shaker (80 rpm, 37°C) to loosen tissue components. Then, add trypsin solution to a final concentration of 10%, and replace the tube on the shaker for 24 min. The trypsinization is stopped by the addition of 10 mL DMEM/ F12 medium supplemented with 10% fetal calf serum. Gently aspirate the tissue suspension through a 10-mL pipet three times, and allow the tissue pieces to settle in the Erlenmeyer that is tipped at an angle. Remove the cell suspension, taking care not to disturb the settled tissue. Filter the cell suspension through Nitex 130 (130 µm) or #100 (140 µm) Cellector sieve. Add 7 mL of culture medium to undissociated pieces, and aspirate them through a 10-mL pipet three times. Allow the tissue pieces to settle, remove the cell suspension and filter through Nitex 130, and combine with the initial filtrate. Repeat this step with 3 mL of culture medium, and combine the resulting cell suspension with the previous filtrates. Centrifuge the total filtrate at 40g for 5 min using a clinical tabletop centrifuge at room temperature. Resuspend the cell pellet in culture medium and proceed to plate cells as indicated in the Nitex bag method (3.2.2.1.).

3.3. Selective Harvesting of Oligodendrocyte Lineage Cells from Primary Mixed Glial Cultures

After 7–9 d of culture, primary cerebral cultures display numerous phase-dark process-bearing cells overlaying a confluent bed layer of phase-light astrocytes (Fig. 2A and *see also* McCarthy and de Vellis, 1980). The majority of process-bearing cells represent oligodendrocyte lineage cells at different stages of maturation (Holmes et al., 1988). A few process-bearing cells are GFAP+. Amoeboid microglia are found to a varying frequency in the top cell layer depending on the sera used (Giulian, 1987). For this reason as well as for eliminating poorly attached astrocytes, a short preshake is used to remove these cells from the cultures before the separation of oligodendrocytes from the underlying astrocytes is initiated.

3.3.1. Sequential Shaking Procedure

The method has worked best with rotary shakers of 0.5-in stroke. The original method used a New Brunswick rotary water bath shaker. It is actually easier simply to place a Lab-line Junior orbit shaker in a dry 37°C incubator. The protocol is as follows: Up to 24 75-cm² flasks of 7- to 9-d-old primary cerebral cultures are selected for their abundance of phase-dark process-bearing cells. The media is replaced with between 10 and 11 mL of fresh DMEM/F12 medium. The caps are tightened and the flasks placed on the shaker platform horizontally. Previously, a layer of packing foam had been placed on the shaker platform to insulate the flasks from the heat generated by the shaker. Each layer of eight flasks is held on the platform by fiber tape. The shaker is set at 200 rpm for the first 6 h. The shake results in the

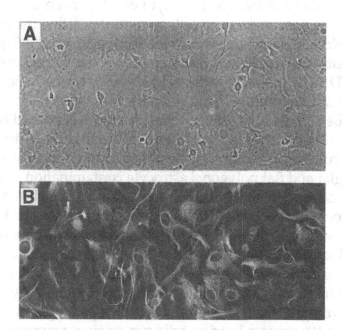

Fig. 3. **(A)** Phase contrast micrograph of primary astrocyte culture viewed just immediately after harvesting oligodendrocytes. Note that a few process-bearing cells still remain in the top layer. Their complete removal requires continuous shaking (*see* Fig. 2D). **(B)** Immunostaining of the same field with GFAP antibody. The majority of cells are GFAP-positive.

detachment of microglia and loose astrocytes into the media. The media is discarded or saved if one wishes to obtain microglia (Giulian, 1987).

Fresh medium, 10 mL/flask, is added, and the flasks are shaken for an additional 18 h. The media from four flasks are aseptically combined into a new 75-cm^2 flask. The flask is placed in a moist 5% CO_2, 95% air incubator. More process-bearing cells can be harvested by adding 10 mL of fresh DMEM/F12 medium for an additional 24-h shake. As previously shown, combine the supernatants of four flasks into a new flask. By examining the cells remaining attached in the primary culture, few process-bearing cells should be seen (Fig. 3A). The bed layer cells are almost all GFAP$^+$ (Fig. 3B). Fresh medium (10 mL) is added if one wishes to utilize the astrocyte bed layer. To prevent the remaining process-bearing cells from repopulating the culture, the flasks are placed on a shaker platform, and set at 100 rpm for continuous shake. Notice that all process-bearing cells are gone (Fig. 2D). The process-bearing cells obtained by the 24- and 48-h shakes adhere rapidly to the culture surface. The flasks are not disturbed for 24 h, at which point the media is poured off and replaced with between 10 and 11 mL of fresh media. After 24–48 h, the cultures are examined by phase contrast microscopy for microglia and flat polygonal astrocyte contamination. It is usually necessary to shake the cultures again to detach the oligodendrocytes selectively, leaving the astrocytes attached to the plastic surface. The oligodendrocyte suspension is poured onto a #500 screen (25-μm pore size), centrifuged in a benchtop clinical centrifuge, and resuspended for counting and plating. These filtration steps tend to retain the few astrocytes and macrophages that had detached and clumped together. A yield of 2.5×10^6 process-bearing cells per original primary culture flask is usually obtained. Seeding density of $50–70 \times 10^3/cm^2$ in DMEM/F12 medium allows good cell growth in either serum-supplemented (Fig. 2B) (McCarthy and de Vellis, 1980) or serum-free medium (Saneto and de Vellis, 1985, or OLDEM, Fig. 2C). Immunostaining of oligodendroglial cultures grown in 10% fetal calf serum or calf serum is shown in Figs. 4 and 5. Immunostaining of oligodendroglial cells for GPDH and GC is shown in Figs. 6 and 7, respectively.

3.4. Purified Astrocyte Cultures

The flasks from which the process-bearing cells were removed are kept at 100 rpm on the shaker for one to several days of continuous shaking in DMEM/F12 with 5% fetal calf serum. The flasks can be used as primary astrocyte cultures or passaged to obtain secondary cultures. To passage the purified astrocytes, discard the medium, and wash the cultures thoroughly—first with 5 mL of versene solution and then with 3 mL/flask of 2–5% trypsin solution that is discarded. Incubate the cultures at 37°C until all cells are dissociated. The cells are resuspended in

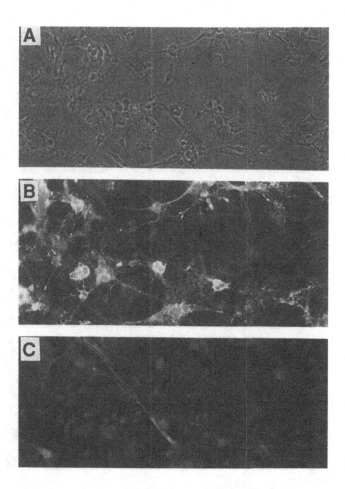

Fig. 4. (A) Phase contrast micrograph of oligodendrocytes grown for 2 wk in 10% fetal calf serum after harvesting from primary glial cultures. (B) and (C) Double immunofluorescence of the same fleld with MBP and trans-ferrin antibodies, respectively. Note that cells form clusters and extend pro-cesses. The MBP immunostaining is mainly localized to the cell body. Most cells are weakly transferrin-positive.

DMEM/F12 medium, spun in a centrifuge, and resuspended in 5 mL of media/flask. Cells are plated at 10^4 cells/cm^2 or at a higher cell density to achieve confluence more rapidly. Purified astrocytes are 95% pure by the criterion of GFAP immunostaining (Fig. 3B) when grown in chemically defined media (Morrison and de Vellis, 1981, 1983).

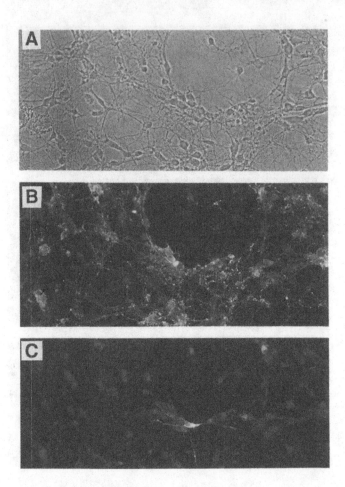

Fig. 5. Phase contrast micrograph of oligodendrocytes harvested from the same primary cultures as in Fig. 4. These oligodendrocytes were maintained for 2 wk in calf serum instead of fetal calf serum. (B) and (C) Double immunofluorescence of the same field with MBP and transferrin antibodies, respectively. Note that cells form a complex network at MBP-positive processes, forming a membranous sheath. Most MBP-positive cells are weakly transferrin-positive. However, strongly transferrin-positive cells can be seen next to MBP-positive clusters.

3.5. Preparation of Pure Oligodendrocyte Cultures from Primary Glial Cultures by the Use of Selective Media

The increased adhesion of oligodendrocytes to their cell substratum in primary cultures older than 10–12 d makes it difficult to isolate these cells by the shaking procedure of McCarthy

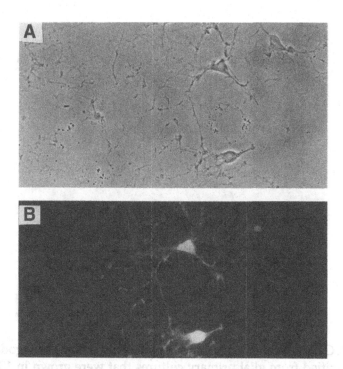

Fig. 6. **(A)** Phase contrast micrograph of oligodendrocytes grown for 7 d in fetal calf serum after harvesting from glial primary cultures. **(B)** Immunofluorescence of the same field with glycerol phosphate dehydrogenase antibody. Note the cytoplasmic localization of GPDH immunostaining.

and de Vellis (Section 3.3.). Increasing the shaking force ruptures the cells, resulting in a low cell viability and yield. An alternative approach is to find a means to kill astrocytes selectively while preserving oligodendrocytes during the passage of the primary glial cultures. Bhat et al. (1981) still used the shaking procedure with the addition of one step. They placed the cell suspension, obtained after shaking, in a balanced salt solution (Hank's balanced salt solution containing 10 mM or 25 mM HEPES, 1% fetal calf serum, or 1% bovine serum albumin, and 10 μm/mL DNase) using a 25 cm² flask. The flask is shaken for 90 min at 100 rpm. In this medium, astrocytes are supposedly osmotically instable. Bhat et al. (1981) reported that this procedure reduced astrocyte contamination in the purified oligodendrocyte cultures. However, this procedure has now been abandoned by the authors (Bhat and Silberberg, 1989).

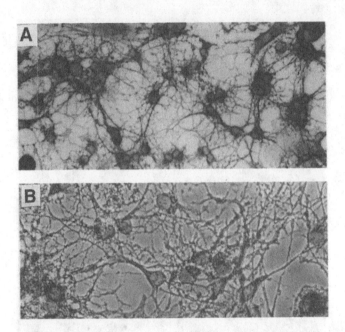

Fig. 7. Galacterocerebroside immunoperoxidase staining of oligodendro-cytes purified from glial primary cultures that were grown in 10% calf serum (A) or OLDEM (B) for 4 d. Note that GC immunostaining is strongly expressed in cell bodies of cells grown in calf serum, but it is evenly distributed between the cell body and processes of cells grown in OLDEM. In OLDEM, the processes are thicker, more ramified, and distributed throughout the culture.

Since there are no antibody-to-astrocyte cell surface antigens yet available, complement cell lysis cannot be used. Espinosa de los Monteros et al. (1986) developed a minimal chemically defined medium in which oligodendrocytes, but not astrocytes, survive after a 48-h exposure. The protocol uses 20-d-old primary cultures prepared in Waymouth medium supplemented with 10% calf serum. Calf serum is preferred over fetal calf serum because it favors oligodendrocyte differentiation rather than Type-2 astrocyte development. The primary cultures are rinsed twice with OSM (DMEM supplemented with 50 U/mL of penicillin, 50 µg/mL of glutamine, 10 mM HEPES, 5 µg/mL of insulin, and 0.8 mg/µL of bovine serum albumin). Then, 3 mL of OSM/100-mm Petri dish are added, and cells are harvested mechanically with a rubber policeman. The cell suspension is placed in a 25-cm² flask at a density of 15 × 10⁶ cells/flask in a total vol of 30 mL OSM. A 7-mm magnetic bar is placed inside the flask. The flask is positioned vertically on the stirring plate at 100 rpm and 37°C for 48 h. After this incubation, cells are pelleted by centrifugation for 10 min at 700 rpm. The pellet is resuspended in Waymouth's medium supplemented with 10% calf serum. Cells are dispersed by using a syringe fitted with a No. 20 needle. Sieve the cells' suspension through a 25-μm pore size nylon screen. Then filter through a double layer of 30-μm nylon screen. Cells are then seeded at 2 × 10⁶ cells/60-mm Petri dishes, previously coated overnight with poly-L-lysine, and washed with Waymouth's medium before use. The cultures are maintained in Waymouth's medium supplemented with 10% calf serum at 37°C and 4.5% CO₂/95.5% air. Cells cannot be plated in OSM because this medium does not support the long-term survival of freshly isolated oligodendrocytes. These cells need a richer medium in order to synthesize myelin components and myelin-like membranes. The medium is changed after 24 h, and twice a week thereafter. By day 13 in culture, 90% of the cells express MBP (Espinosa de los Monteros et al., 1986).

3.5.1. Oligodendrocyte Method Combining Features of the Shaking Procedure and OSM Method

The neonatal rat brain primary cultures are established as described earlier and grown in Waymouth's medium supplemented with 10% calf serum. The separation of oligodendrocytes is initiated by a short but very vigorous shaking (10 min at 400 rpm) of 1- to 5-wk-old cultures on a rotary shaker. The detached cells are pelleted by centrifugation at 700 rpm for 7 min. Cells are resuspended in Waymouth's medium containing 10% calf serum and plated at 1.5 × 10⁵ cells/cm² on polylysine-coated Petri dishes or coverslips for immunocytochemistry. The cultures are kept at 37°C in a humidified atmosphere with 4–5% CO₂/95.5% air. After 4 h, the medium is removed and fresh OSM is added for a 20- to 24-h period. Then, OSM is removed, and serum-free or serum-supplemented medium is added, depending on the experimental requirements. The O-2A lineage cells obtained by this method are more mature than those isolated by the McCarthy and de Vellis method. The latter largely contains progenitor cells. The reasons for obtaining more mature cells with the combined method are in part the use of fetal calf serum, the age of the primary cultures used, and the fact that phase-dark flat progenitors are not easily detached by the short, vigorous shaking whereas the clustered and rounded, more mature cells do come off more easily.

3.6. Serum-Free Chemically Defined Media

Traditionally, cell culture media have included a basal medium composed of amino acids, vitamins, and balanced salts that is chemically defined, to which one adds an animal serum that favors the growth of the desired cells. However, sera are of ill-defined and variable composition. Therefore, Sato and his collaborators undertook to replace serum with a chemically defined supplement of hormones, growth factors, and adhesion molecules selected for their optimal effect on the growth of a particular cell type (*see* Bottenstein, chapter 3).

3.6.1. Serum-Free Media Optimized
for the Proliferation of Oligodendroglial Cells

Saneto and de Vellis (1985) developed a serum-free medium (ODM) that supports the proliferation of oligodendrocyte progenitors (Fig. 8) isolated by the method of McCarthy and de Vellis (1980), described in Section 3.3. Initially, the cells were plated overnight in DMEM/F12 medium containing 10% fetal calf serum to allow the cells to attach to the plastic surface. Subsequently, the cultures are washed two or three times with DMEM/F12 medium without serum, and then ODM is added. The serum-free medium, ODM, consists of DMEM/F12, to which the following supplements are added: fibroblast growth factor, 5 ng/mL; transferrin, 500 mg/mL; insulin, 5 µg/mL; and sodium pyruvate, 100 µg/mL. Stock solutions of transferrin and pyruvate are freshly prepared in DMEM/F12 every 10 d. Insulin is made as a 5 mg/mL stock solution in 0.05N HCl and stored at 4°C up to 1 mo. In ODM, the majority of cells proliferate and display a less differentiated morphology, e.g., fewer processes. The addition of dibutyryl cAMP, hydrocortisone, and thyroid

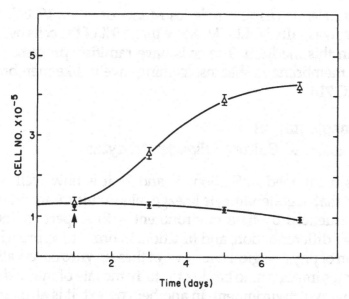

Fig. 8. Growth curve of purified oligodendroglial cells. Cells were maintained in either 10% fetal calf serum-supplemented medium (SSM) or in ODM. Arrow indicates time at which either SSM or ODM was added. Values represent cell number per well (24-well plate) is the mean of six wells ±SEM (Saneto and de Vellis, 1985).

hormones induces morphological and biochemical differentiation. McKinnon et al. (1990) have shown that the addition of platelet-derived growth factor can keep the cells as proliferating progenitors.

Instead of plating cells in serum, it is possible to plate on polylysine-coated dishes or fetuin-coated dishes. The role of adhesion in oligodendrocyte differentiation is important but not yet well characterized (Yim et al., 1986; Cardwell and Rome, 1988a,b).

3.6.2. A Defined Medium
for Mature Oligodendrocytes: OLDEM

Oligodendroglial cells isolated from neonatal glial primary cultures by the method of McCarthy and de Vellis (1980) or Espinosa de los Monteros et al. (1986), as described earlier, are plated in DMEM/F12 supplemented with 5 µg insulin, 16 µm putrescine dihydrochloride, 1.8 g/L of D ± galactose, and 8.6 µm

Sodium Selenite (Espinosa de los Monteros et al., 1988a). This medium was called OLDEM. More than 90% of the cells become MBP⁺ in this medium. The cells have ramified processes that extend membranous sheaths, forming myelin-like membranes (Figs. 2C,7B).

4. Transplantation
of Labeled Culture Oligodendrocytes

As mentioned in Sections 1. and 2., it is now well documented that oligodendrocyte lineage cells in culture are dramatically influenced by their environment with respect to lineage decision, differentiation, and function. In order to approach the question of physiological relevance of these in vitro observations, it becomes important to be able to study the fate of marked cells in the in vivo environment. In another context, it is also important to find out whether transplanted oligodendroglial cells are capable of repairing damage caused by injury or degenerative disorders affecting oligodendrocytes and myelin. Recently, Espinosa de los Monteros et al. (1991) developed experimental conditions for the successful transplantation of developing oligodendroglial cells prelabeled in culture. The fluorescent dye, fast blue (FB), is readily taken up by oligodendrocytes and allows identification of cells even after 4–6 cell divisions, and up to 6 mo after transplantation, the longest time carried out thus far. The majority of grafted cells survive and differentiate in the developing rat brain. Some cells divide and migrate far away from the injection site in both the normal and myelin-deficient (md) rat mutant. No signs of graft rejection were observed up to 6 mo of age in the normal rat. This indicates that the dye does not leak out of the cells in any significant amount.

The FB dye is toxic to oligodendrocytes at high concentrations. Several dye concentrations and time-course experiments were performed to determine the FB concentration that did not alter the development and myelin function of these cells. Stock solution of FB is prepared by adding 100 µL of sterile water to 1 mg of FB dye. The solution is kept in the dark at 4°C. Just before the experiment, 1 µL of FB is added to 1 mL of Waymouth's medium. Twenty-four hours prior to transplantation, oligodendroglial cultures grown in 4-well plates are incubated with the FB containing culture medium for 4–5 h. The cultures are then washed with fresh culture medium at 37°C and kept in 1 mL of medium/well overnight.

The adhering cells are gently scraped from the culture dish and transferred by using a syringe fitted with a spinal needle into a sterile Eppendorf tube. The cells are centrifuged in a microphage for 30 s. The cell pellet is resuspended in 16 μL of Ca^{2+} and Mg^{2+} free Hank's balanced salt solution.

4.1. Transplantation Procedure

Surgical instruments appropriate for the age of the rats are washed with 95% ethanol and air dried. Rat pups, 5–14 d of age, are weighed and anesthetized with nembutal, 30 mg/kg body wt. Animals are placed on a stereotaxic apparatus. The skull is exposed by a short incision of the skin. A 1-mm^2 hole is drilled into the skull. The syringe is inserted to the desired depth to reach the selected site of injection. The cell suspension is injected slowly at a rate of 16 μL/5 min. The syringe is left in place for 5 min to allow diffusion and retention of the cell suspension into the brain parenchyma. The syringe is withdrawn stepwise over a 5-min period. The rat pups are returned to the mother after the anesthetic effect has worn off.

4.2. Characterization of the Grafted Cells

Prior to transplantation, the phenotype of cells is established by carrying out double or triple immunofluorescence with appropriate oligodendrocyte markers (*see* Section 2.) on sister cultures. This determines the phenotype of FB-labeled cells at the time of grafting. To characterize grafted cells, rats are anesthetized with avertin (Aldrich) at a dose of 70 mg/kg body wt. Animals are perfused intracardially for 10 min with sodium phosphate buffer and postfixed by immersion for 8 h. Paraffin sections (6 μm) are prepared, and immunocytochemical analysis of the markers is carried out in a standard manner. The FB-positive cells are easily identifiable (Fig. 9). The FB$^+$ cells expressed the oligodendroglial markers GPDH, GC, and MBP but not GFAP.

Fig. 9. Photomicrograph of a brain paraffin section (6 μm) showing grafted oligodendrocytes. These cells were labeled in culture with the fluorescent dye FB prior to transplantation. A large number of cells remain localized in the area of the injection site. Some cells have migrated away from the injection site.

Acknowledgments

We thank Nancy Wainwright for assisting with the preparation of the manuscript. We also thank Sharon Belkin and Carol Gray of the MRRC Media Unit for assistance in preparing the illustrations. This work was supported by NIH Grant HD-06576 and DOE Contract DE-FC03-87-ER60615.

References

Arenander A. and de Vellis J. (1989) Development of the nervous system, in *Basic Neurochemistry: Molecular. Cellular and Medical Aspects* (Siegel G. J. et al., eds.), Raven, New York, pp. 479–506.

Bhat S., Barbarese E., and Pfeiffer S. E. (1981) Requirement for nonoligodendrocyte cell signals for enhanced myelinogenic gene expression in long-term cultures of purified rat oligodendrocytes. *Proc. Natl. Acad. Sci.* 78, 1283–1287.

Bhat S. and Silbergerg D. H. (1989) Isolation and culture of rat and mouse brain oligodendrocytes, in *A Dissection and Tissue Culture Manual of the Nervous System* (Shahar A., de Vellis J., Vernadakis A., and Haber B., eds.), Alan R. Liss, New York, pp. 145–147.

Bologa L., Aizenman Y., Chiappelli F., and de Vellis J. (1986) Regulation of myelin basic protein in oligodendrocytes by a soluble neuronal factor. *J. Neurosci. Res.* **15**, 521 –528.

Bornstein M. B. and Murray M. R. (1958) Serial observations on patterns of growth, myelin formation, maintenance and degeneration in cultures of new-born rat and kitten cerebellum. *J. Biophys. Biochem. Cytol.* **4**, 499–504.

Breen G. A. M. and de Vellis J. (1974) Regulation of glycerol phosphate dehydrogenase by hydrocortisone in dissociated rat cerebral cell culture. *Dev. Biol.* **41**, 255–266.

Breen G. A. M. and de Vellis J. (1975) Regulation of glycerol phosphate dehydrogenase by hydrocortisone in rat brain explants. *Exp. Cell Res.* **91**, 159–169.

Breen G. A. M., McGinnis J. F., and de Vellis J. (1978) Modulation of the hydrocortisone induction of glycerol phosphate dehydrogenase by N^6, O^2-dibutyryl cyclic AMP, norepinephrine and isobutylmethyl-xanthine in rat brain cell cultures. *J. Biol. Chem.* **253**, 2554–2562.

Bunge M. B., Bunge R. P., and Pappas G. D. (1962) Electron microscopic demonstrations of connections between glia and myelin sheaths in the developing mammalian central nervous system. *J. Cell Biol.* **12**, 448–453.

Cardwell M. C. and Rome L. H. (1988a) Evidence that an RGD-dependent receptor mediates the binding of oligodendrocytes to a novel ligand in a glial-derived matrix. *J. Cell Biol.* **107**, 1541–1549.

Cardwell M. C. and Rome L. H. (1988b) RGD-containing peptides inhibit the synthesis of myelin-like membrane by cultured oligodendrocytes. *J. Cell Biol.* **107**, 1551–1559.

Cameron R. S. and Rakic P. (1991) Glial cell lineage in the cerebral cortex: A review and synthesis. *Glial Special Issue* **4**, 124–137.

del Rio Hortega P. (1921) Postnatal gliogenesis in the mammalian brain. *Int. Rev. Cytol.* **40**, 281–323.

Espinosa de los Monteros A., Roussel G., Gensburger C., Nussbaum J. L., and Labourdette G. (1985) Precursor cells of oligodendrocytes in rat primary cultures. *Dev. Biol.* **108**, 474–480.

Espinosa de los Monteros A., Roussel G., and Nussbaum J.-L. (1986) A procedure for long-term culture of oligodendrocytes. *Dev. Brain Res.* **24**, 117–125.

Espinosa de los Monteros A. (1987) Contribution a l'etude des oligodendrocytes in vitro. Universit Louis Pasteur, Strasbourg, France, Ph.D. Thesis.

Espinosa de los Monteros A. and de Vellis J. (1988) Myelin basic protein and transferrin characterize different sub-populations of oligodendrocytes in rat primary glial cultures. *J. Neurosci. Res.* **21**, 181–187.

Espinosa de los Monteros A., Chiappelli F., Fisher R. S., and de Vellis J. (1988a) Transferrin: An early marker of oligodendrocytes in culture. *Int. J. Develop. Neurosci.* **6**, 167–175.

Espinosa de los Monteros A., Roussel G., Neskovic N. M., and Nussbaum J. L. (1988b) A chemically defined medium for the culture of mature oligodendrocytes. *J. Neurosci. Res.* **19**, 202–211.

Espinosa de los Monteros A., Peña L. A., and de Vellis J. (1989) Does transferrin have a special role in the nervous system? *J. Neurosci. Res. (New York)* **24,** 125–136.

Espinosa de los Monteros A., Zhang M., Gordon M., Aymie M., and de Vellis J. (1991) Transplantation of cultured premyelinating oligodendrocytes into normal and myelin-deficient rat brain. (Submitted for publication).

Gard A. L. and Pfeiffer S. E. (1990) Two proliferative stages of the oligodendrocyte lineage ($A2B5^{+}O4^{-}$ and $O4^{+}GalC^{-}$) under different mitogenic control. *Neuron* **5,** 615–625.

Giulian D. (1987) Ameboid microglia as effectors in inflammation in the central nervous system. *J. Neurosci. Res.* **18,**155–171.

Goldman J. E. and Vaysse P. J.-J. (1991) Tracing glial cell lineages in the mammalian forebrain. *Glia* **4,** 149–156.

Holmes E., Hermanson G., Cole R., and de Vellis J. (1988) Developmental expression of glial specific mRNAs in primary cultures of rat brain visualized by *in situ* hybridization. *J. Neurosci. Res.* **19,** 389–396, 458–465.

Knapp P. E., Bartlett W. P., and Skoff R. P. (1987) Cultured oligodendrocytes mimic in vivo phenotypic characteristics: Cell shape, expression of myelin-specific antigens and membrane production. *Dev. Biol.* **120,** 356–365.

Kumar S. and de Vellis J. (1981) Induction of lactate dehydrogenase by dibutyryl cAMP in primary cultures of central nervous tissue is an oligodendroglial marker. *Dev. Brain Res.* **1,** 303–307.

Kumar S. and de Vellis J. (1987) Glucocorticoid mediated functions in glial cells, in *Glial Cell Receptors* (Kimelberg H. K., ed.), Raven, New York, pp. 243–263.

Kumar S., Cole R., Chiappelli F., and de Vellis J. (1989) Differential regulation of oligodendrocyte markers by glucocorticoids: Post-transcriptional regulation of PLP, MBP and transcriptional regulation of GPDH. *Proc. Natl. Acad. Sci.* **86,** 6807–6811.

Labourdette G., Roussel G., Ghandour M. S., and Nussbaum J. L. (1979) Cultures from rat brain hemispheres enriched in oligodendrocyte-like cells. *Brain Res.* **179,** 199–203.

Labourdette G., Roussel G., and Nussbaum J. L. (1980) Oligodendroglia content of glial cell primary cultures, from newborn rat brain hemispheres, depends on the initial plating density. *Neurosci. Lett.* **18,** 203–209.

Leveille P. J., McGinnis J. F., Maxwell D. S., and de Vellis J. (1980) Immunocytochemical localization of glycerol-3-phosphate dehydrogenase in rat oligodendrocytes. *Brain Res.* **196,** 287–305.

Lu E., Brown W. J., Cole R., and de Vellis J. (1980) Ultrastructural differentiation and synaptogenesis in aggregating rotation cultures of rat cerebral cells. *J. Neurosci. Res.* **5,** 447–463.

Matthieu J.-M., Honegger P., Trapp B. D., Cohen S. R., Webster H. D. F. (1978) Myelination of rat brain aggregating cell cultures. *Neuroscience* **3,** 565–572.

Matthieu J.-M., Roch J.-M., Torch S., Tosic M., Carpano P., Insirello L., Giuffrida Stella A. M., and Honegger P. (1990) Triiodothyronine increases the stability of myelin basic protein mRNA in aggregating brain cell cultures, in *Regulation of Gene Expression in the Nervous System* (Giuffrida Stella A. M., de Vellis J., and Perez-Polo J. R., eds.), Wiley-Liss, New York, pp.109–122.

McCarthy K. D., McGinnis J. F., and de Vellis J. (1978) Alpha-adrenergic receptor modulation of beta-adrenergic, adenosine and prostaglandin E_1 increased adenosine 3':5'-cyclic monophosphate levels in primary cultures of glia. *J. Cyclic Nucleotide Res.* 4, 15–26.

McCarthy, K. D. and de Vellis J. (1980) Preparation of separate astroglial and oligodendroglial cell cultures from rat cerebral tissue. *J. Cell Biol.* 85, 890–902.

McKinnon R. D., Matsui T., Dubois-Dalcq M., and Aaronson S. A. (1990) FBF modulates the PDGF-driven pathway of oligodendrocyte development. *Neuron* 5, 603–614.

McMorris F. A. and Dubois-Dalcq M. (1988) Insulin-like growth factor 1 promotes cell proliferation and oligodendroglial commitment in rat glial progenitor cells developing *in vitro*. *J. Neurosci. Res.* 21, 199–209.

Morrison R. S. and de Vellis J. (1981) Growth of purified astrocytes in a chemically defined medium. *Proc. Natl. Acad. Sci.* 78, 7205–7209.

Morrison R. S. and de Vellis J. (1983) Differentiation of purified astrocytes in a chemically defined medium. *Dev. Brain Res.* 9, 337–345.

Noble M. (1991) Points of controversy in the O-2A lineage: Clocks and Type-2 astrocytes. *Glia* 4, 157–164.

Norton W. T. (1984) *Oligodendroglia*, Plenum, New York.

Peters A., Palay S. L., and deF. Webster H. (1976) *The Fine Structure of the Nervous System: The Neurons and Supportina Cells*, W. B. Saunders, Philadelphia.

Robertson J. D. (1899) On a new method of obtaining a block reaction in certain tissue elements of the central nervous system (Platinum method). *Scott. Med. Surg.* 4, 23–30.

Rome L. H., Bullock P. N., Chiappelli F., Cardwell M. C., Adinolfi A. M., and Swanson D. (1986) Synthesis of a myelin-like membrane by oligodendrocytes in culture. *J. Neurosci. Res.* 15, 49–65.

Saneto R. P. and de Vellis J. (1985) Characterization of cultured rat oligodendrocytes proliferating in a serum-free chemically defined medium. *Proc. Natl. Acad. Sci.* 82, 3509–3513.

Saneto R. P., Altman A., Knobler R. L., Johnson H. M., and de Vellis J. (1986) Interleukin 2 mediates the inhibition of oligodendrocyte progenitor cell proliferation *in vitro*. *Proc. Natl. Acad. Sci.* 83, 9221–9225.

Skoff R. P. and Knapp P. E. (1991) Division of astroblasts and oligodendroblasts in postnatal rodent brain: Evidence for separate astrocyte and oligodendrocyte lineages. *Glia* 4, 165–174.

Szuchet A., Polak P. E., and Yim S. H. (1986) Mature oligodendrocytes cultured in the absence of neurons recapitulate the ontogenic development of myelin membranes. *Dev. Neurosci.* **8,** 208–221.

Yim S. H., Szuchet S., and Polak P. E. (1986) Cultured oligodendrocytes: A role for cell-substratum interaction in phenotypic expression. *J. Biol. Chem.* **261,** 11,808–11,815.

Brain Capillaries

Cell Culture of Capillary Endothelium
Derived from Cerebral Microvessels

Nika V. Ketis

1. Introduction

The microvasculature of the brain is important in the normal development and maintenance of the blood-brain barrier. Since glial cells are in close contact with brain capillaries *in situ*, communication between glial cells and capillary endothelial cells is thought to be important in maintaining homeostasis within the central nervous system. The selective permeability barrier, termed the blood-brain barrier, is maintained by specific membrane transport systems and tight junctions between adjacent endothelial cells. (Patho)physiological stresses such as anoxia, ischemia, metabolic, and/or degenerative disturbances, affect the endothelium, and in turn, affect the selective permeability barrier. Brain capillary endothelial cells in culture would facilitate the study of the mechanisms involved in cerebral vascular injury.

Several methods for the isolation of brain capillary endothelial cells have been described (De Bault et al., 1979; Panula et al., 1978; Phillips et al., 1979). However, it has not always been apparent that these cells retain the properties essential for blood-brain barrier function in vivo. Bowman et al. (1981) did report the presence of tight junctions that were visualized through the use of freeze-fracture techniques in cultures of rat brain capillary

From: *Neuromethods, Vol. 23: Practical Cell Culture Techniques*
Eds: A. Boulton, G. Baker, and W. Walz ©1992 The Humana Press Inc.

endothelial cells. Furthermore, earlier methods did not provide for long-term maintenance of brain capillary endothelial cells. Only recently have culture techniques evolved to the point where pure identifiable cultures of brain capillary cells are available.

2. Cell Culture Techniques

2.1. Source of Tissue

The in vivo culture of blood vessels was one of the first successful forms of tissue culture (Carrel and Burrows, 1910; Feig, 1910). More recently, several groups have developed techniques to isolate capillary endothelial cells from rat epididymal fat pads (Wagner and Matthews, 1975), rat, mouse, rabbit, and bovine brain (Bowman et al., 1981; De Bault et al., 1979; 1981; Spatz et al., 1980; Williams et al., 1980), human dermis (Karasek and Charlton, 1974), bovine, rat, and feline retina (Bruzney and Massicotte, 1979; Frank et al., 1979) and rat, bovine, and human adrenal cortex (Davidson et al., 1980; Del Vecchio et al., 1977). Endothelial cell lines can now be maintained for extended periods of time in culture. The general considerations that must be made when isolating capillary endothelial cells are:

1. Isolate capillaries free from contaminating large blood vessels;
2. Clone out capillary endothelial cells, thus producing a pure culture of endothelial cells; and
3. Develop culture conditions that promote the long-term growth and maintenance of capillary endothelial cells.

The isolation of capillary endothelial cells is performed in at least two steps. The first is accomplished by mincing the tissue (Bowman et al., 1981; Spatz et al., 1980) by disruption with a Dounce homogenizer (Bowman et al., 1981; De Bault et al., 1979; Del Vecchio et al., 1977), or by enzymatic disruption with either trypsin (Davidson et al., 1980), collagenase (Del Vecchio et al., 1977; Folkman et al., 1979; Wagner and Matthews, 1975), or a mixture of collagenase and dispase (Bowman et al., 1981). Reports have indicated that isolated capillaries generated through enzymatic digestion may yield a larger number of viable cells than endothelial cells isolated through mechanical disruption (Williams et al., 1980).

2.2. Sterile Techniques

All procedures are performed to ensure that bacterial and fungal contamination does not occur. It is critical that all tissue culture materials are sterile. The impact of poor laboratory practice has been dramatically reduced by the availability of sterile disposable labware and commercially prepared media and reagents. Proper handling of tissue culture glassware cannot be overstated. This includes such items as pipets used to transfer media. Of particular importance are toxic organic products produced during manufacturing of certain items, inorganic residues from detergent washes, or contaminating metal ions from pipes. Careful attention must be paid to cleaning and sterilization of glassware; several such procedures are available in the literature (Kruse, Jr. et al., 1973; Paul, 1975; Penson and Balducci, 1963).

Dissection and tissue culture manipulation should be carried out in a filtered laminar air ventilated hood that will help to prevent airborn contaminants in cell culture. The laminar flow unit (high-efficiency particle filters, air balance, and air flow) should be checked three times a year. A detailed description of the procedures has been published (McGarrity, 1975). It is noteworthy to mention that the items that are to be used for tissue culture work should be placed in the laminar flow unit 10–15 min prior to use to ensure that particulates are removed from the work area. Furthermore, any unnecessary movement around the tissue culture area should be avoided. Detailed tissue culture techniques are available through a number of reviews and published texts (Freshney, 1983; Jakoby and Pastan, 1979).

2.3. Dissection Precautions Required for Primary Cultures

Brain tissue should not be allowed to dry during the dissection procedure. All the work should be carried out in the laminar flow unit under sterile conditions. The calf brains are soaked in a basin of Buffer A, pH 7.4 at 4°C (Buffer A: Hank's Balanced Salt Solution with 15 mM Hepes (H-H) containing 2 mL/100 mL penicillin-streptomycin (stock - 5000 U/mL penicillin, 5000 mg/mL Streptomycin); 2.5 μg/mL fungizone; 10 U/mL nystatin and 50 µg/mL gentamycin). With fine forceps, gently remove the covering meningeal membrane and blood vessels. Discard this material. Trim off the white matter and place the gray matter into Buffer A containing 0.1% bovine serum albumin (BSA).

3. Isolation of Capillaries

Several techniques to generate brain capillary endothelial cells have been described in the literature (Bowman et al., 1981; De Bault et al., 1979; Mrsulja et al., 1976; Panula et al., 1978; Phillips et al., 1979; Spatz et al., 1980; Williams et al., 1980). The following procedure has successfully produced bovine capillary endothelial cells in culture. The procedure is based on the methods of Spatz et al. (Mrsulja et al., 1976; Spatz et al., 1980).

Once the gray matter has been dissected from the brain, the following steps are taken:

1. The pieces are minced as finely as possible with surgical scissors.
2. The bovine cerebral tissue is then homogenized in 20 vols of Buffer A, containing 0.1% BSA, pH 7.4 with a Dounce homogenizer using ten slow upward and downward strokes (with pestle B, the loose filter pestle). It is important to avoid frothing.
3. This homogenate is then centrifuged at 1500g for 15 min at 4°C.
4. The supernatant is discarded and the resulting pellet is resuspended in Buffer A containing 0.1% BSA. At this point, the slurry can be refrigerated and the procedure continued the following day, or one can proceed directly to Step 5.
5. The slurry is again centrifuged at 1500g for 10 min.
6. Prepare a sucrose step-gradient 90 min in advance. In the bottom of an ultracentrifuge tub place 15 mL of $1.5M$ sucrose in Buffer A, then gently overlay with 15 mL of $1.0M$ sucrose in Buffer A.
7. Discard the supernatant then resuspend the pellet in 60 mL (per brain) of cold Buffer A containing $0.25M$ sucrose, pH 7.4.

8. Layer slurry over the sucrose step-gradient and centrifuge at 58,000g for 30 min.
9. At the end of Step 8, there will be three morphologically distinct fractions (Mrsulja et al., 1976), but only the pellet contains microvessels. Thus, the sucrose is aspirated off and the pellets are resuspended in small vols of Buffer A and pooled. Characterization of the three fractions is provided by Mrsulja et al. (1976).
10. The slurry is then centrifuged at 800g for 10 min at room temperature.
11. The pellets are resuspended in Buffer A and stirred gently at 4°C for 30 min.
12. The slurry is centrifuged at 800g for 10 min at 4°C.
13. One-third of the pellet is resuspended in growth medium and the suspension is plated into 100 mm tissue culture dishes.
14. The rest of the pellet is resuspended in a 1:1 mixture of 1X trypsin/EDTA (Gibco Laboratories, New York, NY) and 0.1% collagenase, and the sample is allowed to sit at room temperature for 5 min.
15. The trypsin/EDTA solution is inactivated by adding 2 mg/ mL of egg white albumin, a trypsin inhibitor, and the suspension is centrifuged at 800g for 10 min.
16. The pellet is resuspended in growth medium and plated into tissue culture dishes (100 mm).
17. The cultures should be plated but not fed for 3 d. Fresh medium should be added on day three, and the cultures should be refed every second day.

This procedure will provide enough material from one bovine brain for 4–8 tissue culture dishes (100 mm). The best bovine capillary endothelial cells are obtained from collagenase/ trypsin-EDTA treatment.

To successfully grow brain capillary endothelial cells, particular attention should be paid to the preparation of medium. Table 1 describes the Taub and Sato K1 basic hormone mix (Taub and Sato, 1980) that is used in the growth medium. The cells are grown in brain capillary endothelial medium (BCE-medium), which is composed of the following components:

Table 1
Taub and Sato K1 Basic Hormone Mix

Component	Final vol. 100 mL of 100× in H-H 1

1. Insulin
    Always add first
	50 mg dissolved in 10 mL of H-H
	    add a few drops (2–3) of HCl concen-
	    trate for insulin to dissolve

2. Prostaglandin E,
    (stock 1 mg/mL;
    store at –20°C
    in 100% ETOH)
	add 2.5 µL

3. T3 (triiodothyronine)
    (stock 0.26M in 100% ETOH
    at –20°C)
	add 100 µL of dilution
	    (dilution 2 µL of stock in
	    10 mL of H-H)

4. TF (transferrin)	add 50 mg dissolved in 10 mL of H-H

5. SS (sodium selenite)
    (stock 10⁻³M, frozen
    at –20°C in H-H)
	add 100 µL

6. HC (hydrocortisone)
    (stock 1.8 mg dissolved
    in 10 mL of 100% ETOH)
    Always add this last
	add 1 mL

Filter 100 mL using a 0.45 µm filter, place 1 mL into 1.5 mL sterile Eppendorf tube, freeze at –70°C. 1–6 indicate order of addition of components.

50% RPMI 1640
25% Dulbecco's Modified Eagle Medium (DMEM)
25% Ham's F10 Nutrient Mixture
0.5 mL/100 mL K1 Basic Hormone Mix
5% fetal calf serum
2.5% Nu serum
1 mL/100 mL penicillin-streptomycin (stock = 5000 U/mL
    penicillin; 5000 ug/mL streptomycin)
2.5 µg/mL fungizone

All products are from Gibco Laboratories (New York, NY) with the exception of the Nu serum (Collaborative Research; Bedford, MA) and K1 Basic Hormone Mix.

4. Primary Cultures of Capillary Endothelial Cells

When the cells obtained from the isolation are first plated, they appear as single cells and small clusters. The cells initially attach to the culture dish, and within 90 min spread out on the vessel's surface. After the first 24 h, the cells begin to proliferate, and by day three, the culture appears to have a heterogeneous population of cells (Fig. 1A), as determined by phase contrast microscopy. However, interdispersed are distinct colonies showing a striking cobblestone appearance (Fig. 1B). Endothelial cells grow in a monolayer of closely apposed, homogeneous, polygonal cells approximately 35×50 μm in size with an oval centrally located nucleus (Figs. 2 and 3).

5. Establishing Post-Primary Cloned Endothelial Cultures, Cloned Cells

Contamination of other cell types can be eliminated by cloning the endothelial cells (DeBault and Cancilla, 1980; Folkman and Haudenschild, 1980; Zetter et al., 1978). This technique is easy and has a high success rate. As such, it has become the method of choice for preparing pure cultures of endothelial cells. Post-primary isolation of brain capillary endothelial cells can be accomplished by first identifying individual colonies on the 100 mm dish by phase contrast microscopy. The colonies are circled with a marker on the base of the 100-mm dish (Fig. 2A). The medium is removed and the cells are washed twice with H-H, pH 7.4. A sterile cloning cylinder is dipped into sterile silicone grease to prevent leakage, and carefully placed with sterile forceps onto the marked circle (Fig. 2B,C) in a 100-mm dish free of H-H, pH 7.4. All the colonies that are identified are encircled by cloning cylinders. Cloning cylinders (penicylinders) can be used to identify and initially isolate single cells from the rest of the

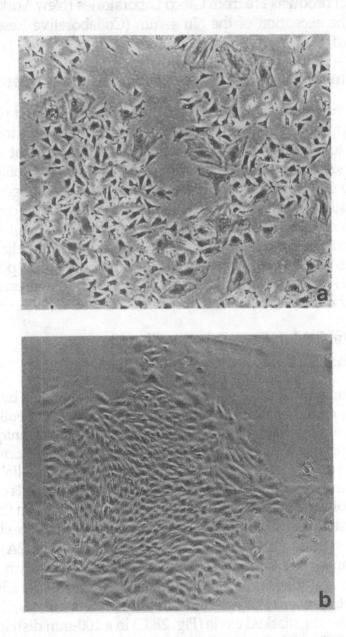

Fig. 1. Phase contrast micrographs of primary cultures of capillary endothelial cells. **(a)** heterogeneous population of cells derived from bovine brain;
(b) interdispersed colonies of endothelial cells.

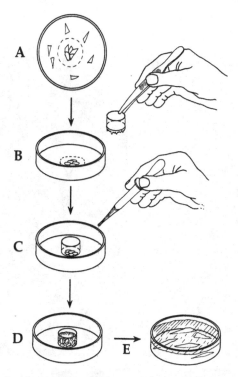

Fig. 2. Schematic of a technique for cloning endothelial cells. (A) distinct colonies showing a striking cobblestone appearance are circled with a marker on base of 100 mm dishes; (B) cloning cylinder bearing silicone grease to prevent leakage is placed with forceps onto the marked circle; (C) into the well of the cylinder is added 1–3 drops of trypsin/EDTA using a Pasteur pipet; (D) the cells are allowed to lift off the vessel's surface; (E) each colony isolate is plated into a tissue culture dish.

population in the tissue culture dish. Cells can be grown to clonal densities within the cylinder prior to harvesting, if so desired. Trypsin/EDTA is added to the well of the cylinder (Fig. 2C) and the cells are allowed to sit at room temperature for 3 min (Fig. 2D). Prior to removing the trypsin-EDTA cell suspension, the marked areas are examined to ensure that the cells have lifted off the vessel's surface. Each colony isolate is then placed into a 60-mm dish or one well of a 24-well-plate containing BCE-medium (Fig. 2E). The cloned cells are allowed to attach and spread (Fig. 3A). The cells are left undisturbed overnight.

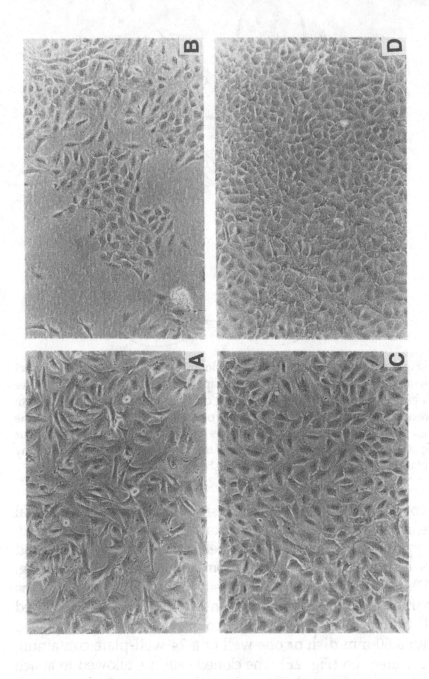

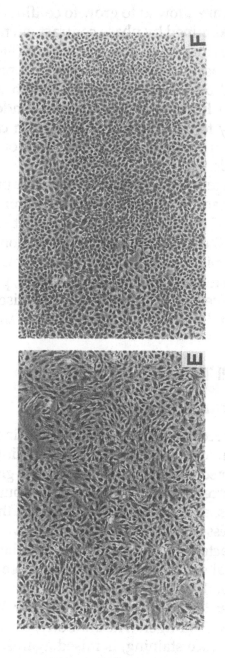

Fig. 3. Phase contrast micrographs of bovine brain capillary endothelial cells (A,C,E) and rat aortic endothelial cells (B,D,F) were plated and allow to grow in T-25 and T-75 tissue culture flasks, respectively. (A,B) Exponentially growing culture; (C,D) near confluent cultures; (E,F) near confluent cultures of (C,D).

The following morning, the cells are washed with H-H, pH 7.4, fresh BCE-medium is added, and the cultures are fed every second day. The cells are allowed to grow to confluence, and the cloned cells are examined by phase contrast microscopy. Pure endothelial clones form confluent monolayers (Fig. 3C,E) of large, polygonal, closely apposed cells. In Fig. 3, panels A, C, and E show cloned bovine brain capillary endothelial cells and panels B, D, and F show cloned rat aortic endothelial cells. It is noteworthy to mention that bovine brain capillary endothelial cells appear somewhat larger than rat or bovine aortic endothelial cells.

To subculture the cell clones, detach the cells to be passaged with trypsin-EDTA by adding this solution to the culture vessel, and allow cells to sit 3–5 min at room temperature until the cells begin to detach in cell sheaths. Triturate the suspension several times with a Pasteur pipet and place it into a 15 mL conical disposable centrifuge tube containing 10 mL of H-H, pH 7.4 at 37°C. Centrifuge the cells at 800g for 5 min and discard the supernatant. Resuspend the cells in culture medium and count them prior to plating.

6. Evaluation of Cell Type

6.1. Immunocytochemical Properties

Most researchers recognize the difficulty in obtaining pure cultures of capillary endothelial cells from brain, epididymal fat pads, retina, and other sources. This is not surprising given that blood vessels are a source of not only endothelial cells but smooth muscle cells, pericytes, and glial cells depending on the tissue origin of the blood vessels. Therefore, it is imperative that the cultures be well characterized. Brain capillary endothelial cells should express general endothelial markers and more specialized properties of neural capillary endothelium.

A general marker for endothelial cells is factor VIII/von Willebrand factor (vWF) antigen. The antibody, used in general immunofluorescence staining, is raised against human factor VIII, but it reacts with factor VIII/vWF-antigens from other species (Fig. 4) (Bowman et al., 1981; Phillips et al., 1979).

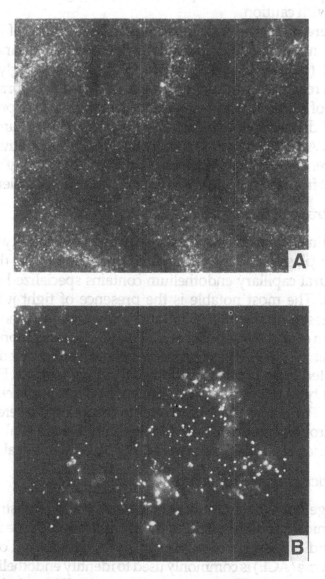

Fig. 4. Immunolocalization of factor VIII antigen in endothelial cells. Endothelial cells were cultured from bovine brain (A) or rat aorta (B). After 7 d in culture, cells were fixed in methanol, washed with phosphate buffered saline, and incubated with 1:25 (A) and 1:50 (B) dilution of rabbit anti-human factor VIII serum K (Behring Diagnostic). The cells were then washed and incubated with rhodamine–conjugated goat anti-rabbit antibodies. There is granular perinuclear fluorescence in endothelial cells.

The pattern of fluorescence that should be seen is typically granular and perinuclear. Other patterns of staining should be interpreted with caution.

There is another marker available for staining of normal human endothelial cells and endothelial tumors. This antibody, BMA120 (Behring Diagnostics) precipitates a 200 kDa glycoprotein and reacts with an antigen located in the cell membrane and cytosol of human endothelial cells. Hemangiomas express the BMA120 defined epitope in mature capillaries and immature endothelium. Angiosarcomas, in some cases, also give a positive staining pattern. The epitope detected by BMA120 is thought to be different from factor VIII-related antigen (Alles and Bosslet, 1986).

6.2. Ultrastructure

Brain capillaries provide the selective permeability barrier between plasma and interstitial fluid. In order to form this barrier, neural capillary endothelium contains specialized characteristics. The most notable is the presence of tight junctions (Simionescu et al., 1975,1976) and a low rate of pinocytosis (Fishman et al., 1985; Jefferies et al., 1984; Olesen and Crone, 1984; Palade et al., 1979). These properties should be reflected in differentiated brain capillary endothelial culture isolates. Figure 5 shows a typical electron micrograph of a bovine aortic endothelial cell. The reader is asked to refer to the following references for micrographs of capillary endothelium: Palade et al. (1979); Simionescu et al. (1975); Spatz et al. (1980); Williams et al. (1980).

6.3. Biochemical Properties

Together with immunocytochemistry and ultrastructure, biochemical properties help in the identification of specific cell types and assessment of culture purity. Angiotensin II converting enzyme (ACE) is commonly used to identify endothelial cells. The presence of ACE does not guarantee a cell is endothelial in origin, but its absence is cause for concern. ACE has been shown to be present in aortic endothelium (Hayes et al., 1978), and endothelium from umbilical vein (Johnson and Erdos, 1977), arteries, and veins (Davison and Karasek, 1981; Johnson, 1980),

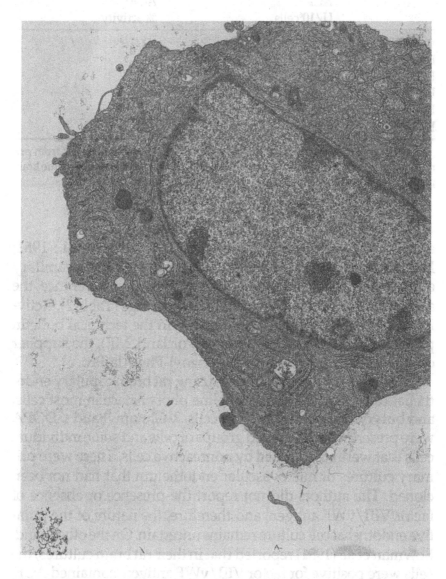

Fig. 5. Electron micrograph showing an aortic endothelial cell from a young calf. The cell appears to have the full complement of organelles expected of bovine endothelial cells. The cells do not contain Weibel-Palade bodies. Refer to Spatz et al. (1980) and Williams et al. (1980) for electron micrographs of brain capillary endothelial cells.

Table 2

Angiotensin II-Converting Enzyme Activity in Endothelial Cells

	ACE* U/10^{6}cells	ACE* % activity
BAEC1°	17,461	100
BAEC P16	10,077	58
BBCE P7	16,440	100
BBCE P8	16,062	98
BBCE P9	10,377	63
BBCE P15	5,729	35

*Angiotensin II-converting enzyme (ACE); BAEC1°, endothelial cells from primary cultures of bovine aortic endothelium; BBCE, passage cultures of bovine brain capillary endothelial cells; P-numeral equals passage number.

and in brain and adrenal capillaries (Bowman et al., 1981; Folkman et al., 1979). Although late passage calf brain capillary endothelial cells maintain a classic cobblestone appearance, the activity of ACE decays with prolonged passage (Table 2). Activity of ACE is determined as described in the technical bulletin provided by Ventrex Laboratories (Portland, ME), the supplier of the radioactive substrate, ^{3}H-benzoyl-Phe-Ala-Pro.

Spatz et al. (1980) described, using rat brain capillary endothelial cells, the presence of alkaline phosphatase in most cells, and butyrylcholinesterase in all cells. λ-Glutamyl and L-DOPA were present in multilayered groups of cells and some individual cells that were surrounded by nonreactive cells. These were primary cultures of microvascular endotheium that had not been cloned. The authors did not report the presence or absence of factor VIII/vWF antigen, and therefore, the nature of the putative endothelial cell culture remains uncertain. On the other hand, Bowman et al. (1981) reported that in their rat brain cultures, the cells were positive for factor VIII/vWF antigen, contained ACE, and did not bind platelets. Electron micrographs revealed typical tight junctions and few pinocytotic vesicles.

To test whether our bovine endothelial cells from brain capillaries have a distinct pattern of protein synthesis relative to bovine aortic endothelium and rat epididymal capillary endothelium, we measured the expression of normal cellular protein synthesis as a function of temperature of incubation and duration of heat treatment (Ketis et al., 1988a).

The fluorographs shown in Fig. 6 display the effect of heat treatment on the rate of protein synthesis in endothelial cells from different origins. The rate of protein synthesis is distinctive in endothelial cells from different origins. Bovine brain capillary endothelial cells (BBCEs) demonstrate a dramatic repression of total cellular protein synthesis relative to control. But unlike passage culture of bovine aortic endothelium (BAECs), BBCEs synthesize a low level of heat shock proteins (HSPs). Two-dimensional gel analysis further substantiates that heat-stressed endothelial cells from different origins are dissimilar in their response to hyperthermia (Fig. 7, *see* page 158). BBCEs respond to conditions of continuous heat treatment (42°C, 2 h) by the induction of HSP 71, 73, and 90 (Fig. 7A). Stressed BBCEs synthesize higher amounts of these proteins than stressed primary BAEC cultures (Fig. 7B), and do not appear to induce HSP90 and 100, which clearly are evident in stressed passage BAECs (Fig. 7D).

Examination of the polypeptides made and released into the growth media of endothelial cells from different origins suggests that the expression of any given polypeptide by the cells is dependent on the (patho)physiological condition. Figure 8 (*see* page 160) shows the protein profile in one-dimensional SDS-polyacrylamide gels of growth media from cultures labeled with ^{3}H-proline for 4 h in the presence and absence of heparin (100 mg/mL). In all four endothelial cells examined, including BBCEs, the thrombospondin/fibronectin ratio in growth media is altered, i.e., is increased relative to control (Ketis et al., 1988b).

The basic functions performed by endothelial cells are similar in blood vessels of all sizes. Despite these similarities, there is reason to believe that there are significant differences between large veins and arteries and those from small capillaries. The blood-brain barrier being facilitated by capillary endothelial cells is only one such example. In neoplasia, arthritis, and psoriasis, there is rapid proliferation of capillaries with relatively little change in large blood vessels. It is then interesting to observe that aortic endothelium (BAECs) vs capillary endothelium

370   Ketis

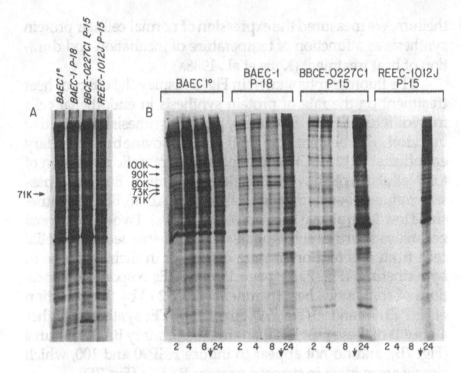

Fig. 6. Effect of temperature on the rate of protein synthesis in endothelial cells from different origins. Endothelial cells from primary (BAEC 1°)
and passage (BAEC) cultures of bovine aortic endothelium, passage cultures of bovine brain capillaries (BBCE), and passage cultures of rat epididymal capillaries (REEC) (numeral equals clone number, P-numeral equals
passage number) were labeled for 1 h with ^{35}[S]methionine at either 37°C
(A) or 43°C (B). Radiolabel was added to the cells 1 h prior to the end of the
treatment time at 43°C, indicated (hours). After 8 h at 43°C, some cells
(marked by ↑) were returned to 37°C for 24 h (the label was present from 23
to 24 h in these cultures). Proteins were analyzed as in Fig. 1 by single-
dimensional SDS-gel electrophoresis and fluorography (from Ketis et al.,
(1988) *Cancer Res.* **48**, 2101; reprinted with permission of publisher).

(BBCEs) from the same species and about the same passage number, appear to differ in their response to thermotolerance and
synthesis of proteins exposed to hyperthermia.

6.4. In Situ *Hybridization*

A powerful technique, *in situ* hybridization, has been
developed (Berger, 1986), that addresses protein-specific transcription within a cell. Certainly, detection of a protein by

immunocytochemistry does not rule out the possibility of protein uptake. Pulse-labeling experiments would address this question, but low protein copy number would mask differential protein synthesis. *In situ* hybridization uses cDNA or anti-sense RNA as a probe to detect specific mRNA coding for a specific protein within a cell. This technique can detect low levels of mRNA in a single cell, and thus the transcription of the gene for a given protein in a given cell can be unequivocally determined.

This technique is analogous to immunocytochemistry, except that the technique detects an mRNA for a protein using a specific cDNA probe, rather than an antibody detecting a specific protein. *In situ* hybridization has been carried out on endothelial cells (Strieter et al., 1989). For a detailed description of *in situ* hybridization, refer to Berger (1986).

7. Cryopreservation

Permanent storage of cells in liquid nitrogen enables the maintenance of viable cells without the need of frequent subculture. Some investigators have described changes in cells upon storage (Berman et al., 1965; Peterson and Stulberg, 1965) but most researchers have not been able to observe detectable changes in cells stored in liquid nitrogen. In fact, no significant change in cells stored in liquid nitrogen for up to 12 yr have been reported (Greene et al., 1967,1975). When cloned cell lines are used, careful bookkeeping is necessary; the clone number and passage number should be noted. The freezing of cells facilitates future experiments with the same clone and passage number. This is particularly important because some cell lines have been shown to change their biochemical properties upon passage or aging in culture (Davison and Karasek, 1981; Johnson et al., 1980; Ketis et al., 1988a).

Materials required for storage of cells are as follows:

1. Dimethylsulfoxide (DMSO), Fisher brand.
2. Sterile freezing vials, Nunc brand.
3. Cells: Should be in log-phase (covering 60–80% of the tissue culture vessel). They should NOT be from a confluent monolayer.

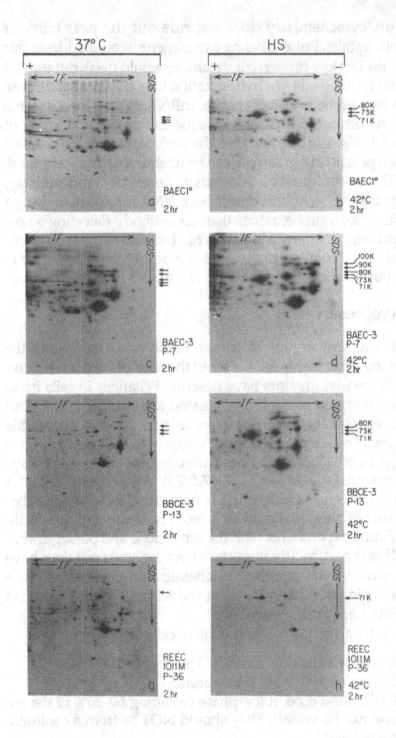

37°C   HS

a. BAEC1° 2hr

b. BAEC1° 42°C 2hr — 80K, 73K, 71K

c. BAEC-3 P-7 2hr

d. BAEC-3 P-7 42°C 2hr — 100K, 90K, 80K, 73K, 71K

e. BBCE-3 P-13 2hr

f. BBCE-3 P-13 42°C 2hr — 80K, 73K, 71K

g. REEC 1011M P-36 2hr

h. REEC 1011M P-36 42°C 2hr — 71K

Freezing Procedure:

1. Treat cells with trypsin-EDTA for 3–5 min.
2. Remove the trypsin-EDTA cell suspension.
3. Add 10 mL of H-H, pH 7.4 at 4°C to (2) and centrifuge cells at 800g for 5 min.
4. Resuspend pellet in 10 mL of H-H, pH 7.4 at 4°C and recentrifuge.
5. Add a 1 mL suspension of freezing medium to the pellet of cells to be stored in liquid nitrogen.

Formula:
10% DMSO
15% Fetal calf serum in growth medium.

The freezing medium is added to the cells slowly over a period of 2 min, thus reducing osmotic shock.

6. The 1-mL aliquots of cell suspension are slowly lowered in temperature. A step-down procedure is followed:
   a. 30 min at 4°C
   b. 1–2 h at –20°C
   c. 1–3 d at –70°C or place overnight in the liquid vapor phase of liquid nitrogen.
   d. submerge vials containing cells in liquid nitrogen,

Fig. 7 *(opposite page).* Two-dimensional gel analysis of normal and stressed endothelial cells from different origins. To label cellular proteins, endothelial cells from different origins were incubated with [³⁵S]methionine for 2 h at either 37°C or 42°C. Cells were solubilized, and the proteins were analyzed by isoelectric focusing (IF) in a pH 4–8 gradient (acid on the right) followed by electrophoresis into an SDS-polyacrylamide gel and fluorography. The stress proteins are indicated by arrows. To control for differences in migration between gel states, equal numbers of cells grown at 37°C were run with each heat-shock condition; A and B, BAEC 1° labeled for 2 h at 37°C or 42°C for 2 h, respectively; C and D, passage BAEC at 37°C or 42°C for 2 h, respectively; E and F, passage BBCE at 37°C or 42°C for 2 h, respectively; G and H, passage REEC at 37°C or 42°C for 2 h, respectively. Note: numeral equals clone number; P-numeral equals passage number. HS, heat shock. (From Ketis et al., (1988) *Cancer Res.* **48,** 2101, 1988; reprinted with permission of publisher).

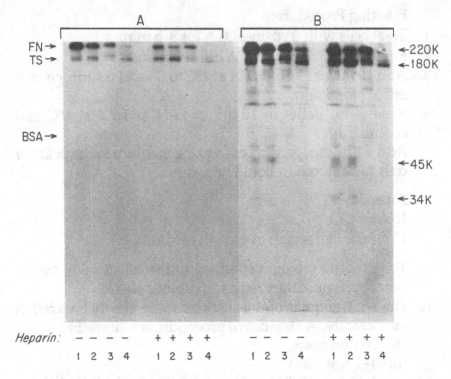

Fig. 8. Effect of heparin on the protein profile of the growth media of endothelial cells from different origins. Endothelial cells were labelled with [³H]-proline for 4 h in the presence or absence of heparin (100 µg/mL). (1) Primary cultures of BAECs; (2) passage cultures of BAECs; (3) passage cultures of rat epididymal capillaries; and (4) passage cultures of bovine brain capillaries. The protein profiles were analyzed by 9% SDS-PAGE and fluorography. Each lane contained equal numbers of cell equivalents. The position of human fibronectin (FN), thrombospondin (TS), and bovine serum albumin (BSA) in the gel are indicated. (A) Short and (B) long exposure of the same gel bearing radioactive polypeptides to X-ray film. Note the change in ratio in the 220,000- and 180,000-mol wt bands in the presence of heparin. (From Ketis et al., (1988) *J. Cell Biol.* **106**, 893; reprinted with permission of publisher.)

<-150°C. Storage of cells at <-150°C prevents ice crystal growth and enzyme activity. Viability will drop if cells are stored at -100°C for extended periods of time.

When needed, thawing of cells should be executed quickly (within 40–60 s) by placing the vials into a 37°C water bath with vigorous agitation. Once the ice has melted, the vial is removed from the water bath and is immersed in 70% ethanol at room temperature. From this point onward, all procedures should be carried out under strictly aseptic conditions in a sterile laminar flow unit. Once thawed, the cells are plated into a T-25 tissue culture vessel. The cells are allowed to attach and spread overnight. The culture medium is changed between 18 and 24 h after thawing. The cultures should be refed every second day. It is noteworthy to mention that recovery of cells will vary and will generally be between 40–80%. The percentage recovery is dependent on cell type, isolation of cells (trypsin-EDTA treatment), and the ability to control the rate of freezing and thawing employed.

8. Conclusion

In contrast to endothelial cell isolations from large blood vessels, the study of neural capillary endothelial cells in culture is still lagging behind. However, progress has been made in the isolation, purification, and growth of brain capillary endothelial cells. The ability to culture brain capillary endothelial cells facilitates the study of the blood-brain barrier. One of the most interesting concepts that has emerged in the study of endothelial cells from different origins is the observation of great diversity among different types of endothelium, and that this diversity is reflected in endothelial cell cultures. Thus, it makes good sense to use endothelial cells from physiologically relevant vessels when studying a vascular problem. Thus, when examining atherosclerosis, cultured aortic endothelial cells are used, whereas when examining encephalopathies, brain edema and tumor angiogenesis, capillary endothelial cells are studied.

The present techniques for isolating, purifying, and growing capillary endothelial cells in culture are not trivial. They require careful manipulation of tissue and isolation of capillaries. Once the cells are established in culture, they can be maintained in culture for long periods. However, caution should be taken when examining passaged brain capillary endothelial cells. It has been shown that prolonged passage of brain capillary endothelial cells alters their biochemical properties. Thus, by using clonal cell lines and freezing the cells, one can examine the same clone and passage number under various (patho) physiological conditions. It is hoped that the reader has received some feeling for the usefulness of brain capillary endothelial cells in culture.

Acknowledgments

I am particularly grateful to Morris J. Karnovsky, Richard L. Hoover, andBenjamin Caleb, under whose agis I learned endothelial cell biology and tissue culture. I would like to thank Brenda McPhail for typing the manuscript, Susan Gray for her technical assistance, and Henry Verstappen for photographic assistance. This work was supported by MRC (MA-1086) and ARC Grants. N. V. Ketis is a recipient of a Queen's National Scholar Award.

References

Alles J. U. and Bosslet K. (1986) Immunohistochemical and immunochemical characterization of a new endothelial cell specific antigen. *Histochem. Cytochem.* **34,** 209–214.

Berger C. N. (1986) *In situ* hybridization of immunoglobin-specific RNA in single cells of the B lymphocyte lineage with radiolabelled DNA probes. *EMBO J.* **5,** 85-93.

Berman L., McLeod M. P., and Powsner E. P. (1965) Monitor cultures of stored frozen human cells. *Lab. Invest.* **14,** 231–234.

Bowman P. D., Betz A. L., Ar D., Wolinsky J. S., Penney J. B., Shivers R. R., and Goldstein G. W. (1981) Primary cultures of capillary endothelium from rat brain. *In Vitro* **17,** 353–362.

Bruzney S. M. and Massicotte S. J. (1979) Retinal Vessels: Proliferation of endothelium in vitro. *Invest. Opth. Vis. Sci.* **18,** 1191–1195.

Carrel A. and Burrows M. T. (1910) Cultivation of adult tissues and organs outside of the body. *J. Am. Med. Assoc.* **55,** 1379–1381.

Davidson P. M., Bensch K., and Karasek M. A. (1980) Isolation and growth of endothelial cells from the microvessels of the newborn human foreskin in cell culture. *J. Invest. Derm.* **75,** 316–321.

Davison P. M. and Karasek M. A. (1981) Human dermal microvascular endothelial cells *in vivo*: effect of cyclic AMP on cellular morphology and proliferation rate. *J. Cell Physiol.* **106,** 253–258.

De Bault L. E., Henriquez E., Hart M. N., and Cancilla P. A. (1981) Cerebral microvessels and derived cells in tissue culture: II Establishment, identification and preliminary characterization of an endothelial cell line. *In Vitro* **17,** 480–494.

De Bault L. E., Kahn L. E., Frommes S. P., and Cancilla P. A. (1979) Cerebral microvessels and derived cells in tissue culture: isolation and primary characterization. *In Vitro* 15, 473–487.

De Bault L. E. and Cancilla P. A. (1980) λ-glutamyl transpeptidase in isolated brain endothelial cell: Induction by glial cells *in vitro*. *Scence* 207, 653–655.

Del Vecchio P. J., Ryan V. S., and Ryan J. W. (1977) Isolation of capillary segments from rat adrenal gland. *J. Cell Biol.* 75, 73a.

Feig C. (1910) Sur la survir d'elements et de systemes cellulaires, en particulair des vaisseaux apres conservation prolongie hors de l'organisme. *C.R. Soc. Biol.* 69, 504–509.

Fishman J. B., Andraham J. V., Connor J., Dickey B. F., and Fine R. T. (1985) Receptor-mediated transcytosis of transferrin across the blood brain barrier. *J. Cell Biol.* 101, 423a.

Folkman J. and Haudenschild C. C. (1980) Angiogenesis *in vitro*. *Nature* 288, 551–556.

Folkman J., Haudenschild C. C., and Zetter B. R. (1979) Long term culture of capillary endothelial cells. *Proc. Natl. Acad. Sci. USA* 76, 5217–5221.

Frank R. N., Kinsey V. E., Frank K. W., Mickus K. P., and Randolph A. (1979) *In vitro* proliferation of endothelial cells from kitten retinal capillaries. *Invest. Opth. Vis. Sci.* 18, 1195–1200.

Freshney R. I., ed. (1983) Culture of Animal Cell: A Journal of Basic Techniques. Alan R. Liss, New York.

Greene A. E., Athreya B. H., Lehr H. B., and Coriell L. L. (1967) Viability of cell cultures following extended preservation in liquid nitrogen. *Proc. Soc. Exp. Biol. Med.* 124, 1302–1307.

Greene A. E., Manduka M., and Coriell L. L. (1975) Viability of cell cultures stored up to 12 years in liquid nitrogen. *Cryobiology* 12, 583.

Hayes L. W., Goguen C. A., Ching S. F., and Slakey L. L. (1978) Angiotensin converting enzyme: Accumulation in medium from cultured endothelial cells. *Biochem. Biophys. Res. Commun.* 82, 1147–1153.

Jakoby W. B. and Pastan I. H., eds. (1979) Cell culture. *Methods in Enzymology vol. LVIII.* Academic, New York, NY.

Jefferies W. A., Brandon M. R., Hunt S. V., Williams A. F., Gatter K. C., and Mason D. Y. (1984) Transferrin receptor on endothelium of brain capillaries. *Nature.* 312, 162–163.

Johnson A. R. and Erdos E. G. (1977) Metabolism of vasoactive peptides by human endothelial cells in culture: Angiotensin I converting enzyme and angiotensinase. *J. Clin. Invest.* 59, 684–695.

Johnson A. R. (1980) Human pulmonary endothelial cells in culture: Activities of cells from arteries and cells from veins. *J. Clin. Invest.* 65, 841–850.

Karasek M. and Charlton M. (1974) Isolation and growth of skin endothelial cells in cell culture. *J. Invest. Derm.* 62, 542.

Ketis N. V., Hoover R. L., and Karnovsky M. J. (1988a) Effects of hyperthermia on cell survival and patterns of protein synthesis in endothelial cells from different origins. *Cancer Res.* 48, 2101–2106.

Ketis N. V., Lawler J., Hoover R. L., and Karnovsky M. J. (1988b) Effects of heat shock on the expression of thrombospondin by endothelial cells in culture. *J. Cell Biol.* **106**, 893–904.

Kruse, Jr., P. F. and Patterson, Jr., M. K., eds. (1973) *Tissue Culture Methods and Applications.* Academic, New York, NY.

McGarrity G. J. (1975) *Tissue Cult. Assoc.* **1**, 181–192.

Mrsulja B. B., Mrsulja B. J., Fujimoto T., Klatzo I., and Spatz M. (1976) Isolation of brain capillaries a simple technique. *Brain Res.* **110**, 361–365.

Olesen S. P. and Crone C. (1984) Serotonin increases microvascular premeability in the brain. *Int. J. Microcirc. Clin. Exp.* **3**, 466–471.

Palade G. E., Simionescu M., and Simionescu N. (1979) Structural aspects of the permeability of the microvascular endothelium. *Acta Physiol. Scand. Suppl.* **463**, 11–32.

Panula P., Joo F., and Rechardt L. (1978) Evidence for the presence of viable endothelial cells in cultures derived from dissociated rat brain. *Experimentia* **34**, 95–97.

Paul J. (1975) Cell and Tissue Culture, 5th ed., Churchill-Livingston, London.

Penson G. and Balducci D., eds., (1963) Tissue cultures in biological research. Elsevier, Amsterdam.

Peterson W. P. and Stulberg C. (1965) *Cytobiology* **1**, 80–92.

Phillips P., Kumar P., Kumar S., and Waghe M. (1979) Isolation and characterization of endothelial cells from rat and cow brain white matter. *J. Anat.* **129**, 261–272.

Simionescu M., Simionescu N., and Palade G. E. (1975) Segmental differentiations of cell junctions in vascular endoethelium. The Micovasculature. *J. Biol. Chem.* **67**, 863–869.

Simionescu M., Simionescu N., and Palade G. E. (1976) Segmental differentiations of cell junctions in the vascular endothelium. Arteries and veins. *J. Cell Biol.* **68**, 705–723.

Spatz M., Bembry J., Dodson R. F., Hervonen H., and Murray M. R. (1980) Endothelial cell culture derived from isolated cerebral microvessels. *Brain Res.* **191**, 577–582.

Strieter R. M., Kunkel S. L., and Showell H. J. (1989) Endothelial cell gene expression of a neutrophil chemotactic factor by TNF-α, LPS and IL-13. *Science* **243**, 1467–1469 .

Taub M. and Sato G. (1980) Growth of functional primary cultures of kidney epithelial cells in defined medium. *J. Cell. Physiol.* **105**, 369–378.

Wagner R. C. and Matthews M. A. (1975) The isolation and culture of capillary endothelium from epididymal fat. *Microvasc. Res.* **10**, 286–297.

Williams S. K., Gillis J. F., Matthews M. A., Wagner R. C. and Bitensky M. W. (1980) Isolation and characterization of brain endothelial cells: Morphology and Enzyme Activity. *J. Neurochem.* **35**, 374–381.

Zetter B. R., Johnson L. K., Shuman M. A., and Gospodarowicz D. (1978) The isolation of vascular endothelial cell lines with altered cell surface and platelet-binding properties. *Cell* **14**, 501–509.

Index